Work and A
A European Pe

D1494050

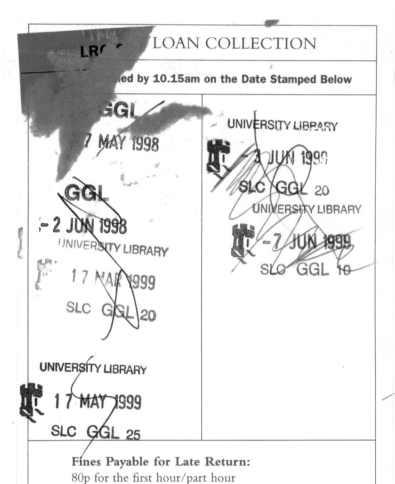

Work and Aging:
A European Perspective

Edited by
Jan Snel

Department of Psychonomics
Faculty of Psychology
University of Amsterdam
The Netherlands

and

Roel Cremer

Institute for Work Integration and Training
IvAS, Heliomare
Wijk aan Zee, The Netherlands

Section Editors:
H. C. G. Kemper

Department of Health Science, Faculty of Human Movement Sciences
Vrije Univeriteit, Amsterdam, The Netherlands

E. Zeef

Department of Psychonomics, Faculty of Psychology
University of Amsterdam, The Netherlands

M. J. Schabracq

Department of Work and Organization Psychology
Faculty of Psychology
University of Amsterdam, The Netherlands

P. T. Kempe

Department of Psychonomics, Faculty of Psychology
University of Amsterdam, The Netherlands

Taylor & Francis
Publishers since 1798

UK   Taylor & Francis Ltd, 4 John St, London WC1N 2ET
USA Taylor & Francis Inc., 1900 Frost Road, Suite 101, Bristol, PA 19007

British Library Cataloguing in Publication Data
A catalogue record for this book is available from the British Library

ISBN 0–7484–0164–4 (Cased)
      0–7484–0165–2 (Paper)

1001230732

Library of Congress Cataloging in Publication Data are available

Cover design by Amanda Barragry

Set in 10/12pt Times
by Santype International Limited, Salisbury, Wilts

Printed in Great Britain by Burgess Science Press, Basingstoke on paper which has a specified pH value on final paper manufacture of not less than 7.5 and is therefore 'acid free'.

Contents

Preface

In the past few years the subject of work and aging has gained much public and professional interest. Many scientific meetings have been organized and many books have been written on the topic, in which experts presented their results, knowledge and views. The impression, however, is that there is still a lack of communication between the groups engaged with scientific, practical and policy-making aspects of work and aging.

In 1993, 'The European Year of the Elderly and Solidarity between Generations', one of our activities was a European Symposium on Work and Aging which was held in Amsterdam on 28–29 January. This symposium was intended not only as a forum for scientists but was also meant for practitioners and policy-makers who are actively involved in this growing field of social interest. The four themes addressed were:

—aging, work and health;
—aging and mental work capacity;
—training and educational programmes; and
—social policy and perspectives.

The last two issues in particular are considered to be of broad social and industrial interest. Much emphasis was laid on topics related to the aging process (biological, physical and mental) in relation to work demands and personal well-being and health. Also, methods and measures were presented in order to evaluate ways of assessing work ability. An important question arose in relation to typical aspects of work which become more demanding when a certain age is reached or when work has to be done with a certain handicap. A related issue those occupations that are risky, in particular for the elderly worker.

The issue of which social policy is adequate was emphasized in many contributions. There is also an urgent need to take goal-directed action for aging workers in the work organization. What is, and what should be, done at the work site in the various European countries, and which methods and measures are available to assess the impact of implemented measures were questions that were frequently asked. An important conclusion is that, at the moment, serious gaps exist in relevant knowledge and that the communication and dissemination of information are poor on the European level, even at the national level.

Recently, the use of age-related vocational education and training has become of increasing interest as a to-be-implemented social policy. It is emphasized that the general programmes should be applied on a continuous basis during the individual's career and be adapted to the situational needs of elderly workers.

Age-related social policy programmes should be more than just 'social'. In the discussion of the different chapters, incentives are suggested for how an organization could benefit from the assets of the aging worker. Finally, training programmes for human resource management, with respect to the elderly and disabled worker in particular, are offered in order to deal effectively with vocational rehabilitation.

A majority of the contributions from the symposium form the basis for this book. In addition, other experts in the field were invited to present their results and views on the issue of the aging worker. The book is meant for scientists, policy-makers and practitioners encountering any combination of issues: work and age-related or handicap-related functional change. We do hope that readers will take time to study the contents and may use the conclusions and recommendations in their future work.

Roel Cremer
Jan Snel

Contributors

B. Baracat
CNRS (ERS 65)
Centre de Recherche en Biologie
    du Comportement
Université Paul Sabatier
118 route de Narbonne
31062 Toulouse Cédex
France

J. G. Boerlijst
Faculty of Management
Twente University
PO Box 217
7500 AE Enschede
The Netherlands

M. Bruyn-Hundt
Faculty of Economical Sciences
    and Econometry
University of Amsterdam
Roetersstraat 11
1018 WB Amsterdam
The Netherlands

K. Brzokoupil
Samhall AB Metodudveckling
PO Box 44
S-146 21 Tullinge
Sweden

A. Carmichael
Age and Cognitive Performance
    Research Centre
University of Manchester
Oxford Road
Manchester M13 9PL UK

B. Cassou
Consultation de Gérontologie
Groupe Hospitalier Ste Périne
49 rue Mirabeau
75016 Paris
France

R. Cremer
Institute for Work Integration
    and Training
IvAS, Heliomare
Relweg 51
1949 EC Wijk aan Zee
The Netherlands

F. Derriennic
INSERM U170
16 avenue P. Vaillant Couturier
94807 Villejuif
Cédex, France

B. de Vries
Ministry of Social Affairs and
Employment
PO Box 90801
2509 LV The Hague
The Netherlands

B. C. H. de Zwart
Study Centre on Work and Health
Academic Medical Centre
University of Amsterdam
Meibergdreef 15
1105 AZ Amsterdam
The Netherlands

J. Espey
Psychologisches Institut der
    Universität Bonn
Psychologische Methodenlehre und
    EDV
Römerstrasse 164
D-5300 Bonn 1
Germany

V. M. A. Gerardu
NV Koninklijke Sphinx
PO Box 1050
6201 BT Maastricht
The Netherlands

W. J. A. Goedhard
Department of Social Medicine
Vrije University
Van der Boechorststraat 7
1081 BT Amsterdam
The Netherlands

G. Holm
Human Resource Development
Danish Technological Institute
PO Box 141
2630 Taastrup
Denmark

J. Ilmarinen
Department of Physiology
Institute of Occupational Health
Laajaniityntie 1
01620 Vantaa
Finland

P. T. Kempe
Department of Psychonomics
Faculty of Psychology
University of Amsterdam
Roetersstraat 15
1018 WB Amsterdam
The Netherlands

H. C. G. Kemper
Department of Health Science
Faculty of Movement Sciences
Vrije University
van der Boechorststraat 7
1081 BT Amsterdam
The Netherlands

A. Kok
Department of Psychonomics
Faculty of Psychology
Roetersstraat 15
1018 WB Amsterdam
The Netherlands

M. H. Lorist
Department of Psychonomics
Faculty of Psychology
University of Amsterdam
Roetersstraat 15
1018 WB Amsterdam
The Netherlands

J. C. Marquié
CNRS (ERS 65)
Centre de Recherche en Biologie
    du Comportement
Université Paul Sabatier
118 route de Narbonne
31062 Toulouse Cédex
France

T. F. Meijman
Section Experimental and Work
    Psychology
Department of Psychology
Grote Kruisstraat 2/1
9712 TS Groningen
The Netherlands

A.-F. Molinié
CREAPT
Centre for Research and Studies on
  Age and Populations at Work
41 rue Gay-Lussac
75005 Paris
France

E. J. Mulock-Houwer
Ministry of Social Affairs and
  Employment
PO Box 90801
2509 LV The Hague
The Netherlands

H. Neuf
Psychologisches Institut der
  Universität Bonn
Römerstrasse 164
D-5300 Bonn 1
Germany

P. Paoli
European Foundation for the
  Improvement of Living and
  Working Conditions
Loughlinstown House
Snakill, Co. Dublin
Ireland

E. Paus
Psychologisches Institut der
  Universität Bonn
Römerstrasse 164
D-5300 Bonn 1
Germany

P. M. A. Rabbitt
Age and Cognitive Performance
  Research Centre
University of Manchester
Oxford Road
Manchester M13 9PL
UK

G. Rudinger
Psychologisches Institut der
  Universität Bonn
Römerstrasse 164
D-5300 Bonn 1
Germany

M. J. Schabracq
Department of Work and
  Organization Psychology
Faculty of Psychology
University of Amsterdam
Roetersstraat 15
1018 WB Amsterdam
The Netherlands

J. Shersby
Commission of the European
  Communities
Directorate General V/C/1 A-1 06 9
rue de la Loi 200
B-1049 Brussels, Belgium

J. Snel
Department of Psychonomics
Faculty of Psychology
University of Amsterdam
Roetersstraat 15
1018 WB Amsterdam
The Netherlands

C. Teiger
CNRS
Laboratoire d'Ergonomie
  Conservatoire
  National des Arts et Métiers
41 rue Gay-Lussac
75005 Paris
France

I. van den Burg
FNV—Federation of Dutch
    Trade Unions
PO Box 8456
1005 AL Amsterdam
The Netherlands

J. van der Velden
OKEG
Hemlaan 3
4837 AR Breda
The Netherlands

F. J. H. van Dijk
Coronel Laboratory
Academical Medical Centre
University of Amsterdam
Meibergdreef 15
1105 AZ Amsterdam
The Netherlands

S. Volkoff
CREAPT
Centre for Research and Studies on
    Age and Populations at Work
41 rue Gay-Lussac
75005 Paris
France

P. Warr
MRC/ESRC Social and
    Applied Psychology Unit
Department of Psychology
University of Sheffield
Sheffield S10 2TN UK

E. Zeef
Department of Psychonomics
Faculty of Psychology
University of Amsterdam
Roetersstraat 15
1018 WB Amsterdam
The Netherlands

1

The European Year of Older People and Solidarity between Generations

J. Shersby

The 'Work and Aging' symposium was convened in 1993, during the 'European Year of Older People and Solidarity between Generations', and was devoted to one of the key issues of that year. The aging of the population is a major challenge for the European Union. The demographic changes speak for themselves: over the past three decades, numbers of people aged 60+ have risen by almost 50 per cent, to one in five of the population. By 2020 the proportion will be on average one in four, and even higher in some member states. Nearly one-third of the population and one-fifth of the labour force are over 50.

The speed and magnitude of these changes have brought the issue firmly on to the EU's political agenda. Interest in questions linked to aging grew throughout the 1980s, particularly on the part of the European Parliament, and, following a proposal by the Union, the Council of Ministers adopted a decision in November 1990 establishing a three-year programme of actions in favour of older people. This programme, which culminated in 1993, the European Year of Older People and Solidarity between Generations, defined three areas for action at European level:

—the definition, at appropriate level, of preventive strategies to meet the economic and social challenges of an aging population;
—the identification of innovative approaches to strengthening solidarity between generations and integration of older people; and
—highlighting the positive contribution of older people to society.

From the outset it was clear that no part of this programme contained a definition of a European policy concerning older people. Legislation in new areas is not on the table. The definition of policies and the organization and financing of services for older people are matters for member states, and responsibilities are exercised not just at the national but also at regional and local level in many cases. Action at the level of the European Union focuses on encouraging debate and exchanges of information and experience and on highlighting examples of good practice. There are three key aspects of that activity:

—*Awareness raising*: the aging of the population affects all of us, whether in our work or in our families. The year aimed to raise awareness of the challenges arising not merely in the more 'traditional' areas of concern such as social services or health care, but right across the board. Projects included activities with schools, initiatives undertaken by police forces looking at safety and security, examination of the needs and wishes of older consumers, and design competitions.

—*Learning*: member states face common challenges. The European year offered not simply a framework for reflecting on the nature of these challenges, but an opportunity to exchange information, to compare experiences and approaches and to look at new and innovative ways of tackling problems across the Union. It provided a framework for mutual learning.

—*Celebration*: not a word frequently associated with aging, but one that was a key part of the European year with its double emphasis on the challenges and opportunities of an aging population. Older people are living longer, healthier lives than ever before. They are increasingly demanding and playing an active and positive role in their communities and in their families, in voluntary work, in social and cultural activities and in undertaking caring responsibilities. The year celebrated aging, and the experience, skills and achievements of senior citizens.

The year 1993 was a European, not a European Union, year. The aim has been to involve and support as many of the interested parties as possible in exploring major issues. Initiatives ranged from seminars and research to television programmes, festivals and competitions. Themes included pensions, services for older people, housing, transport, volunteering, older migrants and many more. Implications for the labour market were a key issue. The conference 'Aging and Work', supported by the Union, examined major questions and compared approaches concerning aging and work. A number of further initiatives have also been taken to mark the year in exploring this area:

a) *Flexible retirement*: A Council recommendation already existed from December 1982 advocating the introduction of greater flexibility and choice for workers deciding when to retire. The recommendation also drew attention to the desirability of more phased or gradual forms of retirement, allowing a smoother transition from work to retirement. At the end of 1992, the Union submitted a report to the Council on the implementation of this recommendation, and this was followed in June 1993 by the adoption by the Council of Ministers of a resolution on flexible retirement. The resolution invites member states, and management and labour, to develop and adapt their employment policies so as to make flexible adjustments in line with changes in demography and the age structure of the labour force. It emphasizes the wish to see older people continue to play an active part in society and, having regard to the situation in each member state, to maintain a link with the labour market.

b) *Analysis*: The EU Observatory on older people and aging, an independent expert group set up to monitor and report on policies affecting older people, produced a report 'Older People in Europe: Social and Economic Policies' which included an assessment of employment policies and older workers in the 12 member states. In addition to this, the Union also funded a new report, by independent experts, on the problems of age discrimination against older workers in the Union. In an opinion survey carried out for the European year ('Eurobarometer: Age and Attitudes') an overwhelming majority of people questioned of all ages believed that discrimination against older age groups existed, particularly in recruitment practices. The report analyses the extent of discrimination in a range of areas including exit policies, recruitment, training and career development.

c) *Exchange of information and experiences*: A major conference was organized jointly by the Union and the French Ministry of Labour in Paris, 22–23 November 1992, to debate public policies and management approaches to older workers. Participants included policymakers, representatives of management and labour, independent experts and personnel managers from large companies.

d) *Highlighting innovative approaches*: A network of innovative projects from across the Union has been set up showing the positive contribution of older people in the field of education and training. The network includes projects in which volunteer senior and retired workers assist younger generations in setting up and running businesses through consultancy, training and advice.

The European year has aimed to provide a forum for debate, exchanges of experience and know-how, the development of networks and raising awareness. This is the essential value of action at European level in the face of the challenge posed by demographic change. The close of the year was marked by the adoption by the Council of Ministers of a declaration of principles setting out common objectives concerning older people and solidarity between generations. While there is much that is different in the traditions and systems in member states, there is also much common ground. For its part the Union's hope is that the European year will have helped to contribute positively to the development of policies and the meeting of the challenges of aging across the Union.

J. Shersby
Commission of the European Communities, Directorate General V

2

Aging at work : a European perspective

P. Paoli

If 1993 was chosen as the 'European Year of the Elderly and Solidarity between Generations', it was not by chance. The aging of the population is a heavy trend throughout Europe. This trend is also affecting the work-force and leading an increasing number of companies to rethink their human resources policies, their training policies, their work organization and the design of workplaces.

The European Foundation has been investigating these issues for a number of years. The present leaflet[1] describes some of the possible answers to the new challenge that aging has put out to companies. It is the result of an active collaboration between the Foundation and ANACT[2] and builds on the conclusions of a seminar organized in 1991 in Paris. The proceedings of this seminar have been published.[3]

A first observation concerns demographic aging. Out of the countries that were represented (Germany, France, Sweden, The Netherlands), the perspectives for the development of the working population's age structure suggest that from about the year 2010 onwards, close to 40 per cent of the population will be aged between 45 and 65. Parallel to this process of aging in the working population, one finds a decrease in the rate of work activity among young people, owing to the rising age of school leaving, in particular. This demographic development will undoubtedly lead to a change in the 'generation structure', linked to the disappearance of certain activities. In other words, the current generations of employees, with their associated experiences, qualifications and occupational orientations, will not be replaced on an identical basis; the new employees will have received different training and will aspire to other forms of work.

At the moment, all of the countries represented are beginning to be concerned with the consequences of aging, not only for social security systems, but also for the structure of available employees in relation to the realities of work.

Up to the present time, the policies implemented have mostly consisted of removing elderly employees from working life by means of various procedures, appropriate to the different countries involved.

Thus, in Sweden and The Netherlands, the development of invalidity pensions has been particularly strong.

In France, early retirement has received the greater emphasis, but in a differentiated manner. Early retirement has been used in certain sectors of activity, such as steel making, where it begins 10 years before the normal retirement age. Faced with this employment policy, a paradoxical consensus appears to have been established in France between the different actors concerned:

—It seems that companies have found in this system a method of carrying out necessary restructuring in order to improve productivity. In addition, this measure has apparently allowed them to deal rapidly with problems of overstaffing.
—Trade unionists, for their part, are said to prefer early retirement to redundancy.
—The public authorities thereby have an option that allows them to regulate employment problems more readily.

These measures, which tend more or less towards the rejection of aging workers, have combined to produce negative consequences which most participants in the colloquium pointed out:

—Demobilization of employees remaining in the company.
—Loss of 'experience-based knowledge', which is hard to replace by theoretical knowledge.
—Imbalance in the age pyramid
—Increased socio-economic costs for society. The Netherlands, for example, where there is a generous social protection system, is beginning to call this into question. It is looking at the possibility of reducing allowances and restricting the criteria for assigning social benefits. Some people are even talking of abolishing them.

At the colloquium, these short-term policies were unanimously criticized, because they seem inadequate in relation to current problems, taking account of the predictable developments in the age structures of the working population. Without rejecting these measures categorically, current thinking is clearly demarcated from them. The approach is now directed towards resisting the systematic exclusion of aging workers, giving priority to the initiation, as early as possible, of a set of actions with more far-reaching perspectives with a view to finding a way of keeping employees at work and in good health. In this context, a number of supplementary proposals have been put forward.

Making companies aware of the problems of aging

The problems connected with aging will not be resolved automatically by demographic development or economic growth. It is therefore necessary to start adopting now a clearly directed policy of actions in favour of aging people. In this framework, companies have a decisive role to play, but most of them are not yet fully aware of this. It seems important, then, to inform them of the consequences of demographic aging, so as to persuade them of the advantages that may derive from keeping their older personnel at work.

When necessary, it can be important to encourage them in this direction—certain employers are reluctant to assign certain tasks to people aged over 50 and to involve them in the most recent developments in the company (technologies with an element of innovation, etc.). This hesitancy owes more to cultural stereotypes about aging than to tangible realities. It is true that natural aging is translated into changes on a physiological level (the motor and visual systems) which can affect work capacity. At a cognitive level, on the other hand, there is no significant change over the agebands that span a working life. It must not be forgotten, however, that aging also implies an accumulation of experience that makes the worker an expert in his or her job, able to find strategies that can make up for his or her shortfall on the functional side.

Lastly, badly designed working conditions and a lack of training are the main factors which turn aging into a handicap rather than the natural aging process in itself. In order to combat these stereotypes, all participants felt that it was more important to provide companies with knowledge about the psychological and the physiological aging process, as well as research findings identifying the ways in which working conditions can affect those processes.

Adapting jobs

There are numerous examples illustrating the mechanisms which exclude aging workers, characterized by unpleasant constraints at work (high speeds, painful postures, high temperatures, etc.). In workshops in the electronics and clothing industries, for example, where there is an accumulation of job demands, female workers find it hard to do their job from the age of 30 onwards. In other words, they are already 'too old' for (and in relation to) these working conditions. With a view to avoiding such processes of premature exclusion, the authors wish to stress the importance of redesigning jobs and adapting them to characteristics of 'aging' employees. These initiatives, carried out for preventive purposes, would make it possible not only to keep employees at work, but also to protect their health. Adapting jobs to employees can be done on two levels, individual and collective:

—On the first level, the challenge is to find among existing jobs those that are best suited to an employee with health problems and a reduced working capacity. This type of post is generally referred to as a 'light' or 'soft' job. To be given such a job is often seen as a form of marginalization in relation to the production process. In addition, it is well-known that these jobs are often the first to be affected if restructuring takes place, and that their number tends to decrease in favour of subcontracting. In the same area of solutions, one finds reconversion workshops that allow employees to remain in a production function which is linked to their previous activity. Many other specific measures adapted to aging people can be envisaged, such as changes in the work timetable and so on. These various actions are important because they are easy to apply in most establishments, and their effects seem to be positive. Moreover, they offer real alternatives to those solutions that would involve the exclusion of employees.

—Another solution, harder for companies to implement for reasons of economic strategy, consists of carrying out a forward-looking policy of improving working conditions. More precisely, this involves deciding during an industrial plan to set up a technical and organizational system which takes as its starting-point the functional capacities of workers and their developments. This type of action operates at a collective level, to the extent that the whole working population is involved. The working population is no longer taken as a variable that can be adjusted to a technical system, as an afterthought, because such a process could lead to the exclusion of workers, or a more rapid aging of the employees concerned. With this solution, the company opts instead for a forward-looking approach which integrates the abilities of employees into the development strategy by positioning them as a variable involved in action.

Health promotion and prevention

The problems of aging are, as we can see, closely linked to those of health. To fight against artificially induced aging means promoting health and preventing certain illnesses caused by work. Some experts suggest three levels of prevention:

—The first level, called the 'primary' level, comprises giving each employee responsibility for his or her own health. Each person must try to maintain his or her functional capacities, by maintaining a certain level of physical activity and controlling lifestyle habits and so on. It would be important, lastly, to be able to define functional age in relation to real age, so as to detect whether there is a problem of premature aging, and to take action arising from this.

—In the framework of a 'secondary' type of prevention, one could monitor the developing state of health of employees who seem to have a high physical and cognitive workload, and of those employees aged over 50, by means of regular examinations. In these examinations, illnesses such as hypertension, arthritis, diabetes or stress-linked diseases could be detected more promptly. In The Netherlands, a company has set up the system for monitoring the development of occupational health, which makes it possible to identify quickly a problem that an employee faces, and also to undertake epidemiological studies related to work organization.

—The third proposed level of intervention consists of putting some extra pressure on companies to recruit a certain percentage of 'handicapped' workers.

Actions relating to health must proceed from an approach which is, above all, individual, depending on the working conditions that employees have to face. Thus, in certain cases, one could envisage particular measures such as a part-time retirement system allowing some people to benefit from a sort of compensatory rest period, while preparing for full-time retirement.

Training

Undoubtedly, consideration of training is given priority, especially by companies. Nevertheless, the training effort undertaken so far by companies is mostly concerned with the age band between 25 and 30, while employees aged over 45 do not exceed 5 per cent of people on training courses. Companies' reluctance to invest in an employee whose life expectancy is reduced largely explains this phenomenon. But it is often forgotten that young people are more likely to leave the company. Other reservations have to do with the prejudices of personnel managers, and of employees themselves: starting from a certain age, it is believed that people find learning difficult. This is not true, as these difficulties derive rather from a lack of initial training, a lack of in-service training and badly designed teaching methods. In particular, the teaching methods should not appear too academic, but should be practical, connected if possible to the work previously done, and requiring active participation from those taking the course. In one German firm two types of training are provided:

—The first is in-service training. This covers a number of varied areas (information technology, safety and so on), thereby encouraging the maintenance of a certain learning ability, as well as keeping knowledge up to date. In this category, there are also progressive training courses for bringing less highly qualified employees up to an appropriate level.

—The second type of more practical training is concerned with the work to be carried out. In this connection, certain companies train employees aged over 50 for the function of trainers, so that they can formalize their own knowledge and pass it on to their younger colleagues. The people who are least suited to group teaching can be involved in tutoring activities or in a system similar to apprenticeship. These actions enhance the standing of more senior employees by recognizing the knowledge that they have. In addition, they show that one can introduce an element of flexibility into professional careers, in order to secure development in the function of aging employees.

Adjustments to work careers

Some people suggest that adjustments should be planned, at the earliest, for the second phase in a work career, around the age of 40. This would involve looking again at the organizational structure of all jobs and the division of work in companies. It would be necessary to go further, and consider a thorough-going revision of a whole set of systems for training, for industrial promotion, for industrial agreements, in the light of labour market conditions.

The basic process of consideration, which is of decisive importance as regards the directions that the company must take, needs to be more fully developed. In fact, it presupposes a real effort on the part of companies to rethink, to change their habits: it requires a new way of looking at work.

This is what happens in a Dutch company, which practises a new type of social management founded on the regular evaluation of competencies, with the aim of promoting employee flexibility. The underlying principle is that 'people who regularly cope with new conditions learn to master change'. Thus, whatever one's level in the hierarchy, whatever one's age, each employee should change jobs about every 5 years. In order to monitor and evaluate careers on an individual basis, an automated system of integrated data has been drawn up. It takes account of information covering both the full range of jobs that have been classified (content, quality of work, working conditions), the employees (competencies, training received, career paths) and the various changes which have taken place. The use of such a tool makes a change in the methods of managing personnel. One is no longer dealing with a quantitative budgetary management of staff in the short term, but with a qualitative management oriented towards the individual monitoring of employees, and consequently an anticipation of problems related to 'aging'.

Creation of a forward-looking management of human resources

The various proposals that we have just presented must not be considered simply as actions in favour of elderly workers, as opposed to actions in favour of younger personnel. These proposals are integrated into a global policy for the company, with the aim of breaking down age barriers. This global policy can only be carried through if the firm is oriented towards a human resource policy integrating a forward-looking management of age factors. As we have already inferred, such management requires instruments that can document the characteristics of work situations, the characteristics of employees and the relationship between the two. Knowledge that is provided by such instruments can represent an aid to decision-making in connection with:

—technical or organizational choices to be made when an industrial project is being planned. These decisions can make it possible to define the criteria on the basis of which a 'consensus' can be established between those managers in charge of personnel, of industrialization, of methods and manufacturing, and which can underlie a process of agreement with worker representatives;
—the development of employee skills, qualifications, training;
—the management of careers and jobs; and
—the management of the age pyramid.

In The Netherlands, certain companies are already believed to be practising such a forward-looking management of human resources. One of these has set up a department entitled 'Health, Safety, Environment', which evaluates the social impact of company policy and modifies decisions thanks to a complete system of information collected through regular interviews with employees.

Most interventions at the colloquium, in illustrating the current state of advancement of questions concerned with aging at work, reveal that we still remain too often at the level of relatively general propositions. Co-operation between research bodies and companies is only beginning to be organized, and study plans are only beginning to take shape:

—In Sweden, for example, a multidisciplinary research programme started at the end of 1989 at the National Institute for Occupational Health.
—In The Netherlands, the Dutch Institute for Health and Welfare asked the Psychological Institute in Hilversum to study the experiences and initiatives found within companies.
—In the Nordic countries, a group set up by the Nordic Council of Ministers in 1971 has been meeting regularly in order to discuss research projects and reach agreement on common actions against the exclusion of older employees.
—In France, the Centre de Recherche et d'Etude sur l'Age et les Populations aux Travail (Centre for Research and Study on Age and Populations at Work) has recently been established. For its part, the Agence National pour l'Amélioration des Conditions de Travail (National Agency for the Improvement of Working Conditions), on the basis of a process of consideration which has been going on for some years, is carrying out information programmes in companies.

—In Germany, the 'Work and Technology' programme has launched a study of 'personnel management and the implications of demographic change'.

Although we are still often only at the level of reflection, certain companies have already embarked on concrete forms of action. Thus Aerospatiale in France has set up training programmes for engineers and project heads so that they may design technical and organizational systems suited to the widest possible population. In Germany, Bayer, to which we have already referred, has drawn up a training programme to guarantee the updating of qualifications for its employees, whatever their age. The Manducher group in France, among a number of detailed actions, is preparing an instrument for identifying the mechanisms of exclusion based on a dynamic study of the characteristics of the population, in relation to the development of working conditions.

These proposals and approaches are the first fruits of innovative policies in human resource management, which ought to make it possible to deal with the aging of the active population. They provide a forecast of co-operative programmes to be established at European level, with a view to sharing the results of those experiences, and helping countries and companies to resolve the common problem of exclusion which faces the population of aging workers.

Notes

1. Written by Sylvie Droit (ANACT), Françoise Guerin (ANACT) and Pascal Paoli (European Foundation).
2. ANACT Agence Nationale pour l'Amélioration des Conditions de Travail, 7 Boulevard Romain Rolland, 92128 Montrouge, France.
3. 'Aging at Work', Proceedings of a European Colloquium, Paris, 12 June 1991, ANACT–European Foundation, 118 pp., 1992.

3

Opinions on work and aging

B. de Vries

Not very long ago, it was regarded as an achievement of society that people were able to end their employment on reaching the age of 65. Because of the unemployment among young people in the late 1970s, it was regarded as socially justified that the older people made way for the younger. It became undesirable for older people to work; and if the older people themselves did not see it that way, other people took them to task.

The symposium on 'Work and Aging' in the European Year of Older People and Solidarity between Generations marks a change of direction in thinking. The wide support given to this symposium demonstrates that attention is being paid to participation in employment by older people. Working and growing older: this does not need to be a problem if we clarify a number of issues so that the policies of the partners and the government can be extended. I shall take advantage of this symposium to discuss such issues.

The demographic development is obvious. The prospects are that the labour market will become less young in the near future; there will be relatively fewer young people on the labour market. Current developments make it difficult to say that the participants in the labour market are 'greying'. At present, employment participation by older people is declining. In the European Union, 80 per cent of men in the age group from 55 to 65 were working in 1970. In The Netherland, the figure has now dropped to just over 40 per cent. The Benelux countries and France score between 5 and 20 per cent lower than the other EU countries. And these figures are much lower than those for the United States or Japan, where the average of men employed between 55 and 65 years of age nowadays is about 50 per cent.

The low employment participation by older people in The Netherlands has a number of specifically Dutch reasons. Employment participation by older women in The Netherlands is lower than in most EU countries. In this country, only 15 per cent of women aged 55 and older are professional employed. The government of The Netherlands imposes quite a strict requirement for retirement at age 65. In comparison with other countries, The Netherlands also has a high level of disablement, with about a quarter of men aged between 55 and 65 receiving disablement benefit.

Many companies and sectors of industry have regulations for voluntary early retirement. Under this scheme, people can stop work at about age 61, with a pension much more favourable than under the statutory security scheme. Two-thirds of those entitled seem to be taking advantage of voluntary early retirement: this was shown by a survey carried out by the Income Department. The result is that 20 per cent of this age group receives voluntary early retirement benefit.

For the organisations of employers and employees, the reason underlying the voluntary early retirement scheme was the desirability of creating leeway for younger people on the labour market, in view of the high levels of unemployment in the younger group. That was 10 years ago. Then there is the background of what are termed the 'Guidelines' for Older People, drawn up by the Ministry of Social Affairs and Employment. These made it possible for older people to be given precedence in the event of enforced redundancy.

A climate developed in which it was expected that the older people 'would be leaving anyway'. Older people were often left out of company training schemes. Apparently, they were not expected to provide a return on the investment in training courses. This incorrect expectation reinforced the downwards spiral.

All these matters contributed to the fact that, in The Netherlands today, about 50 per cent of people aged between 50 and 65 have an income based on some social regulation or other.

It is important to reverse this situation in order to reduce the collective burden of taxes and social charges and to increase the foundations for economic and social security. This foundation is necessary in view of the necessity to make provisions for the increasing number of elderly and aged people. The retention of experience and expertise in the business community is of equal importance. Know-how is an increasingly important aspect in competition.

And, last but not least, a change of climate is necessary to maintain older people's feeling of self-esteem. Participating in work, earning one's own income, is valuable in itself. The Dutch government is undertaking a re-evaluation of being older. The Ministry of Social Affairs and Employment, in particular, has a task in relation to work participation by older employees.

In the light of all this, the 'Guidelines for Older People' on discharge will be withdrawn. This measure fits into a policy in which the Minister of Social Affairs and Employment is the first to sweep away the institutional obstacles to work participation by older people. Measures like those in the fields of disablement and working conditions will also contribute to the objectives we want to attain. In the next few years, we want to reduce the voluntary and involuntary removal of older men from employment. We want to increase the number of older women entering the labour market. In the long run, we will be aided in this by a natural development. The labour market is becoming less young, with relatively fewer younger people becoming available. This will result in an increasing demand for older people.

Nevertheless, an active policy will be necessary. The Organization for Economic Co-operation and Development (OECD) has provided an impulse. In its 1992 document, 'A New Orientation for Social Policy', the organization turned against the growning practice of early retirement in many of the affiliated countries. The recommendations relate to attention for working conditions, for training and work mediation, an adaptation of pension schemes where they promote early retirement.

The development and implementation of policy demand a clear picture of policy on various issues. I shall present a number of them to this symposium in this European Year of Older People and Solidarity between Generations.

Working and growing older: what are the issues?

As mentioned earlier, the present voluntary early retirement regulations are an obstacle to work participation; nor is the prospect of voluntary early retirement an encouragement for additional education. Voluntary early retirement may partly be defended as affording a measure of protection for older people who entered the labour market in those difficult years of reconstruction after the war. They should not be denied voluntary early retirement. One may raise very serious questions about this statement. I do not begrudge anyone voluntary early retirement, but I regard the statement that work in the reconstruction period was much harder as being too hasty and too facile. People are also complaining nowadays of stress and too much pressure of work. Furthermore, this statement needs qualification at the very least. It is striking that the survey by the Income Department, mentioned before, indicates considerable differences in the use of voluntary early retirement schemes between different sectors of industry: in the construction sector, 100 per cent take advantage of it, in the banking and commercial services sector, only one-half do. Research into, and information about, the load on the various age and vocational groups and the endurance they display may contribute to a useful debate on voluntary early retirement as a generic measure.

This debate is additionally important because voluntary early retirement as a measure for the present generation of over 60s also arouses expectations among those who help to pay for the schemes.

There are now prospects for older people to continue working, following what we in The Netherlands term 'age-conscious personnel policy'. The increasing attention paid to this matter is an encouraging development. Options are being developed for older people to work part-time and hence retain part of their salary, and there is a growing awareness that older people should not be forgotten in company training schemes. There is an increasing recognition that the knowledge and experience of older people are important to companies, for continuity and for healthy innovation. Options are being sought for replace older people in bottle-neck posts, instead of forcing them along a route that leads to unemployment or disablement.

And, yet, we should recognize that an age-conscious personnel policy still barely extends beyond immediate care for older people. An age-conscious personnel policy also means more permanent education for younger people, variation in their tasks, extension of their posts and the exchange of posts. In the last case, an age-conscious personnel policy may imply demotion. Transfer to a post less demanding than the previous one need not be perceived as being degrading. This type of personnel policy is, however, not yet general practice among employers, in either the public or the private sector. There is a broad consensus about the need for such a preventive personnel policy. I attach considerable importance, in this European Year of Older People and Solidarity between Generations, to the issue of how this preventive approach can be implemented in the labour organizations.

The future of the baby-boom generation

Recent research seems to suggest that there is no direct connection between the intrinsically biological aging process and a possible decline of professional ability. On the contrary: being able to use the acquired skills or not permanently using them and lack of opportunities at work for training of skills are causes of any decrease of flexibility.

In principle, these data imply positive expectations for work participation by the future generation of older people. They usually entered the labour market at a later age, they work shorter hours, and are better qualified and equipped. This leads to a picture of a more vital future generation of older people.

Their improved circumstances and better intellectual ability, however, also seem to be necessary in view of the sharp increase in work productivity. As already said, one hears complaints about too great a pressure of work. Employment has shifted from industry to the services sector; from work with a heavy physical burden to work with a mainly intellectual burden.

The issue is, from where will the future generation of older people derive its vitality?

There is also an issue surrounding the aspirations of the future generations of older people themselves. It is said that even younger employees expect and want to be able to retire after reaching the age of 55. A further statement is that the younger generations have a different work attitude. They regard work as a means rather than as an end. By this reasoning, it would be less obvious to continue working at a more advanced age. Irrespective of whether one shares this opinion, it is desirable to have a picture of the labour market aspirations of the coming generations of older people and how this differs from reality.

The largest group on the labour market is now the generation of the baby boom. Explicitly or implicitly, these employees are often used as the yardstick. Working conditions, productivity requirements—many things are aligned to the capacities and requirements of this generation, or, if possible, to a picture of younger, dynamic people.

One might debate the question of whether this yardstick can be retained. The largest group on the labour market is getting orlder. Can an age-conscious personnel policy suffice in the future, or should we also examine developments in technology, organization and management for their social consequences?

Cutting across all the issues is the necessity of differentiation and gradation. We can no longer speak of *the* older person on the labour market, just as we cannot speak of *the* younger person. There are vital older people and 'worn-out' younger people. The specific situation of generations, branches of industry and professional groups has to become more open to our understanding.

In this address, a few key issues were raised. Clarification of these issues should be regarded as important for two reasons: first, in order to offer an advantage to the social partners in the various branches of industry in their policy for viewing work participation by older people as work made to fit. And, second, because the public sector will derive from it a clearer picture of the general conditions under which it can operate.

The European Year of Older People and Solidarity between Generations has made a contribution to the ultimate objective. This objective is that keeping on working becomes an attractive and obvious option. This will not come about automatically. The European symposium on Work and Aging, spring 1993, provided a good opportunity for discussing the issues. I anticipate that research, debate and international exchanges on important points will provide clarity. Let us hope that in a few years' time, your contribution will be a success in making older people *and* their environment say: 'Working and growing older? No problem.'

Minister de Vries of Social Affairs and Employment
The opening address was delivered on the minister's behalf by E.J. Mulock
Houwer, Director General of Labour.

4

When I'm sixty-four

I. van den Burg

'Nobody has the eternal life', as Dutch people say sometimes. Getting older is a phenomenon that's inevitable for every one of us. Sometimes even young people care about that. I remember from a time when I wasn't even half of my present age, a beautiful song from the Beatles—also much younger then ... 'When I get older, losing my hair, many years from now'. The singer wonders, with grave concern, 'will you still need me, will you still feed me, when I'm sixty-four?'.

This wondering might express the concern of elderly workers, and could be directed to employers as well: will you still need me, will you still feed me, when I'm sixty-four? It's not so peculiar, however, that this Beatles generation, the 'baby-boom-generation', in the picture since the election of Clinton, worries about its future as elderly workers and/or pensioners. Demographic developments force us to be aware of the forthcoming problems of labour supply and income guarantees.

The need for greater participation in the labour market and anxiety about the growing costs of social security benefits for older generations (including future ones) are not the only reasons for this. The individual's need and wish to stay accepted as a full member of society and of the working community should also not be underestimated.

For the seniors of the generation that endured the war period and had a difficult job rebuilding the post-war economy, the emphasis lay on reaching the 'finish' of their working life as soon and as healthy as possible. For new generations of seniors, who started less young, are better educated and have better working conditions and medical and health and safety care, other aspects count: they don't feel as old when they reach their 50s or even 60s! But, this is a general picture. At the same time, we notice that technological changes, high productivity standards, and so on, put a heavy burden on the workers of the 1990s.

Stress is almost the number-one professional disease at the moment. This same conclusion has been drawn by the Dutch Minister of Labour. Nevertheless, I don't agree with the direction of his conclusion, that this equality should lead to equally insufficient protection. At present, 'work and aging' is a popular subject in many discussion circles, not least among the social partners: employers' and workers' organizations in The Netherlands.

The Dutch Federation of Trade Unions (FNV), the largest one in The Netherlands, recently adopted a policy document about work and aging: 'getting older, while working' is the translation. The title expresses the underlying intention: improving the possibilities for elderly workers to participate and keep on participating in the working process untill they reach pensionable age. Noticing that actually only one out of five employees is 50 years or older, and that participation rates are extremely low, and knowing that in the majority of cases it is not the voluntary choice of these non-participants to leave their jobs, but the consequence of health, disability problems or unemployment, the FNV wanted to promote a social policy actively involving elderly workers in the labour process.

In general terms, this policy refers to four distinct phases of the labour process:

(1) the phase of entering the labour process: the start or often 'false start' of engagement;
(2) the period throughout engagement;
(3) the end of engagement; and
(4) the phase of incapacity for work for disability reasons, and subsequent unemployment.

In each of these stages, specific problems occur for elderly workers. Such problems are: younger people are preferred in recruitment procedures; education and training and retraining programmes have maximum age limits; careeer perspectives are missing beyond a certain age.

There is a tradition of 'social' schemes of dismissal of elderly workers in the case of enterprise reorginizations or mergers—albeit with favourable financial arrangements—but nevertheless resulting in a practically zero chance at the labour market. The latter situation does not apply to otherwise unemployed or partially disabled elderly workers and is even worse for the female and migrant elderly worker. To tackle these problems the FNV has worked out policies in different fields: legislation, collective bargaining and social policy of personnel mangement.

Legislation

In terms of legislation or, as it is more generally known, the public sphere, there are several fields of legislation and public activity in which measures can be taken and changed, which are meant to stimulate work participation.

Two evident fields are social security regulations and labour market policy. The FNV advocates a more active approach to social security, especially with regard to disability, that simulates reintegration and maintenance on the job— adapted or not, and sometimes part time or on 'half speed'. Essential elements for that approach must be found in the preliminary phase, in the prevention of disability and motivation of elderly workers, through health and safety pro- grammes, and improvement of the quality of work. These aspects are also part of our activities in the two other spheres mentioned earlier.

Collective bargaining

In collective agreements, arrangements can be made for health and safety and quality of work aspects in general, but also specifically for 'senior workers'. For instance, provisions through which workers above a special age are exempted from nightwork, or—another frequent feature of collective contracts—have access to extra holidays, so called 'senior' days. At the presen- tation of the policy document on work and aging, Johan Stekelenburg, the FNV's president, even launched the proposal of a four-day working week for 50 + workers, on a voluntary basis and without loss of pay or pension rights.

A more structural approach towards adaptation of the workload for elderly workers

A regional bus-transport enterprise in the northern part of Holland has intro- duced a scheme already, which will gradually reach a growing category of age cohorts, starting with the eldest. A very topical debate in the sphere of collec- tive bargaining, and one that is especially pushed by employees, is the dis- cussion about the 'VUT', a 'pre-pension' scheme, which is not financed through 'payment in advance' by the future pensioners as in normal pension funds, but through a 'solidarity' contribution from the total work-force falling under the collective concern contract.

This VUT system started in a period of high unemployment and was meant in the first place as a measure to provide job chances for younger (unemployed) people, who could go into the jobs that these older workers left. Gradually it became very popular among the seniors. On average, more than 80 per cent —and in many cases even more—of employees who are entitled to use this VUT on early retirement, which is voluntary (the letter V stands for 'voluntary'), take the opportunity. In general, they start to be 'VUT-ers' at the age of 62 or 63, sometimes even before the age of 60, and in most cases with a benefit of 80–85 per cent of the last earned salary.

At present, this is seen as a well-established right, and is heavily defended, especially by workers who are close to the age of entitlement to the provision. We can't deny that employers have noticed with concern the growing costs of the yearly/monthly contributions to these schemes, and are opening all new negotiations these days with a fierce attack on these VUT schemes. The trade unions have tried to keep the future of this provision in their own hands. This is at least as relevant as the extent to which members are willing to pay for their older colleagues, although the trade unions are not unwilling to negotiate the future of this provision and to some extent replace this financial arangement by measures for elderly workers that keep them inside the workplace, provided that other provisions will lessen the burden of those last years of engagement. Possible options are 'part-time VUT', a 3- or 4-day working week, extra holidays, and so on, together with the introduction and improvement of the employer's social and personnel policy, with respect to work and aging.

Social policy of personnel management

The third field of great importance is the social policy at the enterprise level, the sphere of personnel management in the broadest sense. Good-quality jobs and work environment are conditions *sine qua non* for equal and adequate opportunities for elderly workers to stay active, motivated and healthy. Not only the personnel department but management as a whole is responsible for this social policy. Through the Dutch legislation on workers' participation in medium-sized and large firms there is also a role for work councils to promote and monitor such active policy. To attain that goal there are several instruments available, such as the social year report that enterprises should draw up annually, which give work councils the opportunity to monitor the social policy of enterprises. The key formula for a good social policy towards work and aging is an age-dynamic personnel management comprising two aspects, which are interdependent and indispensable.

(1) A general personnel management and social policy that takes into account the qualities and restrictions of workers, which can vary according to age-specific factors. The Dutch Health and Safety legislation, the Arbo-legislation, instruct employers to take into account such individual factors.

(2) Age-dynamic personnel management demands for a specific policy towards senior workers. There is a tension between these two aspects, in that the senior policy should not be an excuse for a shortage of good general social policy, in which all ages are integrated.

Not only age is such a distinctive factor. Gender, colour, and specific factors related to responsibilities in the private sphere (especially in the case of young parents) should also have the comprehensive attention of the management. Special policies for elderly workers are necessary where general and preventive policies are insufficient or inadequate. As for the age-specific measures, there should be a carefully balanced approach to avoid age discrimination. Voluntariness and opportunities for real choices are crucial in this respect. This fits into a modern approach of collective arrangements, based on solidarity, but with individual tailoring, dependent on individually distinctive circumstances and preferences. What is seen to be discriminatory by one person can be felt as a very desirable privilege by another.

There is such a debate in The Netherlands at present about a regulation concerning dismissal in the case of large reorganizations or mergers. Employers and trade unions negotiating about the social consequences of such radical personnel reductions for years had the opportunity to make use of an exception to a public regulation on dismissals to lay off a disproportionate number of elderly workers—who through these negotiations (and partially through general unemployment schemes), would have a reasonable income guarantee in the future. The Dutch Minister of Labour suddenly announced the withdrawal of this regulation, argueing that a social policy in which elderly workers could stay integrated in the labour process was preferable.

The Dutch Federation of Trade Unions protested against this withdrawal at such short notice, but did not disagree that a good social policy is preferable. The measure supports our approach, which is being worked into the recommendations of the central employers and trade union organizations. As a Dutch proverb says, you must not throw away your old shoes before you have good fine new ones. For the present generation of senior workers, such an age-dynamic social policy has not been at hand. You can't fill such a gap from one moment to another, especially not in the difficult climate of lay-offs and reorganizations. In these latter cases, voluntariness should be the solution. For, as well as the opposite, it is unacceptable that elderly colleagues should be pressed to disappear, to save jobs for the younger ones. When they prefer staying, they must have full opportunities to do that.

In conclusion, this contribution has given an impression of the FNV approach to the policy on work and aging in different fields and in different phases of employment. Some dilemmas were presented, occurring especially in this era of new orientations and priorities. The hope is that it will be clear that, as is the case with all special policies towards special target groups, if well done, the need for special measures diminishes. General preventive and age, gender, colour, etc., dynamic policies must include, integrate and take account of the distinctive needs of individual employees. And, as far as special measures are advisable, these should not be too general and obligatory to prevent them from coming into the danger zone of discrimination and restrictions for individual preferences.

The negative spiral of attitudes towards elderly workers should be broken. Elderly women and men should be treated as full-fledged participants in the labour market and working communities. This will not be reached through ruthless demolition of won privileges—as employer organizations and, in their footsteps, the Dutch Ministry of Labour advocate—but through carefully building up new attitudes and policies.

With respect to this future, I prefer to be optimistic and constructive, and not to worry too much about what the situation will be 'when I'm sixty-four'.

I. van den Burg,
Dutch Federation of Trade Unions

Part I
Aging, work and health

Introduction

Low birth rates in combination with a longer life expectancy today change the build-up of our population. Since 1950, the population in The Netherlands has increased from 10 to 15 million, but the percentage of older males and females (45–64 years) increased by 60 per cent, while the percentage increase in younger males and females was only 45 per cent. Nowadays, there are considerably more older than younger workers in comparison with 1950. In the year 2010, 20 per cent of the Dutch population will be more than 60 years old and will not be part of the work-force.

Recently, life expectancy of Dutch people was estimated to increase between 1950 and 2010 by about 5 years in the average male and by about 8 years in the average female (Ruwaard and Kramers 1993). Although males will reach a mean age of 75 years and females of 81.5 years, and both will live considerably longer, it is not to be expected that they shall also live longer in good health. Recent epidemiological data estimate that the total number of years that we live in good health for both sexes is 60 years. This means that men suffer from chronic diseases for as long as 15 years and females even for 20 years. The general trend that health care elongates our lives, but is not able to give us more years in good health, has an important impact on the future: a smaller proportion of our population has to work for an increasing proportion of people on pension or with work disability owing to chronic diseases.

Therefore, there is a great need for strategies that can change this trend by diminishing the costs for health care and reduce the economic work power of the population. In this context, it is important to know to what extent work plays a role in the cause of the chronic diseases and health problems in elderly workers and retired people. To be able to develop efficient preventive measures, it is necessary to know the causes of the increase of chronic diseases.

People spend a considerable part of their total life at work and, for the majority, work is also important for their well-being. However, the physical load that goes along with daily work represents a potential harmful component of health. Physical load has to be in balance with the physical capacity of the worker. Both underloading and overloading deteriorate the worker's health. In the case of underloading (hypokinesia), chronic degenerative diseases such as cardiovascular, pulmonary diseases and some types of cancer may occur. Overloading (hyperkinesia) increases the incidence of musculoskeletal injuries (back pain, overuse injuries).

In view of this imbalance of workload and physical capacity of the elderly worker, the authors of Part I evaluate the current research on the interaction between age and workload from the perspective of health.

In the first paper, Kemper describes the aging process of physiological capacities, showing that from age 25 onwards these capacities decline with an average of 1 per cent per annum. This results in elderly workers that have only 60 per cent of their young adult capacity left at the age of 60. Nevertheless, in most workplaces the same work has to be done at both ages.

In Chapter 6, Ilmarinen poses the question of why the combination of growing old and working is so difficult to fit. Aging and work are combined with lifestyle and disease. A solution to the problem of the interaction between age and work is to prevent the decline in physical capacity and adjust the work to the physical capacities. With a work ability index the worker's ability related to work demands was evaluated against clinical assessments of health status in a large cross-sectional sample of Finnish workers. Finally, a description is given of the so-called 'FinnAge programme', which is aimed at redesigning working conditions and at health promotion of aging workers.

Teiger, in Chapter 7, presents an interdisciplinary approach to the problem of aging and work. She discusses three questions in this respect: (1) aging with regard to work, (2) aging through work, and (3) aging in work. The importance of starting with ergonomic measures, training and retirement schemes is underlined.

The question of the impact of working conditions on the development of age-related and/or age-dependent diseases was investigated by Cassou and Derriennic, Chapter 8, in two prospective longitudinal surveys in France: a study in retired subjects on the relation of job mobility and heavy physical work with their health evaluation, and a cross-sectional result in an active population about the relation between age and health-related symptoms such as low back pain, carrying heavy loads and decision latitude in the job.

Goedhard's contribution deals with the possible relationship between aging, the baroreflex function and job stress in a group of health workers between 20 and 64 years.

The last chapter is a review of the effects of shiftwork in elderly workers. De Zwart and Meijman hypothesize that shiftworkers suffer more from health complaints and sleep disturbances and this worsened health will be more prevalent in older than in younger shiftworkers. In this epoch, more and more work is organized over 24 hours. In the European Union, 17.6 per cent of all workers work regularly during the night, although there is a considerable variation over countries in this respect (the highest in Ireland with 32.6 per cent and the lowest in Denmark with 9.8 per cent).

It is self-evident and very important to know more of the (self)selection and the adjustments of the aging workers and the consequences of shiftwork for health.

Reference

Ruwaard, D. and Kramers, P. G. N. (eds). (1993) *Public health, future expectancies, health status of the Dutch population in the period 1950–2010* (in Dutch), Rijks Instituut voor Volksgezondheid en Milieuhygiene (RIVM), Sdu, The Hague.

5

Physical work and the Physiological consequences for the aging worker

H. C. G. Kemper

Summary. In this chapter the age-related changes of different physiological character-istics such as maximal aerobic power, maximal muscle force and skeletal mass, as found in the literature, are described. From age 25 onwards, physiological capacities lessen by an average of 1 per cent per year. Of great importance is the balance between work capacity and workload. It is suggested that an important measure against an age-related imbalance between these two factors is physical exercise and/or measures taken to adapt workload to the worker's capacity. The impact of physical training on several physiological systems is described. An illustration of this type of research is given by presenting results on the effects of taking ergonomic measures on the work capacity of refuse collectors.

Keywords: work capacity, work load, physical training ergonomic measures

Introduction

There is abundant evidence that aging makes elderly people weaker, slower, less powerful and also causes a deterioration in their balance and reaction time. Nevertheless, it is still difffficult to make a clear distinction between the effects of aging, deconditioning and disease with respect to decreased per-formance and health status. Since most of the aging studies have cross-sectional designs and therefore are prone to confounding, they cannot give a clear answer about the true effects of aging.

Taking this uncertainty for granted, elderly workers demonstrate lower physical capacity than younger workers. If physical load in the working situ-ation remains constant over the years and, at the same time, work capacity as a worker-related factor decreases, this can introduce an imbalance between

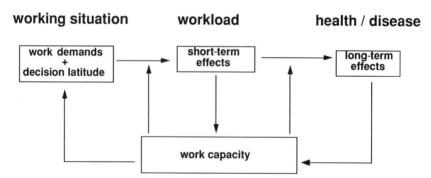

Figure 5.1. The workload model of Van Dijk et al. (1990) giving the interactions between the working situation (including work demands and decision latitude) and the work capacity of the worker in the short and long term.

workload and capacity and, in the long term, can lead to disease and inability to work. This workload model from van Dijk et al. (1990) is illustrated in figure 5.1.

In principle, two possibilities can prevent the imbalance between workload and the capacity of the worker: decreasing the physical load by ergonomic measures at the workplace or increasing or retarding the decrease in physical capacity of the workers by employee fitness programmes.

More important is the question of what are the beneficial effects of regular exercise on the aging process, not only as employee fitness training but also during free time in recreational sports. In recent years, evidence of the health benefits of exercise has been gathered from numerous training studies, although these studies have been continued only over relative short periods, that is weeks or months. However, the current, prevalent perception that strenuous exercise also has long-term beneficial effects on the work ability index and health is largely based on emotional reactions to exercise and wishful thinking (Holloszy 1983).

The scientific evidence that exercise promotes long-term benefits is surprisingly scarce and it is therefore necessary to design longitudinal studies in elderly workers to determine whether exercise can prevent, slow or reverse the aging process, and thus to have a beneficial impact on the well-being of the elderly worker.

In the last part of this chapter an example will be given of an investigation to measure workload and work capacity in refuse collectors, using ergonomic principles in order to create a healthy work situation for elderly workers.

Hypotheses on the role of exercise and life expectancy

Although vigorous exercise nowadays is being encouraged as an important component of a health maintenance programme, this was not the case 50 years ago. The negative attitude towards exercise was largely a consequence of the rate-of-living theory formulated by Pearl (1928) and primarily based on the work of Rubner (1908). This theory stated that the greater the rate of energy expenditure and oxygen utilization, the shorter the life-span (Sacher 1979). Subsequently, the stress hypothesis of Selye and Prioreschi (1960) had a further negative influence regarding the effects of exercise. It said that vigorous exercise represents a stress with long-term harmful effects similar to those of infections, trauma and nervous tension.

Also, the observations that males have a shorter life expectancy than females, whereas males are generally more active than females (Kemper et al. 1989), and that houseflies live longer the less energy they spend (Sohal and Buchan 1981) seemed to support the rate-of-living theory.

These days, these hypotheses are not generally thought any longer to have relevance to the long-term effects of regularly performed strenuous exercise. No evidence has accumulated to support the concept that increasing daily energy expenditure by daily physical activity causes the body to 'wear out' more rapidly or shortens the life-span (Holloszy 1983). On the contrary, it now seems well established that, in contrast to motorcars that wear out more rapidly the more they are used, the cells, tissues and organs of humans develop an adaptive increase in functional capacity and power in response to increased use, which runs counter to the deleterious changes that occur with age. While there is now much supportive evidence for the concept that 'if you don't use it, you will lose it' (in Dutch: 'rust roest' or 'to rest is to rust'), it does not necessarily follow that strenuous exercise has an effect on the aging process itself. Holloszy (1983) states that: (a) death is not usually caused by loss of functioning; and (b) although no evidence has accumulated in support of the rate-of-living theory, this does not necessarily mean that it is incorrect.

Information regarding the effect of exercise on the aging process is difficult to obtain, particularly in humans. Fitness testing in training programmes attracts a highly selective population: whereas in younger adults 50 per cent of a population may agree to participate, in middle and old age the proportions of volunteers can be as low as 10 per cent (Shephard and Sidney 1978).

The possible mechanisms of aging are reviewed by Birren and Schaie in their *Handbook of the Psychology of Aging* (1990). The general findings are that biological systems decline as the result of cell death (Shock 1977). This death results in a decline in capacity and functions, which correlates directly with the progressive loss of body tissue.

Physiological performance

Average changes in physiological performance over age

One of the most obvious manifestations of human aging is a decline in the ability to exercise and a failure to return to normal levels of functioning as quickly as the young (MacHeath 1984). The reduction in adaptability of general, physiological functions of men in our industrialized world, is indicated in figure 5.2 (Smith and Serfass 1981). The two lines indicate the differences between sedentary and active subjects.

The values exemplify average changes in men over time. In general, peak performance is reached at about age 30. After that time ('over the hill'), functional capacity declines gradually, varying between 0.75 and 1.0 per cent per year, depending on the individual and the organ system. An identification of the proportions of the aging curve attributable to genetics and to the environmental contribution to aging, such as regular physical activity, is not yet clear and requires continued research.

Whether regular exercise affects aging in humans is extremely difficult to determine, since humans are a long-lived species with large, genetically determined interindividual differences in longevity. It is also a practical impossibility to control for a wide range of dietary and other environmental factors that may affect longevity in a freely living aging population (Skinner et al. 1982).

Changes in the skeletal system

The support system of the body is the skeletal system, which loses bone at different rates depending on sex and individual factors. Bone mass increases

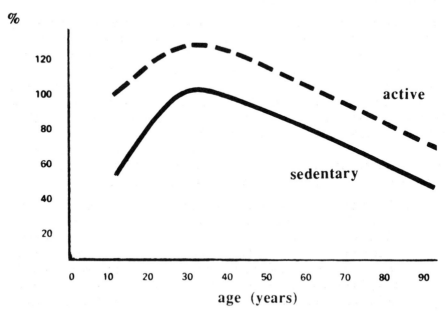

Figure 5.2. The percentage of general physiological performance over age. The solid line is the general change in sedentary humans and the broken line is hypothesized for active humans (after Smith and Serfass 1981).

during growth in both sexes and reaches peak values between 25 and 30 years of age. The average female and male starts to lose bone at age 35 to 40 and, after the menopause, it is faster in women because of lower oestrogen levels. By the time a man is 70 years old, he has lost 15 per cent of his bone and a woman 30 per cent.

Osteoporosis is a general and major bone mineral disorder in the older adult with an estimated prevalence of 10 per cent. This bone loss is characterized by a decreased bone mineral mass and an enlarged medullar cavity. Owing to the decreased mass and increased cell death, osteoporotic bone is weaker than normal bone. Osteoporosis is therefore often defined by the presence of fractures which occur spontaneously or as the result of only mild trauma. The fractures most commonly associated with osteoporosis are fractures of the wrist (radius, ulna) and hip (femoral neck and vertebral columns) (Arnold et al. 1966).

The higher the peak bone mass that can be reached in youth, the longer it takes to attain the critical bone mass at old age. Mechanical load of contracting skeletal muscles and gravity in weight-bearing activities, as well as nutrition (calcium and vitamin D intake), are environmental factors that are of importance for the aetiology of osteoporosis. In elderly people, preventive measures aimed at these factors can reduce the incidence of fractures.

Changes in the neuromuscular system

Moving is generally seen as including the continuous perception of signals by means of sensory organs, the processing of this information in the central nervous system, which may in turn lead to the generation of impulses in the motor neurones and to the contraction of muscles. With aging, this total system will function at an increasingly lower level caused by a decrease in functioning of the components; either as a consequence of the aging process or because of inactivity, diseases, undernutrition or a combination of these factors. The effects are, for instance, manifested as a lengthening of reaction time, an increase of performance execution time, especially of complex movements, and decreased motor control (Mortimer et al. 1982, Joseph 1988).

The aging nervous system demonstrates two major changes:

(a) Cells within the brain die at a constant rate, until about age 60. This cell death results in a decreased number of nerve cells in the brain. The exact functional consequences of this are not yet clear. The rate of cell death does not appear to have any effect on one's intellectual capacity or the decision-making process as one grows older. The only exception is that in the extremely old individual, decreased nervous tissue functioning is directly related to the blood flow to brain tissue.

(b) The nerve conduction velocity, or the speed at which the nerve carries messages from one point to another, is also decreased (Shock 1977), as is the impulse of firing rate of nerves. Both changes result in a slightly slower response to various environmental activities by older individuals.

One of the most clear phenomena with aging is the loss in muscle force. The data in figure 5.3 are cross-sectional and depict mean values of different muscle groups collected by Hettinger (1961) and published by Åstrand and Rodahl (1986).

It is interesting to note that the rate of decline with age is not the same for all muscle groups (Larson 1982). The force of the lower leg muscles, for instance, seems to decrease at a faster rate with age than that of the lower arm muscles. Probably, arm muscles are used more and more intensively than leg muscles at older age. This decrease in force is associated with a decrease im muscle mass. Young (1986) reports that the relation between muscle force and cross-sectional area is equal in younger subjects and in old females. However, he also reports that this is lower for old-aged male subjects. These last findings may be related to the fact that no use is made of what he calls 'age-matched normative data'; in other words, reliable reference values are missing.

The characteristic related to 'speed' of contraction is usually defined as the time needed to reach the peak of the developed force during a twitch contraction, mostly isometrically. This time needed to reach this peak is called time to peak tension (TPT). As can also be seen, it is characterized by the velocity of contraction. Several studies showed that TPT increases with increasing age (e.g. Newton and Yemm 1986, Davies et al. 1986, Petrella et al. 1989). In other words, the development of maximum force takes a longer time in the elderly

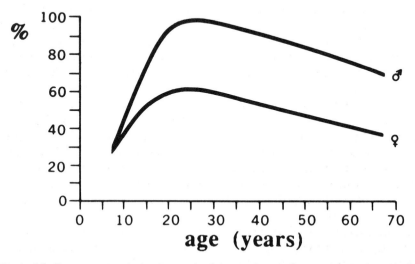

Figure 5.3. Percentage variation in maximal isometric muscle strength over age in both sexes taking the age of 25 years in males as the reference (100 per cent) (after Åstrand and Rodahl 1986).

compared with the young. The force of isokinetic contractions is shown to be lower at older age, especially at high velocities (Cunningham et al. 1987, Harries and Bassey 1990). These data imply that at the same force of contraction, older subjects develop less velocity than the young ones.

In this paragraph, aerobic endurance in muscle performance will be emphasized. This means that the force exerted is not maximal but in the order of 60–50 per cent. According to Larson (1982), isometric as well as isotonic endurance, expressed as percentage of the maximal force, are not affected by age or may even be slightly increased. If the absolute force is taken, it will be obvious that endurance will be decreased. However, Laforest et al. (1990) showed that dynamic fatigue did not differ between young and old subjects.

Changes in the energy-delivering system

Of the many changes that occur with aging, the degeneration of the cardio-respiratory system is the one that causes considerable morbidity and mortality. In the USA, at age 60 cardiovascular disease has an incidence (per 10^5) of 250 in females and 750 in males, while at age 70 the incidence is 700 and 1750, respectively. Cardiovascular disease is the most common cause of death in the elderly population (Bove 1990).

Despite advancing age, in resting conditions many of the physiological variables remain well regulated: thus, body temperature, blood glucose levels and blood volume are similar in elderly and young adults. Only resting oxygen intake ($\dot{V}O_2$) decreases slightly with age, which is related to loss of active tissue. The $\dot{V}O_2$ is a function of cardiac output and the rate of extraction of oxygen from the blood by working tissues. The decline of $\dot{V}O_2$ is much less than the decline of cardiac output; thus, the arterio–venous difference (a $-$ v O_2 difference) increases with age. Since cardiac output is the product of heart rate (HR) and stroke volume (SV) and resting heart rate is not greatly altered, the fall of cardiac output must reflect a decrease of stroke volume. Blood pressure increases owing to alterations in the compliance or stiffness of the aorta and other vessels (Safar 1990). On average, the systolic blood pressure (Psyst) increases between ages 30 and 70 by 10 to 15 mmHg respectively and diastolic blood pressure (Pdiast) by 5 to 10 mmHg. The increase of mean arterial pressure and the decrease of cardiac output indicate an increase of total peripheral resistance. This increase may be caused by: (a) a reduction in muscle mass; (b) increased vasoconstrictor sensitivity; (c) loss of microvascular channels in the peripheral vascular bed; (d) stiffening of the vessels; and (e) fatty deposits on the walls of the vessels.

During submaximal effort, the regulation of homeostasis is less efficient in the elderly than in the young: the rate of adjustment to an exercise stimulus and the subsequent recovery both being impaired. Thus the time required for heart rate (HR) and blood pressure to attain steady state is prolonged. Submaximal cardiac output is maintained with advancing age, since cardiac dilatation and increased SV compensate for the diminished HR. As a result of the increased arterial pressure, stroke work by the heart increases with aging (Lakatta 1985). The oxygen cost of a given submaximal workload remains the same (Shephard and Sidney 1978).

As for respiration, there is an age-related decline of respiratory function. Vital capacity decreases by 40–50 per cent between the ages of 30 and 70, while the lung residual volume increases by 40 per cent. Perhaps the most significant cause of the decline is the diminishing static elastic recoil force of the lungs and the resistance to deformation of the chest wall by a general disintegration of the supporting fibrous network (Reddan 1981). In the elderly, for a given amount of submaximal work, ventilation may be higher, because more energy is required to sustain any given level of ventilation (VE) and the ventilatory equivalent (VE/VO$_2$) is higher. The latter may be due to the fact that older people compensate for the lower gas exchange efficiency (Davies 1972).

The best single indication of physical working capacity is the maximal oxygen uptake ($\dot{V}O_2$ max). Which is the main factor accounting for the limiting maximal performance or $\dot{V}O_2$ max is the subject of considerable controversy (Rowell 1974, Wagner et al. 1986). So far, in young, healthy subjects, it is unknown which part of the oxygen transporting system is the weakest: heart, lung or muscle metabolism. In sedentary older people effort is often limited by other than circulatory factors, such as motivation, electrocardiographic abnormalities, fear of overexertion or muscle weakness.

Notwithstanding the fact that there may be a larger error in the prediction of $\dot{V}O_2$ max, the research literature shows a mixture of directly measured and predictive tests. The $\dot{V}O_2$ max of adult men and women declines progressively with age: this has been documented by both cross-sectional and longitudinal studies. At age 65 $\dot{V}O_2$ max is reduced to 70 per cent of the average value at age 25 in both males and females (Åstrand and Rodahl 1986).

Data from cross-sectional studies suggest a decline in adulthood in $\dot{V}O_2$ max at a rate of 0.40–0.45 ml/kg min. per year in males and 0.30 ml/kg min. per year in females. The rate of loss tends to be larger in sedentary compared with active individuals. This age-related decline in $\dot{V}O_2$ max is inevitable and appears to be caused by the decline in maximal cardiac output. The maximal cardiac output, which is the product of two components, stroke volume and heart rate, declines by about 30 per cent between ages 30 and 70. The reduced maximal heart rate is a consistent finding, and is ascribed to decreased end-organ sensitivity to catecholamines. To offset the decline in maximal heart rate, maximal stroke volume may increase. As for the maximal arterio–venous oxygen difference, a decline in the elderly is found, but not always: a decline in muscle mass and a lower capillary/fibre ratio would contribute to lower a − vO$_2$ differences, but a reduced cardiac output to a greater a − vO$_2$ difference. Pulmonary function does not appear to limit $\dot{V}O_2$ max as in young adults, although in elderly people this function may be less efficient, while ventilating during strenuous exercise.

Effects of exercise training in the elderly

If one increases the amount, intensity and/or frequency of regular exercise in a systematic way, this induces structural and functional adaptation processes in the individual. This overload principle implies a common process. The kind of exercise, however, determines the effects that are specific for each organ: e.g. strength training, characterized by high intensive, low repetitive exercises improves muscle contraction (i.e. strength and speed) and endurance training, characterized by low intensive, high repetitive exercises, enhances the oxygen transport system (Kemper 1990).

Training effects on bones and joints

Training effects on bone density is investigated extensively by different methodologies:

—cross-sectional differences between active and inactive people (Talmage et al. 1986) and right to left comparisons within subjects (Montoye et al. 1980);
—effects of immobilization, such as bedrest or casting after injuries (Issekutz et al. 1966); and
—longitudinal intervention studies (Smith et al. 1989, Dalsky et al. 1988).

From the last studies, the conclusion is that physical exercise prevents bone loss and may even increase bone mineral content in elderly people when the training consists of weight-bearing activities with a frequency of at least three times a week over a period of 1–3 years. Also a number of changes occur in the joint components: the cartilage, tendons, ligaments and synovial fluid. Exercise can prevent loss of mobility and instability in the joints of elderly people (Chapman et al. 1972) and, although the activity patterns diminish in elderly people (MacHeath 1984), it is uncertain whether aging or disuse is the primary cause.

Training of the neuromuscular system

Studies have showed that training may improve the muscle functions' force and endurance, even in very aged subjects. Moritani (1981) found different training adaptations in the arm muscles of older men compared with young men. In the absence of any significant muscle hypertrophy, it was suggested that in the older subjects, the effect of muscle training may rest entirely on the neural factors which could be improved by training and thus result in higher levels of muscle activation. The importance of the central nervous system above muscular factors as such can also be concluded from the study of Stelmach et al. (1989) and Spirduso's (1982) statement that 'age-related psychomotor slowing is almost universally attributed to delays in central processing rather than in peripheral components'.

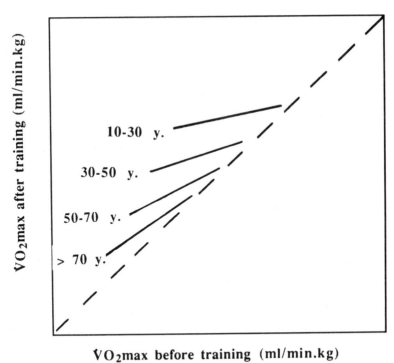

Figure 5.4. Trainability of $\dot{V}O_2$ max (ml/min/kg) in young and old populations.

Taking these points together, they suggest that for some motor abilities, the age-related decrease in muscular functions may be of more importance than for others. It also suggests that there is a critical level for the capacity of one or more motor functions below which a considerable decrease in motor performance is detected and above which the decrease is not easily detected. This level may depend on the specific motor abilities under study. Young (1986) demonstrated that the muscle force of leg extensors can be critical for the ability to stand up from a chair or to enable a subject to walk at a speed of 3.85 km/h across a Pelican crossing (Bassey et al. 1988).

Training of the cardiorespiratory system

Relatively few studies report directly measurement of the changes in $\dot{V}O_2$ max resulting from training in the elderly (Seals et al. 1984). Elderly males and females are capable of demonstrating a training effect in response to endurance training, regardless of previous physical activity patterns and current training status. Whether older individuals respond to physical training to the same degree as their younger counterparts is not certain: complicated matters such as initial fitness level and artificially low pre-test score levels caused by a lack of motivation and/or fear can create big changes in the post-test scores. Also, elderly people may not be capable of engaging in vigorous exercise early in training owing to orthopaedic limitations and unfamiliarity with excessive exertion (Stamford 1988).

Nevertheless, there is now convincing evidence that old people who train at a relatively high intensity can induce a substantial increase in their directly measured $\dot{V}O_2$ max over the first few months of conditioning (Shephard 1978). The differences in trainability of $\dot{V}O_2$ max appear to depend on at least two factors. The first factor is age: the older the people, the lower the training effect. The differences in trainability with age are summarized in figure 5.4 by giving the relation between the change in $\dot{V}O_2$ max and the initial pre-training level of different groups. The figure shows that trainability, indicated by the slope of the lines, of subjects >70 years and 50–70 years, is lower than that of younger subjects (10–30 and 30–50 years) at the same pre-training level.

The second factor is the pre-training level: the lower the pre-training level, the higher the training effect for a given age group. This is illustrated in figure 5.5 by nine training studies with elderly males and females to improve their $\dot{V}O_2$ max. In these studies, mean ages varied between 63 and 78 years and the training programmes were continued between 9 and 26 weeks. The mean

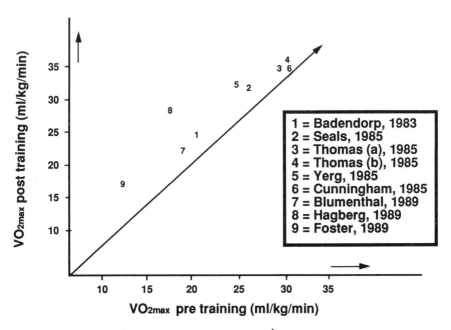

Figure 5.5. The relation between training effect on $\dot{V}O_2$ max and pre-training level of nine training studies with elderly people.

effects of each study (numbered from 1 to 9) are plotted as $\dot{V}O_2$ max per kilogram body mass with pre- and post-training values on the X and Y axis, respectively. All studies showed significant improvements of the initial $\dot{V}O_2$ max from 10 to 30 per cent. It is also clear that the greatest training effects are found in populations with the lowest pre-training levels.

The consequences for the aging worker

In the framework of Van Dijk's model (Van Dijk et al. 1990) (see figure 5.1), the physical load in the work situation, as a work-related factor, remains in general the same for each worker during his or her career. On the other hand, the physical capacity of the aging worker, as a worker-related factor, declines. This results at older ages in an imbalance between physical load and physical capacity, leaving less work reserve at older ages. The short-term effects reflect a higher physical strain: older workers use a higher percentage of their maximal muscle force, maximal oxygen uptake and/or the maximal compression force of their skeleton. In the long term this can induce physical adaptations that can be harmful to health and can cause work-related diseases and even work inability.

A key question is designing healthy work situations for elderly workers is: what percentage of physical capacity is acceptable as daily workload? In order to answer this question, the ratio between physical capacity and daily workload should be measured. This has to be done at the workplace itself or, when this is not possible, in a mock-up situation. Although guidelines have been constructed and based on physiological, biomechanical, psychological and epidemiological data, the scientific basis, however, remains poor.

If we restrict the answer to the key question, that is to the energetic aspects of physical workload, norms vary between 30 per cent (Åstrand and Rodahl 1986) and 50 per cent $\dot{V}O_2$ max (Ilmarinen 1992). The same uncertainty holds for the norms of biomechanical load on the musculoskeletal system: the National Institute of Occupational Safety and Health (NIOSH 1981) advised maximal acceptable limits of lifting loads of 40 kg or lower, depending on circumstances such as rate, distance and posture. Recently, this advice has been updated, taking also into consideration asymmetrical lifting situations, contact of hands with the load and trunk rotation. This has resulted in a recommended weight limit of 25 kg.

Prospective intervention studies that compare workers who do or do not surpass the limits are scarce. The long-term effects of overloading over 5 or more years are, to our knowledge, unknown. Moreover, each work situation is unique concerning the kind of physical work, the duration, the frequency and the recovery phases. This uniqueness complicates the application of general guidelines on specific work situations.

Apart from aging, sex and lifestyle are also factors that influence the work capacity of workers. A safe workplace can therefore only be attained when the ratio between workload and work capacity can be determined very specifically: a distinction has to be made between, for example, arm, leg or combined arm and leg work and between static or dynamic work. It is also important to measure as directly as possible. In the case of energetic load it is better to measure oxygen uptake rather than heart rate (Kemper et al. 1990). In the case of biomechanical load it is more precise to measure compression forces by objective recording methods than to estimate them from observational methods (De Looze 1992).

An application of these principles has been made in a study aimed at assessing the workload of three different collecting systems by refuse collectors in The Netherlands (Stassen et al. 1993). In refuse collectors a relatively high incidence of health complaints is found, resulting in frequently occurring sick leave and which, in the long run, will lead to a high percentage of permanent disability. Above the age of 45, hardly any refuse collector is able to continue his job because of health complaints. In a series of experiments, both energetic and biomechanical load were measured in refuse collectors working with metal dustbins (6 kg), polythene bags and later with small, two-wheeled (120 and 240 L) and large, four-wheeled containers (1100 L). The consequences of this workload for aging were studied by dividing the subjects into three age groups. Energetic load was measured directly in mock-up situations by continuous measurement of oxygen uptake and indirectly by whole day heart rate registration and individual calibration of the heart rate/oxygen uptake relationship. Biomechanical load was measured by means of observations at the workplace of all tasks during the whole working day and in the laboratory by measuring the net moments and forces acting on the lower back with a two-dimensional dynamic linked segment model (De Looze 1992). The results pointed out that by an ergonomic measure to replace metal containers with polythene bags, the same physical load could be doubled without influencing the oxygen uptake $\dot{V}O_2$ of the refuse collectors (Kemper et al. 1990). However, in both situations the ratio between physical load and physical capacity was well above the norm of 30 per cent of $\dot{V}O_2$ max during an 8-hour working day. The energetic load with large containers showed the lowest values compared with minicontainers and polythene bags.

These norms were also applied to the different age groups of refuse collectors: $\dot{V}O_2$ max declined with age from 3.8 L/min (< 30 years) to 3.6 (30–39 years) and 3.0 (> 40 years), respectively. As a consequence, the loading time for 90 per cent of the population of refuse collectors with polythene bags was only 4–5 hours/day in the youngest and 2–3 hours/day in the oldest group. In this few hours' time, refuse collectors older than 39 years are allowed to collect 4–5 tons of refuse.

Collection with polythene bags was also unacceptable, because of the overload on the lumbar spine. In almost all situations, the compression forces (peak load) of collecting with polythene bags surpass the NIOSH norm of

3400 N on the lumbar spine and, in addition, the frequency of lifting, carrying and throwing these bags was very high (on average 800 times per day). The combination of high peak loads and frequently submaximal loads (van Dieën and Toussiant 1992) makes this collection method unacceptable. Peak load and frequency were considerably lower with minicontainers and also with large containers, provided that the latter could be transported over flat surfaces and were operated by two collectors at the same time.

Therefore, the general recommendations from this study were:

(1) Replacing the polythene bags collecting method by the method with small and large wheeled containers.
(2) Diminishing the workload of elderly workers during their career, taking into account the age-dependent loss of their work capacity.
(3) To introduce a system with rotation of heavy and light tasks over the day (i.e. sweeping, driving and collecting) (Kuijer et al. 1993). Such a system appears to be the best solution for physically heavy tasks.

References

This chapter is partly based on 'Exercise and the physiological consequences of the aging process', in J. J. F. Schroots (ed.) *Aging Health and Competence* (Elsevier, Amsterdam), Chapter 6, 1993.

Arnold, J., Bartley, M., Tout, S. and Jenkins, D. 1966. Skeletal changes in aging and disease, *Clin. Orthop.*, **49**: 17–38.

Åstrand, P.-O. and Rodahl. K. 1986 *Textbook of Work Physiology, Physiological Bases of Exercise* (McGraw-Hill, New York).

Bassey, E. J., Bendall, M. J. and Pearson J. M. 1988. Muscle strength in the triceps surae and objectively measured customary walking activity in men and women over 65 years of age, *Clinical Science*, **74**: 85–89.

Birren, J. E. and Schaie, K. W. (eds) 1990. *Handbook of the Psychology of Aging* (Academic Press, New York).

Bove, A. A. 1990. Exercise and aging: effects on the cardiovascular system. In G. P. H. Hermans and W. C. Mosterd (eds), *Sports, Medicine and Health*, (Exerpta Medica, Amsterdam), 107–118.

Chapman, E. A., de Vries, H. A. and Swezey, K. 1972. Joint stiffness: effects of exercise on young and old men, *J. Gerontol.*, **27**: 218–221.

Cunningham, D. A., Morrison, D. Rice, C. L. and Cooke C. 1987. Aging and isokinetic plantar flexion, *Eur. J. Appl. Physiol.*, **56**: 24–29.

Dalsky, G. P., Stocke, K. S., Ehsani, A. A., Slatopresky, E., Lee, W. C. and Birge, S. J. 1988. Weight bearing exercise training and lumbar bone mineral content in postmenopausal women, *Ann. of Int. Med.*, **108**(8): 824–828.

Davies, C. T. M., 1972. The oxygen transporting system in relation to age, *Clin. Sci.*, **42**: 1–13.

Davies, C. T. M., Thomas, D. O. and White, M. J. 1986. Mechanical properties of young and elderly human muscle. In P.-O. Åstrand and G. Grimby (eds), *Physical Activity in Health and Disease* (Almquist and Wiksell International, Stockholm), 219–226.

Dieën, J. H. van and Toussaint, H. M. 1992. Prediction of vertebral endplate fratures in different loading protocols. In *Scientific Conference on prevention of work-related musculo-skeletal disorders* (Premus, Solna, Sweden).

Dijk, F. J. H. van, Dormolen, M., Kompier, M. A. J. and Meijman, T. E. 1990. Herwaarderingsmodel belasting belastbaarheid, *T. Soc. Gezondheidszorg*, **68**: 3–10.

Grimby, G. and Saltin, B. 1971. Physiological effects of physical training, *Scand. J. Rehab. Med.*, **3**: 6–19.

Harries, U. J. and Bassey, E. J. 1990. Torque–velocity relationship for the knee extensors in women in their 3rd and 7th decades, *Eur. J. Appl. Physiol.*, **60**: 187–190.

Hettinger, T. H. 1961. *Physiology of Strength* (Thomas, Springfield, IL).

Holloszy, J. O. 1983. Exercise health and aging: a need for more information, *Med. Sci. Sports Exercise*, **15**(1): 1–5.

Hollweg, R. A. T. 1992. De trainbaarheid van de maximale zuurstofopname bij oudere populaties. Scriptie vakgroep Gezondheidkunde, Vrije Universiteit Amsterdam.

Ilmarinen, J. 1992. Job design for the aged with regard to decline in their maximal aerobic capacity, part II: the scientific basis for the guide, *Int. J. Ind. Ergonomics*, **10**: 65–77.

Issekutz, B., Blizzard, J. J., Birkhead, N. C. and Rodahl, K. 1966. Effect of prolonged bedrest on urinary calcium output, *J. Appl. Physiol.*, **21**: 1013–1020.

Joseph, J. A. (ed.) 1988. Central determinants of age-related declineness in motor function, *Ann. N. Y. Acad. Sci.*, **515**: 1–429.

Kemper, H. C. G., Verschuur, R. and de Mey, L. 1989. Longitudinal changes of aerobic fitness in youth ages 12 to 23, *Ped. Ex. Sci.*, **1**: 257–270.

Kemper, H. C. G. 1990. Exercise and training in childhood and adolescence. In R. P. Welsh and R. J. Shephard (eds) *Therapy in Sports Medicine 2* (Decker, Toronto), 11–18.

Kemper, H. C. G., Aalst, R., Leeghwater, A. Maas, S. and Knibbe, J. J. 1990. The Physical and physiological workload of refuse collectors, *Ergonomics*, **33**(12): 1471–1486.

Kuijer, P. P. F. M., Visser, B. and Kemper, H. C. G. 1993 Effect van taakroulatie op de fysieke belasting bij reinigers/beladers/chauffeurs, Ergocare, Amsterdam.

Laforest, S., St-Pierre, D. M. M., Cyr, J. and Gayton, D. 1990. Effects of age and regular exercise on muscle strength and endurance, *Eur. J. Appl. Physiol.*, **60**: 104–111.

Lakatta, E. G. 1985. Age-related changes in the heart, *Geriatric Medicine Today*, **4**: 90–97.

Larson, L. 1982. Aging in mammalian skeletal muscle. In F. J. Pirrozolo and G. J. Maletto (eds). *The Aging Motor System* (Praeger, New York), 60–97.

Looze, M. P. de 1992. Mechanics and energetics of repetitive lifting. Thesis Vrije Universiteit Amsterdam, Copyprint 2000 Enschede.

MacHeath, J. A. 1984. *Activity, Health and Fitness in Old Age* (Croom Helm, London).

Montoye, H. J., Smith, E. L., Fardon, D. F. and Howley, E. T. 1980. Bone mineral in senior tennis players, *Scand. J. Sports Sci.*, **2**: 26–32.

Moritani, T. 1981. Training adaptations in the muscles of older men. In E. L. Smith and R. C. Serfass (eds) *Exercise and aging* (Enslow, Hillside, NJ), 149–166.

Mortimer, J. A., Pirozzolo, F. J. and Maletta, G. J. (eds) 1982. *The Aging Motor System* (Praeger, New York).

Newton, J. P. and Yemm, R. 1986. Changes in the contractile properties of the human first dorsal interosseus muscle with age, *Gerontology*, **32**: 98–104.

NIOSH. 1981. *Work Practices Guide for Manual Lifting* (National Institute for Occupational Safety and Health: Taft Industrie, Cincinatti, 81–227.

Pearl, R. 1928. *The Rate of Living* (Alfred Knopf, New York).

Petrella, R. J., Cunningham, D. A, Vandervoort, A. A. and Paterson, D. H. 1989. Comparison of twitch potentiation in the gastrocnemius of young and elderly men, *Eur. J. Appl. Physiol.*, **58**: 395–399.

Reddan, W. G. 1981. Respiratory system and aging. In E. C. Smith and R. C. Serfass (eds) *Exercise and Aging* (Enslow, Hillside, NJ), 89–107.

Rowell, L. B. 1974. Human cardiovascular adjustments of exercise and thermal stress, *Physiol. Rev.*, **54**: 75–112.

Rubner, M. 1908. *Das Problem der Lebensdauer und seine Beziehungen zur Wachstum und Ernährung* (Oldenbourg, München).

Sacher, G. A. 1979. Theory in gerontology, part 1. *Ann. Rev. Gerontol. Geriatr.*, **1**: 3–25

Safar, M. 1990. Aging and its effects on the cardiovascular system, *Drugs*, **39** (suppl.): 1–18.

Saltin, B. 1986. Physiological adaptation to physical conditioning. In P.-O. Åstrand and G. Grimby (eds) *Physical Activity in Health and Disease, Acta Med. Scand. Suppl.*, **711**: 11–25: (Almquist, Wiksell, Stockholm).

Seals, D. R., Hagberg, J. M., Hurley, B. F., Ehsani, A. A. and Holloszy, J. O. 1984. Endurance training in older men and women, *J. Appl. Physiol.*, **57**: 1024–1029.

Selye, H. and Prioreschi, P. 1960. Stress theory of aging. In N. W. Shock (ed.) *Aging, some Social and Biological Aspects, Am. Ass. Adv. of Sci.* (Washington), 261.

Shephard, R. J. 1978. *Physical Activity and Aging* (Croom Helm, London).

Shephard, R. J. and Sidney, K. H. 1978. Exercise and aging, *Exerc. Sport Sci. Reviews*, **6**: 1–59.

Shock, N. W. 1977. System integration. In C. Finck and L. Hayflick (eds) *Handbook of the Biology of Aging* (Van Nostrand Reinhold, New York), 634–655.

Skinner, J. S., Tipton, Ch. M. and Vailas, A. C. 1982, Exercise, physical training, and the aging process. In A. Viidik (ed.) *Lectures on Gerontology. Vol: On biology of aging. Part B* (Academic Press, New York), 407–439.

Smith, E. L., Gilligan, C., McAdam, M., Ensign, C. P. and Smith, P. E. 1989. Deterring bone loss by exercise intervention in pre-menopausal and post-menopausal women, *Calcif. Tissue Int.*, **44**: 312–321.

Smith, E. L. and Serfass, R. C. 1981 *Exercise and Aging, the Scientific Basis* (Enslow, Hillside).

Sohal, R. S. and Buchan, P. B. 1981. Relationship between fluorescent age pigment physiological age and physical activity in the housefly, musca domestica, *Mech. of Aging and Development*, **15**: 243–249.

Spirduso, W. W. 1982. Physical fitness in relation to motor aging. In J. A. Mortimer, F. J. Pirozzolo and G. J. Maletta (eds) *The aging motor system* (Praeger, New York), 120–151.

Stamford, B. A. 1988. Exercise and the elderly, *Exerc. Sport Sc. Rev.*, **16**: 341–379.

Stassen, A. R. A., Markslag, A. M. T., Frings-Dresen, M., Kemper, H. C. G., de Looze, M. P. and Toussaint, H. M. 1993. Arbeidsbelasting van huisvuilbeladers bij reinigingsdiensten (SDU uitg, Den Haag).

Stelmach, G. E., Philips, J., DiFabio, R. P. and Taesdale, N. 1989. Age, functional postural reflexes, and voluntary sway, *J. Gerontology*, **44**: B100–B106.

Talmage, R. V., Stinnett, S. S., Landwehr, J. T., Vincent, L. M. and McCartney, W. H. 1986. Age-related loss of bone mineral density in non-athletic and athletic women, *Bone and Mineral*, **1**: 115–125.

Wagner, P. D., Hoppeler, H. and Saltin, B. 1991. Determinants of maximal oxygen uptake. In R. G. Crystal and J. B. West (eds) *The Lung* (Raven Press, New York), 1585–1593.

Young, A. 1986. Exercise physiology in geriatric patients. In P.-O. Åstrand and G. Grimby (eds) *Physical Activity in Health and Disease* (Almquist and Wiksell International, Stockholm), 227–232.

6

Aging, work and health

Juhani Ilmarinen

Summary. Essentially, the general problem of aging and working is the imbalance of functional capacities and work demands. Evidence points out that the critical period of this imbalance lies around 60 years of age. Three groups of risk factors that may add to this imbalance have been identified: physical demands, environmental factors and factors related to work organization. A crucial point for assessment of the imminent imbalance is the way work capacity is measured. The Work Ability Index has been developed to measure work capacity. The newly started FinnAge-programme has two aims: redesign of work and promoting the worker's health and functional capacity. In order to attain these goals the programme is divided into three parts: research, which is in progress, and training and education. The last two parts are in preparation.

Keywords: aging, health, functional capacities, work ability index, work demands

Introduction

Most European countries are facing an increase in the aging work-force. The problems are, in principle, the same everywhere: as aging advances, health, working capacity and physical condition deteriorate. The trend is that older workers are leaving the labour force, on average, well before their official retirement age. The high rates of unemployment can encourage older workers to retire early and to make room for the young unemployed. However, it is not so simple to replace an older experienced worker by a young one. Also, those jobs where older workers have been laid off are not necessarily suitable for younger workers.

A key question is why growing old and working is becoming so difficult to fit. We are living longer than ever before and working life should be better today than it was 50 years ago. There is probably no simple answer to it, but a very complex set of reasons, which are affecting decisions at individual, enterprise and governmental levels. Positive examples, however, are now urgently needed. Aging and work can be fitted successfully. As stated recently in *The Lancet* (1993), we should not dwell on the problems of growing older, but should, rather, emphasize the benefits of aging.

Interactions

Aging and work are combined with health and lifestyle (figure 6.1). Lifestyle, work and aging influence the severity of disease. In the absence of disease, lifestyle and work affect indirectly the rate of aging. The presence of disease modulates lifestyle and work, and may influence aging as well, and both disease and aging modify lifestyle. The comprehensive model in figure 6.1 emphasizes that action to help older workers should be multifactorial. The model also suggests that in research on older workers a multidisciplinary approach is valuable. The interaction of work and aging is least known, and in such research lifestyle and disease should be controlled.

Problem and solution

In principle, the problem between aging and work is the conflict between functional capacities and work demands. Aging is often combined with a decline of functional capacity. On the other hand, work demands do not systematically decline with age; rather, the contrary. With more experience, the demands of work, at least in mentally demanding jobs, tend to increase with age.

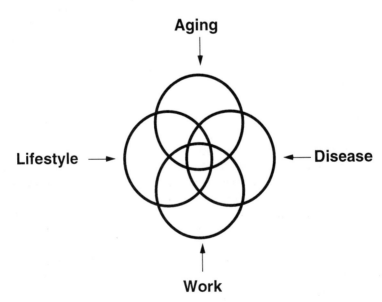

Figure 6.1. Interaction of lifestyle, disease, work and true biological aging.

Problem

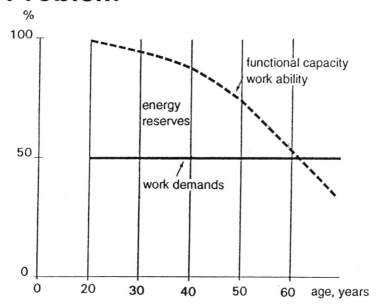

Figure 6.2. Relation between functional capacity and work demands with aging.

The problem is illustrated in figure 6.2. The curves of capacity and work demands are crossing with advancing age. The critical age for crossing, however, is individual and demonstrates the large variability of older workers with regard to their physical, mental and social capacities for work. Workforce participation rates of older workers, however, are getting lower in many countries and indicate that the critical period is stabilizing between the ages of 50 and 60 years (WHO 1993). The problem arises earlier in physically rather than in mentally demanding jobs, because physical work capacity starts to decline after the age of 45 years.

The solution to the problem requires two actions to be taken. The first is to prevent the premature decline of the capacities needed for work. This can be done, for example, by physical exercise, if physical work capacity needs to be kept at a high level. It should be noticed, as indicated in figure 6.3, that an age-related decline in physical work capacity is, in all circumstances, normal, but an accelerated decline should be prevented. The second action includes measures for decreasing or adjusting work demands. This can be done in several ways and job redesign is often needed. In physically demanding jobs, less lifting and carrying per shift among older workers can be compensated, for example by using their skills and experience more for training younger workers or for developing work processes. Also, new work schedules can be effective for adjusting the work demands for older workers. The model given in figure 6.3 should be tailored to company level. According to this model,

Juhani Ilmarinen

Solution

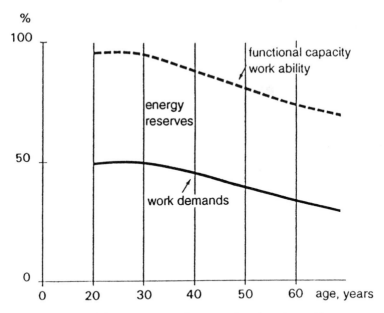

Figure 6.3. New concept for supporting the aging worker.

older workers at the age of 60 years are capable of coping with working life as well as the younger workers at the age of 30 years. Individual and flexible solutions, however, are needed and their economic benefits should be clarified in the long run.

Health

Aging is associated with an increase in the prevalence and incidence rates of diseases (Verbrugge 1984, Pisa and Uemura 1989, Aromaa et al. 1989). Although there is a lack of comprehensive morbidity statistics to determine morbidity trends in various populations, it is obvious that between one-third and two-thirds of workers aged 50 years and older have at least one diagnosed disease, mainly musculoskeletal or cardiovascular disease. So, the aging work-force can be characterized by an increase in chronic diseases. It should be pointed out, however, that such a trend is typical in all populations and is associated with the aging process.

The increase in prevalence rates of diseases with age can be seen both among men and women, as well as in different jobs. The increase of prevalence rates of musculoskeletal and cardiovascular diseases has been illustrated in a 4-year follow-up study covering three work categories in Finland (Tuomi et al. 1991). Among the men, the prevalence rate of musculoskeletal diseases

increased from 32 per cent to 45 per cent between the ages of 51 and 55 years.) ⌄
Among men doing physical work, 50 per cent of workers had at least one
diagnosed musculoskeletal disease at the age of 55 years. Also, men doing
mixed work (physically and mentally loading jobs, such as in transport work)
had high disease rates. It should be noted that musculoskeletal disorders were
not uncommon in mentally demanding jobs (nearly 40 per cent of men) (figure
6.4).

Among the women, similar figures were found: in physically demanding
jobs, however, the prevalence rates of musculoskeletal disorders increased up
to 60 per cent at the age of 55 years (figure 6.5).

The prevalence rates of cardiovascular diseases were at a lower level than
the rates of musculoskeletal diseases both at the age of 51 and 55 years among
women and men. The rates were highest in physically demanding jobs, but
(cardiovascular diseases, especially hypertension, were prevalent in one-quarter ⋀
of men and women doing mental work and aged 55 years (Tuomi et al. 1991).) ⌄

It is obvious that a disease does not have to be an obstacle for continuing in
working life. Moreover, the presence of diseases should be considered as a
criterion for taking action on work redesign.

If the work organization has been constructed for young, healthy men and
women, older workers who experience a minor negative impact of lower

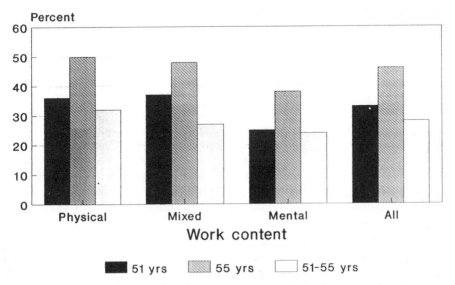

Morbidity
Musculoskel./ Men

Figure 6.4. Prevalence rates and incidence rate of musculoskeletal diseases of men by work content.

Morbidity
Musculoskel./Women

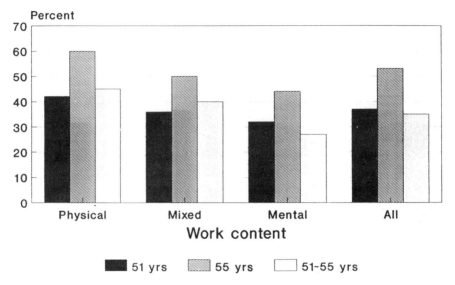

Figure 6.5. Prevalence rates and incidence rate of musculoskeletal diseases of women by work content.

health status on working capacity will be discriminated against. Therefore, one of the main tasks for solving the problem of the aging work-force is that working should also be adapted for the majority of older workers, who have a lower health status than the younger co-workers.

Functional capacities

Human functional capacities change with age. A schematic diagram of an individual's physical, mental and social capacity illustrates that different capacities have simple and complex levels (Heikkinen et al. 1984). For example, basic physical capacities are needed for activities in daily life, but unrestricted physical capacity is needed for the higher demands of physical work.

Surprisingly, physical capacities can start to decline earlier and faster than expected. Cardiovascular capacity ($\dot{V}O_2$ max) has been shown to decline by 20–25 per cent during the four years after the age of 45 years (Ilmarinen et al. 1991c). The rate of decline of $\dot{V}O_2$ max is individual, as can be seen from figures 6.6 and 6.7 (see left panels of the figures). The fast rate of decline is explained mainly by the lack of physical exercise. On the other hand, the effects of physical exercise on $\dot{V}O_2$ max have been demonstrated among

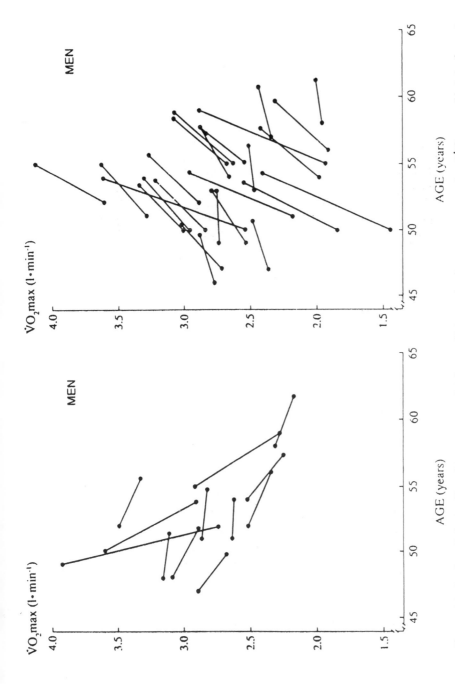

Figure 6.6. Decline (left panel) and improvement (right panel) in maximal oxygen consumption ($\dot{V}O_2$ max 1/min) of men in the four years after the age of 45 years.

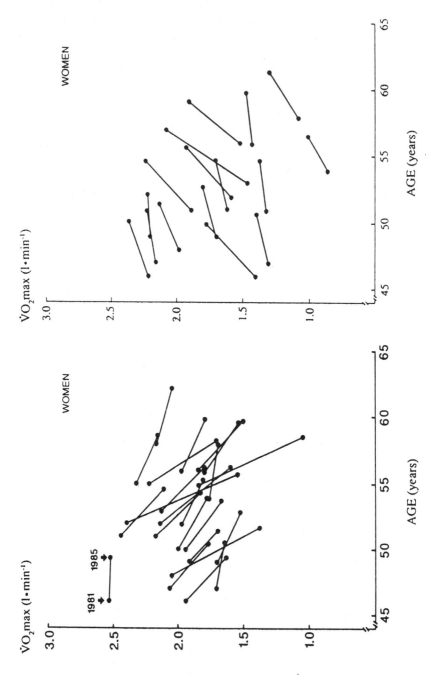

Figure 6.7. Decline (left panel) and improvement (right panel) in maximal oxygen consumption ($\dot{V}O_2$ max, 1/min) of women in the four years after the age of 45 years.

workers over 45 years of age and working in the respective occupations (see figures 6.6 and 6.7, right panel). The rate of increase in $\dot{V}O_2$ max was related to the type and intensity of physical exercise. Women exercised by regular walking and men used more intensive aerobic activities. The results indicate that without regular physical activities cardiovascular capacity can decline remarkably between 45 and 60 years of age, but this decline can be fully compensated for or prevented, for example by regular walking (Ilmarinen et al. 1991c).

The changes in musculoskeletal capacity between 51 and 55 years of age illustrate a similar trend (Nygård et al. 1991b). Among men, the strength of trunk flexion decreased by 22 per cent in four years. The decrease was about the same in physically, mixed or mentally demanding occupations (table 6.1).

Among women, the disability index, based on 11 tests, declined by 26 per cent on average, over four years. The decline was as high in physically as in mentally demanding occupations, but those doing mixed work, like nurses, showed a more modest decline (table 6.2).

Because both cardiovascular and musculoskeletal capacities tend to decline earlier in life and by a faster rate than would be desirable, special attention should be given to those older workers exposed to physical demands during daily routines. The premature decline of physical capacities can be prevented

Table 6.1. Musculoskeletal capacity changes between 51–55 years. %

Men	N	Disability index	Trunk flexion	Hand grip
Physical	17	+14	−21***	−9
Mixed	12	−9	−26***	+2
Mental	10	−10	−20**	−3
All	39	+1	−22***	−4

** p < .01, *** p < .001

Table 6.2. Musculoskeletal capacity changes between 51–55 years. %

Women	N	Disability index	Trunk flexion	Hand grip
Physical	11	−30	−14	5
Mixed	9	−11	−12*	+8
Mental	24	−29*	−6	−7*
All	44	−26*	−9*	−4

* p < .05

by exercise. On the other hand, the decrease of physical load with age is justified for several reasons: (a) a normal age-related decline of capacities should be acceptable, (b) musculoskeletal disorders increase with age, and (c) workload itself has no training effect on physical capacities.

It should be emphasized that the differences in physical work capacity between men and women are significant throughout the work career. In consequence, the physical load should be in absolute units about 30 per cent and in relative units (related to body weight) about 20 per cent lower for women than for men of the same age (Ilmarinen 1992).

Mental capacity includes simple and complex functions; among them cognitive processes, socio-emotional processes and creative behaviour, which are needed increasingly in working life. Fortunately, mental capacity does not show a similar premature decline to physical capacities. It does not mean, however, that we should overlook the role of mental capacities among aging workers. Our experiences have shown that although the older workers doing mentally demanding work have higher scores in short-term memory than those doing mixed or physical work, the change over four years for different work groups is about the same. In complex short-term memory the decline was about 6 per cent both in mentally and physically demanding jobs (table 6.3). The test score of a simple, short-term memory test indicated a 6 per cent decline in mental working occupations whereas no decline, rather an increase, was found in a physically working group over a period of four years (Suvanto et al. 1991).

The results suggest that doing mentally demanding work and loading the memory pattern at work do not prevent the possible decline of short-term memory with age. Perhaps a one-sided and continuous loading of memory pattern in several occupations is not optimal for maintaining such a function unchanged during aging. Our findings from teachers and nurses support the loading theory. Another example was found among the bus drivers. In a 4-year follow-up, visual speed, a function which is needed continuously for daily driving in city areas, decreased the most among bus drivers between 51 and 55 years of age (Suvanto et al. 1991). Before a definite conclusion can be

Table 6.3. Short-term memory

Work	N	Short-term memory		Complex short-term memory	
		Age 51 yrs score	Age 55 yrs %-change	Age 51 yrs score	Age 55 yrs %-change
Physical	28	5.1	+10*	4.1	−7
Mixed	21	5.6	−1	4.3	−6
Mental	34	6.3	−6*	5.1	−6
All	83	57	+1	4.5	−6

*p < 0.05

drawn, the data from a 10-year follow-up of the same subjects need to be analysed.

Generally, functional capacities are the basis of working capacity. Independently of occupation, some physical, mental and social capacities for daily work are needed. The quantity and quality of functional capacities needed are determined by specific work demands. It is obvious that there are important interactions between the different capacities. In other words, a remarkable decline in physical capacities affects also the mental and social capacities. On the other hand, a remarkable improvement in physical capacities can also improve cognitive abilities in older workers (Clarkson-Smith and Hartley 1989, Bashore 1990).

Work ∨

Work itself, including work content, work organization and work environment, has a central role for successful aging in working life. Three groups of risk factors, which add to the deterioration of the work ability of aging workers, have been identified. The results are based on the 4-year follow-up study of 6257 aging workers in Finland, covering the physical, mixed and mental occupations (Ilmarinen et al. 1991a).

The first group of risk factors for work ability includes physical demands that are too high (table 6.4). Traditional items can be found on the list: static muscular work, lifting and carrying as well as awkward work postures all produced a significant decrease in work ability index (see next chapter) during a 4-year follow-up. It is obvious that owing to the decline of physical capacities and increase in the prevalence of musculoskeletal disorders, such physical demands are not only loading but also harmful for older workers. In auxiliary work, installation work, home care work and in nursing work physical load is still common (Ilmarinen et al. 1991b).

The second group of risk factors includes environmental factors (table 6.5). Outdoor work and the risk of work accidents as well as exposure to heat and

Table 6.4. Factors of work, which add to the deterioration of the work ability of aging workers*

Physical demands that are too high
Static muscular work
Use of muscular strength
Lifting and carrying
Sudden peak loads
Repetitive movements
Simultaneously bent and twisted work postures

* Significant factors explaining the decrease of work ability index during four years among 6257 workers aged 45 to 58 years.

Table 6.5. Factors of work, which add to the
deterioration of the work ability of aging
workers*

Stressful and dangerous work environment
Dirty and wet workplace Risk of work accidents Hot workplaces Cold workplaces Changes in temperature during workday

* Significant factors explaining the decrease
of work ability index during four years
among 6257 workers aged 45 to 58 years.

cold have been found to deteriorate work ability. Tolerance for heat and cold decreases with age, which explains their importance to the older worker.

The third group is a collection of factors typical for poorly organized work (table 6.6). Many of the factors listed are associated with the duties of foremen. Role conflicts cover questions of responsibility. It is often unclear how responsibilities change with aging and experience. Is the older worker more responsible than a younger one, when, for example, work accidents or production losses happen? How is the foreman supervising or tackling the work? Are mistakes and failures also accepted for an experienced worker? Are there enough degrees of freedom for rescheduling one's own work to cope with time pressure? With experience, possibilities to develop one's own work should increase and older workers need regular professional training to improve their competence. Also, the need for acknowledgement and respect is important for the older worker.

Table 6.6. Factors of work, which add to the
deterioration of the work ability of aging
workers*

Poorly organized work
Role conflicts Supervision and tackling of work Fear of failure and mistake Lack of freedom of choice Time pressure Lack of influence on own work Lack of professional development Lack of acknowledgement and appreciation

* Significant factors explaining the decrease
of work ability index during four years
among 6257 workers aged 45 to 58 years.

Surprisingly, organizational factors were as harmful as physical or environmental factors for the work ability of aging workers. The more risk factors that an older worker is facing daily, the higher is the risk that his/her work ability declines remarkably in the four years after the age of 50 years. On the other hand, if any one of the risk factors is influenced; positive results can be expected in relation to work ability.

Work ability index

Measuring work capacities or abilities is a complex issue. The physical, mental and social capacities of the individual should be related to the respective demands of work. It is not only the measurable human capacities or demands of work which are important, however, but also the motivation and experience of the worker, which should be taken into consideration. Additionally, health and the decline in health status and its perceived consequences for managing work have an important role among aging workers. In spite of the complexity of measuring work ability, a need for an overall method was evident.

*Table 6.7. Work ability index**

Item	Scale	Explanation
1. Subjective estimation of present work ability compared with the lifetime best	1–10	0 = very poor 10 = very good
2. Subjective work ability in relation both to physical and to mental demands of the work	2–10	2 = very poor 10 = very good
3. Number of diagnosed diseases	1–7	1 = 5 or more diseases 2 = 4 diseases 3 = 3 diseases 4 = 2 diseases 5 = 1 disease 7 = no disease
4. Subjective estimation of work impairment due to disease	1–6	1 = fully impaired 6 = no impairment
5. Sickness absence during past year	1–5	1 = 100 days or more 2 = 25–99 days 3 = 10–24 days 4 = 1–9 days 5 = 0 days
6. Own prognosis of work ability after two years	1, 4, 7	1 = hardly able to work 4 = not sure 7 = fairly sure
7. Psychological resources (enjoying daily tasks, activity and life spirit, optimistic about the future)	1–4	1 = very poor 4 = very good

* Range 7–49 with poor work ability = 7–27, moderate work ability = 28–43, and good work ability = 44–49.

Using the first cross-sectional data of the follow-up questionnaire study of 6257 workers aged 45 to 58 years, an index was constructed describing older workers' capabilities related to work demands. After correlation and regression analysis, seven items were chosen and the index was called Work Ability Index (WAI). Work ability was evaluated against the clinical assessment of health status and work ability (Eskelinen et al. 1991). The associations between functional capacities and work ability have also been studied (Nygård et al. 1991b). The work ability index has been described in detail recently (Ilmarinen and Tuomi 1993).

The seven items of the work ability index are illustrated in table 6.7. The range of the summate index was 7–49, which was classified into the following three groups: (a) poor work ability (score 7–27), (b) moderate (score 28–43), and (c) good work ability (score 43–49). The cut-off points for poor and good work ability were chosen from the 15th percentile of the index distribution of the total population before the follow-up.

The WAI was found to predict the incidence of work disability for a group of 51-year-old men and women. One-third of the individuals in the group with poor work ability according to the index were granted a work disability pension during the 4-year follow-up (table 6.8).

Poor work ability at the age of 55 years seemed to be job-related. Among the women aged 55 years, the highest prevalence rates of poor work ability were found in auxiliary (21.0 per cent) and home care workers (20.0 per cent) and the lowest among dentists (3.9 per cent) and in administrative work (4.6 per cent). Poor work ability among men at the age of 55 years was most common among workers dumping place (37.5 per cent) and in auxiliary work (23.9 per cent). The lowest rates were found among doctors (0.0 per cent), in administrative work (7.3 per cent) and among teachers (8.0 per cent) (Ilmarinen and Tuomi 1993).

Risk ratios of poor work ability have been reported by age, gender and work (Ilmarinen and Tuomi 1992). The results suggested that men experience poor work ability about four years earlier in physically demanding jobs than

Table 6.8. Work ability index of 51-year-old municipal workers and the work disability pension rate during the follow-up period from 51 to 55 years of age by gender

| Work ability index at the age of 51 years | Work disability pension between 51–55 years of age | | | |
| | Men | | Women | |
	Number	Percent	Number	Percent
Poor	119	37.8	109	33.3
Average	120	8.4	85	4.4
Good	39	0.8	7	1.5
Poor, average and good combined	242	11.5	201	7.5
Information lacking	56	12.4	45	9.9

in mentally demanding ones. The risk ratio of poor work ability among men and women in mentally demanding jobs was different. The risk ratio was lower for men than for women before the age of 55 years, but it was higher for men than for women after this age. The differences in workload, in individual factors and in early retirement explain this finding. The WAI has proved to be a valuable method both for research purposes and for daily praxis in occupational health service. The WAI has been translated into English and used in research projects in several countries.

FinnAge—respect for the aging

FinnAge: action programme to promote the health, work ability and well-being of the aging worker

Among the OECD countries and especially in several European countries, the proportion of 45-to 64-year-old individuals in the work-force is increasing remarkably. Also, the number of people of work age (15–64 years) in relation to the number of people retired will change over the next few decades (Ilmarinen et al. 1991a). Finland, Sweden, Luxemburg, Switzerland and Japan will represent the oldest countries among the OECD: more than 40 per cent of their work-force will be older than 45 years by the year 2000.

In Finland, a longitudinal research project was started in 1981 with the overall objective of determining whether the job-dependent retirement ages were still justified. Since 1986 in the private sector and 1989 in governmental and municipal sectors, a law requiring an early retirement system was passed. According to this new concept, an individual disability pension can be granted if impairments, disabilities, handicaps, aging, heavy workload, a poor work environment or unreasonableness to continue own work can be established. Based on the results and experiences of the 4-year follow-up study of aging workers (see Ilmarinen 1991) an action programme called FinnAge was started in 1991. The Finnish Institute of Occupational Health in Helsinki has taken aging workers as one of the highest-priority topics for the 1990s. The aim of the FinnAge programme is to (a) redesign working conditions for aging workers and, (b) promote the health and functional capacities of aging workers.

The FinnAge programme has three parts: research, training and education. The first phase of FinnAge (1991–1995) is research-oriented. It now has 28 research projects running in private, governmental and municipal sectors in Finland. Besides the intervention projects at work sites, new educational material for training purposes is in preparation. A handbook *Aging and Work* is a collection of current gerontological and occupational health knowledge of the topic. Later, training packages for different user groups (managers, foremen, occupational health and safety personnel, older workers, and so on) will be produced. Information on the programme is disseminated continuously.

The programme's own bulletin and own publication series have been established. In the middle of the first phase of FinnAge, a total of 13 research reports and one symposium proceedings concerning aging and work (Ilmarinen 1993) have been published. FinnAge is also collaborating with other parties, especially in the University of Jyväskylä. Altogether, about 30 manpower and about $1 million per year will be invested in the FinnAge-Action programme. The major costs will be covered by the Finnish Institute of Occupational Health and the Finnish Work Environment Fund.

References

Aromaa, A., Heliövaara, M., Impivaara O. et al. 1989. Health, function limitations and need for care in Finland. Basic results from the Mini-Finland Health Survey (Social Insurance Institution, Helsinki/Turku), 792 pp.

Bashore, T. R. 1990. Age, physical fitness, and mental processing speed. In M. P. Lawton (ed.) *Annual Review of Gerontology and Geriatrics, Volume 9: Clinical and Applied Gerontology* (Springer, New York).

Clarkson-Smith, L. and Hartley, A. A. 1989. Relationships between physical exercise and cognitive abilities in older adults, *Psychology and aging*, 4(2): 183–189.

Eskelinen, L., Kohvakka, A., Merisalo, T., Hurri, H. and Wägar, G. 1991. Relationship between the self-assessment and clinical assessment of health status and work ability, *Scand. J. Work Environ. Health*, 17(suppl. 1): 40–47.

Heikkinen, E., Arajärvi, R.-L., Era, P., et al. (1984). Functional capacity of men born in 1906–10, 1926–30, and 1946–50: a basic report, *Scand. J. Scoc. Med. Suppl.*, 33: 193.

Ilmarinen, J. (ed.) 1991. The aging worker, *Scand. J. Work Environ. Health*, 17(suppl. 1): 1–141.

Ilmarinen, J., Tuomi, K., Eskelinen, L., Nygård, C.-H., Huuhtanen, P. and Klockars, M. 1991a. Background and objectives of the Finnish research project on aging workers in municipal occupations, *Scand. J. Work Environ. Health*, 17(suppl. 1): 7–11.

Ilmarinen, J., Suurnäkki, T., Nygård, C.-H. and Landau, K. 1991b. Classification of municipal occupations, *Scand. J. Work Environ. Health*, 17(suppl. 1): 12–29.

Ilmarinen, J., Louhevaara, V., Korhonen, O., Nygård, C.-H., Hakola, T. and Suvanto, S. 1991c. Changes in maximal cardiorespiratory capacity among aging municipal employees, *Scand. J. Work Environ. Health*, 17(suppl. 1): 99–109.

Ilmarinen, J. 1992. Job design for the aged with regard to decline in their maximal aerobic capacity: part II–the scientific basis for the guide, *International Journal of Industrial Ergonomics*, 10: 65–77.

Ilmarinen, J. and Tuomi, K. 1992. Work ability of aging workers, *Scand. J. Work Environ. Health*, 18(suppl. 2): 8–10.

Ilmarinen, J. (ed.) 1993. *Aging and Work*. International Scientific Symposium on Aging and Work, 28–30 May 1992, Haikko, Finland. Proceedings 4 (Finnish Institute of Occupational Health, Helsinki), 254 pp.

Ilmarinen, J. and Tuomi, K. 1993. Work ability index for aging workers. In J. Ilmarinen (ed.) *Aging and work*. Proceedings 4. Helsinki: (Institute of Occupational Health, Helsinki), 42–151.

Lancet (Editorial) 1993. Ageing at work: consequences for industry and individual. Vol. 340, 9 Jan: 87–88.

Nygård, C.-H., Eskelinen, L., Suvanto, S., Tuomi, K. and Ilmarinen, J. 1991a. Associations between functional capacity and work ability among elderly municipal employees, *Scand. J. Work Environ. Health*, **17**(suppl. 1): 122–127.

Nygård, C.-H., Luopajärvi, T. and Ilmarinen, J. 1991b. Musculoskeletal capacity and its changes among aging municipal employees in different work categories, *Scand. J. Work Environ. Health*, **17**(suppl. 1): 110–117.

Pisa, Z. and Uemura, K. 1989. International differences in developing improvements in cardiovascular health, *Annals of Medicine*, **21**(3): 193–197.

Suvanto, S., Huuhtanen, P., Nygård, C.-H. and Ilmarinen, J. 1991. Performance efficiency and its changes among aging municipal employees, *Scand. J. Work Environ. Health*, **17**(suppl. 1): 118–121.

Tuomi, K., Ilmarinen, J., Eskelinen, L., Järvinen, E., Toikkanen, J. and Klockars, M. 1991. Prevalence and incidence rates of diseases and work ability in different work categories of municipal employees, *Scand. J. Work Environ. Health*, **17**(suppl. 1): 67–74.

Verbrugge, L. M. 1984. Longer life but worsening health? Trends in health and mortality of middle-aged and older persons, *Milbank Memorial Fund Quarterly*, **62**: 475–519.

7

We are all aging workers: for an interdisciplinary approach to aging at work

C. Teiger

Summary. Aging at work is considered as a *broad process*, which should be taken account of throughout working life, but should also be perceived as a *result*, partly as a consequence of the effects of harmful work constraints. The understanding of this complex issue, which enables effective action to be taken from the ergonomic viewpoint, involves an interdisciplinary approach that links results from various origins without superposition. Some examples are given, taken from recent research. In the present demographic, social, economic and technical context 'aging at work' should and may be perceived as a *dynamic* and *positive phenomenon*, encompassed within the human being's development, so that the beautiful Swedish slogan, 'to age is to grow', can become reality, in working life at least.

Keywords: aging, work, health, performance, ergonomics, multidisciplinarity, training, job requirements, design, age stratification.

Introduction

The relation between work and the aging of operators can be covered by three questions:

(1) In what way may transformations with age be incompatible with the job requirements laid down by the technique and work organization? This concerns the question of aging *with regard to work*.

(2) In what way does work, and its conditions of execution, accelerate and exaggerate the usual aging process in operators? This concerns the question of aging *through work*.

(3) What ergonomic actions have to be used from the work design stage in order to maintain the mental and psychic health of operators without risk, and to develop their skills throughout working life? This concerns aging *in work*.

The three questions are important in ergonomics for at least three reasons:

(1) Ergonomics needs to know the diversity of the characteristics of the working population in order to ensure that this population is not reduced to a working model of the human operator—that is, unique and stable through time. Yet, age is a very important factor of inter-individual diversity and intra-individual variability, which are ignored or greatly underestimated by the traditional work organization.

(2) In industrially developed countries, the proportion of the working population aged over 40 is increasing and will continue to do so over the next decade. Even if early retirement schemes are maintained, this evolution will not be prevented. Moreover, the use of such schemes varies considerably in the different industrialized countries, as shown by the employment rate of those aged 55 to 59 years of age (table 7.1), which ranges from 6.1 per cent in Western Germany to 82.7 per cent in Sweden. In the year 2010, in France, for instance, more than half of the working population will be aged over 40 (50.4 per cent compared with 42.3 per cent at present (table 7.2); in Canada, half of the

Table 7.1. Employment rate for 50 to 59 years of age (year 1989).

	%		%
West Germany ('87)	6.1	Finland	59.1
France	56.1	Japan	71.6
Norway	73.2	Holland	44.6
United Kingdom	64.6	Sweden	82.7

Source: OECD.

Table 7.2. Distribution of workers by age (France 1987).

	Under 25	25–39	40–54	55 and over	Total
1990	14.3	43.4	33.1	9.2	100
1995	12.8	42.3	36.9	8.0	100
2000	11.2	41.3	40.0	7.5	100
2005	11.0	39.3	39.7	10.0	100
2010	10.6	39.0	40.4	10.0	100

Source: INSEFE.

working population will be aged over 45, compared with one out of three at present (Tétreault 1992).

(3) The rapid technological changes which have taken place in companies over the past few years have often been accompanied by the exclusion of certain categories of workers considered, a priori, to be 'handicapped by situation' because of their low qualification level, but especially owing to their age, despite a certain improvement which has taken place in France, at least, over the past few years (Molinié and Volkoff 1993): For instance in 1987, the maximum proportion of users of computers (microcomputer or terminal) in the age range 25 to 44 years old, dropped over the age of 45. Since that time, there has been a great expansion in the use of computers, which has concerned all ages, to the point where, in 1991 the proportion of employees aged between 50 and 54 who used a computer in their work reached the same level as that noted in 1987 for the 25 to 44 years category. But in 1991, just as in 1987, work on a computer was less and less frequent from 45 years onwards. It appeared that employees who used a computer in their work in 1987 continued to do so in 1991, regardless of their age, but that the chances of becoming a new user clearly diminished for older generations. The use of computers therefore appears to depend to a great extent on an employee's age at the when computers are introduced and on the conditions under which this technology is introduced.

Technological changes take place quickly. Yet, the skills of experienced operators remain vital for the operation of the production system. From now on, therefore, it is necessary to organize the training of operators and the design of new production means in order to take into account the aging of the working population with the aim of organizing current situations for older workers. This should be done especially with the aim of prevention, considering that aging in work should be taken into account from the start of working life and that we all are aging workers. Certainly, this question is not new (Teiger 1989) and was the subject of decisive efforts made by English researchers, during and after the Second World War. These scientists opened the way to research into age-related work characteristics through both experimental and statistical research and field studies in companies (cf. the survey in Welford 1958). This research clearly highlighted the differing age distributions between sectors of economic activity and even among several professions in the same sector. Such questions were then expressed on a European scale, mostly under the auspices of the OECD (Organization for Economic Co-operation and Development 1965). However, the present context makes it dramatically urgent and suggests strongly that the terms must be posed differently by taking into account constant phenomena and the significant transformation which have taken place over the past 30 years in demographic, socio-economic, technological and scientific fields.

So, in order to be mastered, the question of the aging of workers in regard to work, through work and in work requires an interdisciplinary approach (see, for example, Laville et al. 1991), since these manifestations of aging/work relations and their underlying processes are varied. And, for that, it is necessary to understand the relations between age, the specific work activity, health and employment at both the individual and the collective level.

We shall illustrate the interest of organizing the results from various origins by using the example of the relation of time pressure with work and aging. In effect, the work situation is a special place for the confrontation between two temporal logics. The first one concerns the human system subjected to instantaneous biological rhythms (ultradian and circadian) and their slow and continuous transformation through the course of time provoked by aging. The second logic affects the production system, which imposes a temporal constraint on both the duration of the activity and its position in the nychthemeron (working hours) and on the instantaneous activity rhythm (work rate). The conflicts between those two logics probably play a major part in the difficulties observed in aging in work.

Aging with regard to work

Thanks to work demography, it is possible to define those production centres where professions are particularly young or old or close to the age structure of the working population. By following such a population through time, they enable us to propose preliminary explanations or concerns for ergonomics. For instance, figure 7.1 shows three examples of the evolution of age structures for three industrial sectors over six years (1962 to 1968) in France. It illustrates two types of evolution, according to hypotheses on the typology of age distribution as developed by Smith (1969). Hypothesis A (age) states that it is the work conditions which determine the difference is in the age distribution of workers. This A hypothesis illustrates these differences in a much stronger way than hypothesis H (history), a simple historical explanation. This is further emphasized by the persistence of these structures through time:

(1) The clothing sector, with an overrepresentation of young age classes and an under representation of older classes, remains young and stable. It is Smith's A hypothesis that is then plausible: that is, the job requirements impose a clearly defined age structure, as if 'one cannot age' in this profession. This raises the intriguing problem of the reasons and mechanisms that lead to such a situation.

(2) The tobacco and matches sector showed a rise in ages between 1962 and 1968 which can be explained by Smith's H (history) hypothesis. High recruitment at a given time, then stabilization, means relative aging in the job. In this industrial sector, workers are protected by a special status, which guarantees employment and takes age and

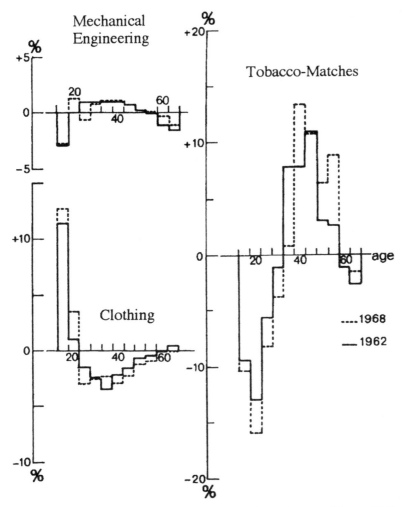

Figure 7.1. Age structure of female workers in three industrial sectors (Teiger 1989: data from 1962 and 1968 are expressed as differences between the percentage of workers in each age group, in one sector, and the average percentage of all workers in this age group = baseline).

service with the company into account. The age pyramid of the staff shows a normal age distribution, although the work conditions and the type of production are similar to those of other sectors which provoke elimination with age, for example in the sectors of electrical construction or clothing. In such a case, the problem of compatibility emerges between increasing age and work demands.

(3) The engineering industries sector clearly corresponds to the average evolution of the population of workers over this period.

The phenomena described persist through time, as shown by a more recent study done on the basis of the 1982 census (Molinié 1984). The same type of approach may be applied to different levels of the production system: such as the company (figure 7.2), at level of workshops in the same company (Teiger 1989), workstations in the same workshop (Teiger and Villatte 1983) or the type of tasks (figure 7.3 Volkoff 1990).

An example is the evolution of ages according to the type of task in a automobile construction company in France between 1979 and 1984 (Desriaux and Teiger 1988). There was a significant statistical relation between the average age and the five types of task selected in the company, regardless of the observation period in question. The average age is lowest in the qualified tasks like

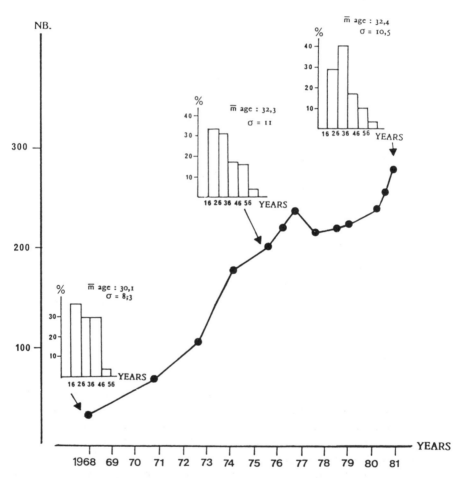

Figure 7.2. The evolution of the work-force and of the age distribution of female workers in a surgical equipment company from the beginning of production (Duraffourg et al. 1983 in Teiger 1989: the general age structure of female workers is young and a retrospective study shows that this age structure was stable for 13 years)

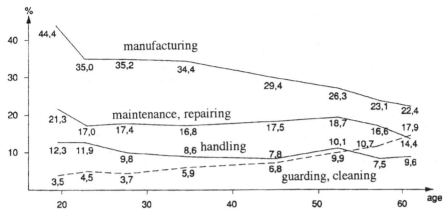

Figure 7.3. Distribution of workers of different ages, in France, according to four main tasks, Volkoff 1990).

maintenance of production installations and in the production-line tasks, which are subjected to time pressure. The average age is the highest in administrative tasks, preparation tasks (supply of production lines) or unqualified maintenance tasks. Over the 5-year period, transfer of workers from one type of task to another was significantly linked to age. For example, those who move to unqualified maintenance tasks had a higher average age than those who remained in their original type of task. It is concluded that time constraints or time pressure bring about a selection that is all the earlier in terms of age, since it is related to certain other stressors such as posture, physical load, accuracy and task complexity.

With this level of knowledge, although the role of work conditions can be made clear, the work characteristics that bring about this selection cannot be defined. By associating the demographic study with an ergonomic study, however, it is possible to single out the work conditions that either cause the selection of workers in terms of age or that, on the contrary, enable the job to be kept and thus to infer the explanatory mechanisms.

Aging through work

By associating a precise analysis of the work activity and its determinants (job requirements, work conditions) with an estimate of the cost in terms of workload (felt and objectified) and the effects on health, ergonomic studies of the clinical type enable the factors in question (positive or negative) to be hierarchized. Globally speaking, the time constraints (rates and hours) seem to play a crucial role in discrimination as regards the age of workers. It is not a simple relation, however, and the results of studies carried out in those work situations, where these types of constraints dominate, show at least the three following phenomena.

Differential aging in job occupation: a combination of several factors

As regards the age-to-work ratio, a simple dichotomy cannot be based on the existence or absence of time constraints or even on their degree of intensity, but it is most often the combination of several factors of workload and job requirements that determines the job's occupation age.

Teiger (1989) gives some examples for France:

(1) In a company that makes safety gloves, the age structure of sewing-machine operators is compared with that of more versatile workers who check the gloves and perform other operations. All the workers are paid according to output, but the sewing-machine operators have very high production rates and have to do an accurate job. It is of interest to note that this double demand for speed and perception-motor accuracy, associated with a wage based on output, leads to sewing-machine operators leaving this job at a very young age (age limit, 25; average age, 21 years; sd 3.9), while the versatile workers have varied tasks which present different levels of precision, enabling compensation of the rate demands from one task to the other. They stay in the job up to a higher age (average age 30.6; sd 11.9).

(2) In an automobile assembly line (table 7.3) all workers are subject to the same time constraint characteristics. Those on the upholstery assembly line have an average age of 33 years owing to the fact that they work inside the vehicle and their posture is very unbalanced. Those who work on the body assembly line, with a great physical load, are, on average, 39.3 years old. From the viewpoint of age, the association of instantaneous work rates and unbalanced posture is more discriminating that the time × physical load.

(3) In shiftwork and nightwork, which most workers give up around the age of 40–45 years, the association of increased task complexity (for example, owing to modernization of machines, such as in the glass industry) introduces age discrimination at a secondary level, as regards the allocation of the workers in this population who are already 'preselected' in terms of age. The most automated machines are generally operated by the youngest workers and the oldest machines by operators of all ages, with a predominant age of 45 to 50 years, although there is no formal difference in qualification.

Relation between age and job requirements through health problems

At least one part of the imbalance between age and job requirements is a consequence of health problems, including, in a broad sense of the term, workload and infrapathology. The short-term, medium-term and long-term effects on health can be identified (Teiger and Laville 1981).

Table 7.3. Relations between work conditions, age and health at two worksites of an automobile plant (upholstery and body-assembly): (a) the estimated causes of workload, (b) recovery time of basic heart rate of the workers after 4 minutes of pedalling on an ergocycle at 300 W/min (Marcelin and Valentin 1969, in Teiger 1989).

| | Automobile construction workshops | | | |
| | Upholstery assembly (mean age = 33 yrs) | | Body assembly (mean age = 39.3 yrs) | |
	N	%	N	%
(a) Estimated causes of job load				
painful posture	19	0	2	12
monotony	1	4	1	4
need for special physical conformation	0	0	1	4
need for special ability	2	8	1	4
work too varied, associated with a too fast rate	0	0	1	4
many operations	1	4	0	0
unhealthy jobs	4	16	0	0
too many changes of location	1	4	0	0
cognitive difficulties	1	4	0	0
(b) Recovery time of heart rate				
>1 minute	8	32	4	16
from 1'30 to 2'	1	4	12	48
from 2'30 to 3'	3	12	4	16
from 3'30 to 4'	1	4	1	4
>4 minutes	12	48	4	16

Short-term effects on health

The short-term effects (workload) of work rates, associated with severe postural constraints in automobile assembly, are demonstrated, for example, by the heart rate recovery time following physical effort which differs considerably from one workshop to another according to age (table 7.3b).

As for working hours, short-term effects of time pressure, in terms of age, are mainly manifested in sleep problems, both quantitative and qualitative, which worsen with late working hours and nightwork, and by gastro-intestinal and nervous disorders owing to the desynchronization of biological rhythms (Quéinnec et al. 1992).

Medium and long-term effects

The long-term effects of the time pressure of work rates can be revealed by epidemiological studies. A study by Vézina and colleagues (1989) over the 1975–1985 period of 800 workers in the clothing industry in Québec, spread over three age periods (45–50 years, 50–64 years, 65–70 years), showed that, for those who are paid according to output i.e. subject to very fragmented work at a very high rate, there is a positive linear relation between the number of years worked and the prevalence of serious functional incapacity. The risk of osteoarticulatory and arthritic disorders is nine times higher in female operators doing piece-work than in those paid by the hour and also higher for all other causes of permanent severe incapacity (table 7.4).

As regards the medium and long-term effects of the time constraints of working hours, the question is more complicated owing to the 'healthy worker effect'. The most important effects therefore show up in the population of workers who gave up or had to give up shiftwork or nightwork, often for health reasons. This phenomenon, which was highlighted some time ago (Aanonsen 1959), is still of significance. Paradoxically, workers who have been exposed to these sorts of working hours in the past have a greater probability of having a high morbidity rate, which increases with age, than those who are

Table 7.4. *Distribution of the causes of permanent severe incapacity according to the remuneration method (Vezina et al. 1989).*

Causes	Payment Hourly	Piece-work	Total
Arthritic and osteoarticulatory disorder*	1	30	31
Cardiovascular disorders	4	18	22
Mental disorders	1	6	7
Others	7	34	41
Total**	13	88	101

*OR (odds ratio) = 9.3 (check for age, section work and nicotinism);
**Excluding three missing values

exposed to it at present (Bourget-Devouassoux and Volkoff 1991). Despite this selection of the most resistant, the deterioration of the health of shiftworkers has increased more over the years than that of day-workers. The difference between the state of health of the two categories of workers increased with age, in particularly from 40 onwards (Haider et al. 1980).

At a more serious level, early mortality may be considered to be a general effect of the stress of working hours. Permanent nightshift, by disturbing biological rhythms, increases the potential effects of all other constraints of the work situation by lowering the defences of the system. In occupational groups in the printing industry, the cumulative percentages by age group–accidents excluded–show that, between 30 and 50 years, permanent nightshift workers, compared with pressmen, have the highest mortality rate, while correctors, working during the day and evening, have the lowest rate (Teiger et al. 1981).

Thanks to epidemiological and demographical studies, it is presently acknowledged that, in a general way, work conditions suffered throughout working life have an impact on the physical, mental and social health of workers after retirement (David and Bigaouette 1986, Derriennic 1990, Ilmarinen 1990, Cribier, 1990). This is the more true, in particular, if we consider life expectancy without incapacity (Chanlat 1983, Colvez and Blanchet 1983). A difference of eight years for life expectancy at 35 between executives and manual workers (Desplanques 1991) is explained not only by differences in social class and living standards, but also by the high load of work conditions.

Age selection in work: a dynamic but hidden process

Aging in work, in situations where constraints are high, leads to a *de facto* selection (where workers are not protected by a status which guarantees job stability). This selection leads to mobility inside the company (reallocation to another job, at a given time, at the request of the person concerned or by order) or to external mobility, (voluntary exclusion from the company—the worker 'overwhelmed' by the incompatibility of job requirements and his functional condition—or by involuntary dismissal). Teiger and Villatte 1983 show that the proportion of day- and shift-workers in a chemical plant clearly inverses around 40 years of age: the number of shiftworkers becomes less and less as they move to day jobs. This type of forced mobility is generally associated with a declassification of the worker and the distribution of workers at different ages according to task characteristics. As shown in figure 7.3 (Volkoff 1990) 'manufacturing' tasks are most often done by youngsters, while, in contrast, tasks such as 'care-taking, sweeping and cleaning' are more common among elderly workers. Two other tasks, 'handling' and 'maintenance, turning and repair' tend to decrease with age but not very significantly. It clearly shows that those tasks where the proportion of workers increases with age are care-taking and cleaning, mainly to the detriment of production tasks.

These selection phenomena, which become 'visible' at a certain time leading to 'pivot periods', should be interpreted in dynamic terms. Selection through health is progressively taking place in the company.

Even in the context of the present economic crisis, it is a reasonable assumption that workers who leave the company, officially for economic reasons, are those who suffer most from the effects of work conditions and do so in an invisible way, that is with no objectified professional pathology. From the start of the 1980s, a Finnish study (Koskela *1981*) showed the influence of the economic situation on the selection of workers by health. The selection is higher during periods of recession than in periods of expansion. The study also shows that where transfer is impossible—or low—the morbidity of workers increases with age: where it is high, the morbidity curve drops to around 40–44 years of age. Stoppage of the increasing morbidity of company workers is then compensated by an increase in morbidity in the group of workers of the same age who left the company (Koskela *1982*). In other words, this selection is apparently not the result of a conscious policy of the company, which actually protects those with a legally recognized professional pathology from being dismissed (Davezies et al. 1991).

Aging in work

Aging process: functional capacities, operative strategies, organizational conditions

Hypotheses concerning the explanation of selection phenomena through work conditions, in terms of age, are based on experimental studies. There are many studies that highlight the age-related slowing of reaction and decision processes, which can be attributed to structural changes of the nervous system and functional modifications of the cognitive processes of information processing and decision making. Evidently, speed in itself is probably not a criterion that is sufficient for assessment of the capacities of an individual. All the more so, since most of the experiments mainly concerned task performance in short-term experimental situations on tasks that were more or less complex but which had no relation, or only a very remote one, with everyday life, particularly with professional activities. In particular, the performance of experimental tasks rarely makes use of acquired everyday experience. In contrast to this, in work situations, Davies and Sparrow (1985) have shown that, among maintenance engineers for example, complex task performance of older engineers, owing to their greater experience, was better than that of young engineers, while younger individuals were more productive in simple tasks.

More information on the aging process comes from experimental work, which addresses the specific difficulties encountered by the oldest workers and the types of mistakes they make. In particular, those who try to understand how the compensation process shows up and how functional capacities are reorganized with age, are able to perform an activity with 'means' that are different from those used at a younger age, such as, in particular, the mobilization of one's experience (for example Czaja et al. 1989). The pioneer work in this research area is that of Szafran (1955, reported by Welford 1958). Particularly in England, during the 1950s, a reversal of thinking took place in the scientific, psychological approach to aging. It stresses strategies rather than simple performance. More recent work in psychology and ergonomics, although difficult to implement, shows that different strategies are employed with age. Such strategies are: prior structuring of information for decision making, the anticipated search for this information, reorganization of perception-motor co-ordination and control of the activity at sensory, motor, postural or cognitive levels (Teiger 1975, Salthouse 1984, Marquié et al. 1988; Paumès 1992).

Consequently, this may mean that, when time pressure is high and anticipation is impossible, elderly operators are often severely penalized and are gradually eliminated from these tasks. Inversely, if the rigidity of the time factor is reduced, and the task and work organization enable various strategies to be used, then the performance level of the aging operators can be maintained. It should be emphasized that these transformations are just as favourable for young operators, since also at their age, selection, although not as tough as for the aging, does exist and does not concern their skills but their speed of working.

Generally, statistical data show that the rigid time pressure has increased even more over the past few years (in France between 1984 and 1991), particularly for workers up to 45 years of age and especially in women. This difference in terms of sex is maintained. In the same way, the time pressure of working hours, which diminished somewhat in the youngest employees between 1984 and 1991, is maintained or even rises between 40 and 50 years of age (Molinié and Volkoff 1993). This problem is all the more crucial, since the 'indicators of the end of exposure' or the symptoms of impending burn-out for various stressors clearly indicate that, in the case of employees aged around 50, it is these time factors (working hours and work speed) that lead or led to the highest burn-out or giving up (table 7.5, Volkoff 1990). In a much wider sense, it raises the question of professional mobility. The comparison of indicators that concern the entire population of present workers and ex-workers who worked for the greatest part of their working life, illustrates the essential role of professional mobility. Stated in other words: a very high proportion of 'end of exposure', as a consequence of some stressor, occurs by leaving the worker category. This is, however, outside the scope of our subject.

Taking into account the aging process

All these experimental approaches have another interest from the viewpoint of ergonomics. They can open up perspectives for the transformation of work situations. On the basis of observations of the actual status of the 'aging at

Table 7.5. Value of 'indicators of impending burn-out' to job stressors
for male employees of the 1938 generation (Volkoff 1990).

	All employees	Present workers	Ex-workers
Shiftwork	0.56	0.46	0.72
Nightwork	0.67	0.63	0.69
Rates	0.59	0.51	0.94
Postures	0.37	0.29	0.71
Heavy loads	0.61	0.56	0.80
Noise	0.43	0.35	0.71
Heat	0.56	0.52	0.73

work' problem (characteristics of work situations, type and intensity of stressors, situation of employees at work-sites according to their age, technological
developments) and scientific knowledge of the aging process, there appear to
be a few promising routes. From the ergonomics view point, these tracks
would enable us to take action on the present and future transformation of
work situations in order to take account of the aging phenomenon in two
forms: namely, workers presently 'aging' and the foreseeable aging of others.

It is a matter of bringing together the characteristics of populations with the
technical-organizational characteristics so that work conditions are adapted to
the fact that the human being changes with age (Cremer 1991). The import
point to focus on is the functional age and not the chronological age of the
working population (Dirken 1972).

Certainly, these concerns are not new (Welford 1958, OECD 1965, Laville et
al. 1975, to mention a few). However the research effort should continue, since
in concrete work situations, the proposals already made in the past are far
from being integrated and should be taken into account in view of the current
changing context (Gaullier and Thomas 1990, Laville and Teiger 1990, Matthews 1990, Cremer 1991, Teiger and Marquié 1991, Tétreault 1992, Volkoff et
al. 1992, Molinié and Volkoff 1993). Two fields, in particular, should be
explored: the design of new production means and training.

Design of new production means

As regards the design of production means, on the basis of all these data we
can then come back to prospective ergonomic-demographic approaches. In
this respect a recent study in the French industry is significant. On the basis of
the current situation concerning automobile assembly lines and with the help
of occupational physicians, Sailly and Volkoff (1990) classified workers according to their aptitude: (a) to make physical effort and maintain constraining
postures; and (b) to follow high output rates. This would be monitored for the
running 5 years while, in the meantime, the age/aptitude ratio was increased.
They note that taking into account those workers who retire and the fact that

youngsters are not being recruited, this population would age by 5 years, reducing the number of the most resistant. The authors claim that they could roughly estimate the extent of this reduction. On the basis of this projection of characteristics of the current population of assembly-line workers, the study shows that the next assembly lines should be organized so that its foreseeable evolution is taken into account. This means that there should be a reduction in the number of jobs on the line with constraining postures and requiring great physical effort, and an increase in the number of jobs off the line, where the time pressure is less strict and which can easily be reorganized to reduce postural constraints and physical load.

Training

Concerning training, the situation is paradoxical. On the one hand, it is obvious that the rapid technological changes of the past few years (which will probably continue) require training that is adapted to all ages and to all workers, but especially to those who need to be reconverted. Most of the time, the latter, when they are 'aging' workers, are victims of a serious prejudice according to which their learning capacity is greatly reduced beyond the age of 40, all the more so when the tasks have important symbolic or status components. The well-known and rather widespread consequences are that workers are then excluded from 'modern' jobs, as we have already noted, especially from access to the continuous training given in companies. In this field, women are even more penalized than men: on average, men have two to three times more continuous training than women but everyone is penalized by age in terms of access to training (figure 7.4, Gadrey 1983, in Teiger 1990). Nevertheless, work on this subject has shown for some time that, as long as there is respect of and patience with long learning periods and the pedagogical principles enable elderly workers to use their experience, know-how and expertise, and where the temporal conditions leave sufficient freedom to use varied procedures, workers can 'learn' up to a very advanced age (Belbin 1965, Pacaud

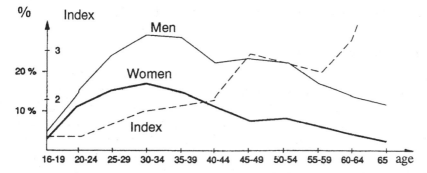

Figure 7.4. The global access rate and index of disparity according to age and sex in access to training at the initiative of the employer (Gadrey 1983, in Teiger 1990).

1971, Salthouse 1984, Marquié et al. 1988, Teiger 1990). However, the learning process and its evolution with age are still poorly known and need more scientific attention.

Conclusion: towards a multidisciplinary approach

The problems of the relation of age and work conditions require a multidisciplinary approach in order to envisage effective ergonomic actions on the design of work facilities. As such, thanks to work demography associated with an ergonomics analysis of work situations, it is possible to assess the importance of the effects of selection, produced by combinations of stressors. Epidemiology brings forward the medium and long-term effects. Though experimental physiological and psychological approaches explain the general causes, the ergonomics approach indicates the specific difficulties of aging operators and offers an aid to the design of work facilities. However, these approaches have to be organized together if they are really to produce knowledge and actions. It would be advisable for each approach to clarify a specific aspect of the problems, but also to pose questions for other disciplinary fields. It is our opinion that each approach should not only develop its own coherence, but, through its results, interrogate the other disciplines. In this way, the complexity of the problem could be revealed and explained and we could deal with it in terms of both forecasting and prevention.

In such a perspective, the viewpoints of demographics and history (Minois 1987, Bourdelais 1993) have to be taken into account along with socio-economic data (Matthews 1990, Gaullier 1992). As regards the present context of demographic aging, it should be mentioned that some people contest the concept of 'aging of the population'. At the start of the century, this concept was developed based on the scientific model of that time and on obsolete age classifications. As a result of this, the concept conveys a gloomy connotation and a non-evolutionary representation, leading to a sort of collective resignation and to a discourse of catastrophe.

If we take into account the state of health of persons and not their chronological age, we note that 'the age of being old' has risen considerably in industrially developed countries. Pessimism concerning demographic aging is therefore unjustified and the term 'future—inevitable' should be changed into 'future—potential', which encourages dynamism rather than resignation. Still, there is a persistent gap between the facts and the social representations (Bourdelais 1993). All the more so when this concerns the 'aging' working population.

As regards the socio-economic context, specialists emphasize that models of production, work organization and, especially, the workplace itself, are undergoing extensive change. The matter of the transformation of the industrial system, based on Fordist and Taylorist organization, into a post-Fordist system corresponding to new models of international competition and encouraging innovation and adaptation to new production conditions, is at the heart of the debate. Certain positive, major consequences for the future elderly workers are sometimes already envisaged (Matthews 1990). The most important fact is perhaps that full employment is no longer the norm and will maybe never return again. For these reasons, professional careers become more and more flexible and discontinuous. They alternate between periods of work, training and unemployment and put an end to the model of the traditional but rigid working life-cycle that encompassed three specific stages which 'normally' followed each other. These stages were: training, working life and retirement (Gaullier 1992). It is, therefore, especially the adaptative capacities which should be developed and in which initial and continuous training will play a major role.

In practice, ergonomics is faced with a major issue when it deals with the diversity of populations at work and, more particularly, when dealing with the aging working population. Highlighting the specific characteristics of aging workers means not only being able to take these features into account in the design of work facilities, but also to attribute to them a particular status with risks of exclusion and even harsh selection. These risks are all the greater since the tasks of producing objects or services will be subject to severe and repetitive time pressure, affecting simple physiological functions and cognitive processes, as we have seen from the results presented. However, we can also consider this aging population to belong to the entire working population, and thus to have characteristics in common. But, aging is a continuous process also; it is a normal feature of the working population. So, the phenomenon of being 'worn down' at work starts from the outset of working life. This evolution in thinking took place in the 1960s, particularly in France, as a result of the work of Pacaud (1965), who compared genetics and gerontology emphasizing that these two disciplines are concerned with the process of change and not with invariance and posed the provocative question: 'Do we age since birth?', consequently the work conditions that are particularly critical for the aging are also critical for young workers, although to a lesser extent.

Current scientific trends which integrate aging and the general issue of a human being's development throughout the life-span, perceive aging as a dynamic process and not as a discrete succession of stable conditions, as was traditionally the case with the linear and unequivocal model, which simplistically stated: growth—maturity/stability—decline. In addition to this new conception, pluralist models of cognitive development (Lautrey 1990) should be proposed that enable renewal of the representation of aging and progress of knowledge. In Sweden, for example, a large present-day, nation-wide campaign is entitled 'to age is to grow'....

In this way, conditions of development could be defined and skills and abilities of all sorts could be maintained. This would be a great improvement on the definition of thresholds of unsuitability in terms of age, which, in companies for a long time was considered, hidden or openly, as a category of some handicap.

In work situations, age is only an indicator of difficulties to which the *entire* population is subjected and any lessening of these difficulties will be beneficial for *everyone*. So, taking away difficulties will prevent the effects of 'wearing out' in youngsters and enable those who are older to continue to age at work, without any risk for health, and with continuous development of their skills.

References

Aanonsen, A 1959. Medical problems of shift work, *International Medical Surgery*, **28**: 422–427.
Belbin, R. M. 1965. *Méthodes de formation. Emploi des travailleurs âgés*, no. 2 (OCDE, Paris).
Bourdelais, P. 1993. *L'âge de la vieillesse* (Odile Jacob, Paris).
Bourget-Devouassoux, J. and Volkoff, S. 1991. Bilan de santé des carrières d'ouvriers, *Economie et Statistique*, **242**: 83–93.
Chanlat, J. F. 1983. Usure différentielle au travail, classes sociales et santé: un aperçu des études épidémiologiques contemporaines, *Le Mouvement Social*, 153–169.
Colvez, A. and Blanchet, M. 1983. Potential gains in life expectancy free of disability: a tool for health planning, *Int. Journal of Epidemiology*, **12**: 224–229.
Cremer, R. 1991. Maintaining a working life: fitting work to the capacities of the older worker. In Y. Quéinnec and F. Daniellou (eds) *Designing for Everyone*, Vol. 2 (Taylor & Francis, London), 1600–1603.
Cribier, F. 1991. Les générations se suivent et ne se ressemblent pas: deux cohortes de nouveaux retraités parisiens de 1972 et 1984, *Annales de Vaucresson*, **30–31**: 181–197.
Czaja, S. J., Hammond, K., Blascovitch, J. J. and Sweder, H. 1989. Age related differences in learning to use a text-editing system, *Behaviour and Information Technology*, **8** (4): 309–319.
Davezies, P., Durand J. D., Luzy, A. and Prost, G. 1991. Study of differential aging related to working conditions in a metal plant. In Y. Queinnec and F. Daniellou (eds) *Designing for Everyone* (Taylor & Francis, London), 1598–1600.
David, H. and Bigaouette, M 1986. Le poids de l'inaptitude au travail dans les prises de retraite d'une, grande municipalité, *Sociologie et Sociétés*, **18** (2): 47–60.
Davies, D. R. and Sparrow, P. R. 1985. Age and work behaviour. In N. Charness (ed) *Aging and Human Performance* (John Wiley, New York), 293–332.
Derriennic, F 1990. Les expositions professionnelles influencent-elles l'état de santé après la retraite? In H. David (ed.) Actes du Colloque Le vieillissement au travail, *une question de jugement*. Mars 1989 (IRAT, Montréal), 36–39.
Desplanques, G. 1991 Les cadres vivent plus vieux. INSEE Première, no. 158.
Desriaux, F. and Teiger C. 1988. L'âge, facteur de sélection au poste de travail, *Gérontologie et Société*, **45**: 33–45.
Dirken, S. M. 1972. *Functional Age of Industrial Workers* (Wolters Noordhoff, Groningen).
Gadrey, N. 1983. Hommes et femmes devant la formation professionnelle continue, *Education Permanente*, **68**: 33–42.
Gaullier, X. 1992. Le risque vieillesse, impossible paradigme, *Sociétés contemporaines*, **10**: 23–45.
Gaullier, X. and Thomas, C. 1990. *Modernisation et gestion des Ages* (La Documentation Francaise, Paris).

Gérontologie et Société 1989. Travail et vieillissement, no. 5 spécial.

Haider, M., Kundi, M. and Koller, M. 1980. Methodological issues and problems in shift-work research. U.S. Dept. of Health and Human Services. NIOSH Proceedings.

Ilmarinen, J. 1990. La relation entre les conditions de travail, l'incapacité et la retraite chez les travailleurs municpaux. In H. David (ed.) Le vieillissement au travail, Mars 1989 (IRAT, Montreal), 25–29.

Koskela, R. S. 1981. Occupational mortality and morbidity in relation to selective turnover, *Scandinavian J. of Work Environmental Health*, **8** suppl. 1: 34–39.

Lautrey, J. 1990. Esquisse d'un modéle pluraliste du développement. In M. Reuchlin, Lautrey, J. and Marendoz, F. (eds) *Cognition, l'individuel et l'universel* (PUF, Paris), 185–216.

Laville, A., Teiger, C. and Wisner, A. 1975. *Age et contraintes de travail* (Naturalia et Biologia, Paris).

Laville, A. and Teiger, C. 1990. Evolution of production techniques and aging of workers. In L. Berlinguet and D. Berthelette (eds) *Work with display units '89* (Elsevier, North Holland) 291–299.

Laville, A., Teiger, C. and Volkoff, S. 1991. The aging process and work: the benefits of a multidisciplinary approach. In Y. Queinnec and F. Daniellou (eds) *Designing for Everyone* (Taylor and Francis, London), 1610–1612.

Marquié, J., Saumès, D., Quéinnec, Y. and Thon, B. 1988. Age, material structuration and time of day effects on visual information processing. Proceedings of the 10th Congress of International Ergonomics Association, **2**: 493–495.

Matthews, J. 1990. Vers un avenir post-fordiste por les travailleurs plus âgés en Australie. In H. David (ed) Actes du Colluque, Le vieillissement au travail, une question de jugement, Mars 1989 (IRAT, Montréal), 76–82.

Minois, G. 1987. *Histoire de la vieillesse—De l'Antiquité la Renaissance* (Fayard, Parts).

Molinié, A. F. 1984. L'âge des ouvriers de l'industrie. Dossiers statistiques du Travail et de l'Emploi (Ministère du Travail, Paris), **9**: 39–64.

Molinié, A. F. and Volkoff, S. 1993. Conditions de travail: des difficultés à prévoir pour les plus de 40 ans, *Données Sociales*: 195–201.

Organization for Economic Co-operation and Development. 1965. L'emploi des travailleurs âgés. Séminaires internationaux: rapport final.

Pacaud, S. 1965. Psychologie génétique et gérontologie—Le commencement de la fin: vieillissons-nous dés la naissance? *Physchologie francaise*, **4** (10): 300–335.

Pacaud, S. 1971. Quelques cas concrets illustrant les difficultés ou les facilités que l'âge entraîne dans la formation professionelle des travailleurs, *L'information psychologique*, **44**: 92–105.

Paumès, D. 1972. Expression du vieillissement en situation professionnelle: un cas de modification des stratégies de travail. Actes du Séminaire Vieillissement et Transport, 3 Décembre 1992 (INRETS-LESCO/CNRS, Bron), 59–63.

Quéinnec, Y., Teiger, C. and Terssac, G. de 1992. *Repères pour négocier le travail posté*, 2nd ed (Octarès, Toulouse).

Sailly, M. and Volkoff, S. 1990. Vieillissement de la main d'oeuvre et adéquation prévisionnelle des postes. Le cas des ouvriers de montage dans l'industrie automobile, *Formation et Emploi*, **29**: 75–81.

Salthouse, T. A. 1984. Effects of age and skill in typing, *J. of Experimental Psychology*, **113** (3): 345–371.

Smith, J. M. 1969. Age and occupation: a classification of occupations by their age structure, *J. of Gerontology*, **24** (4): 412–418.

Teiger, C. 1975. Caractéristiques des tâches et âge des travailleurs. In A. Laville, Teiger, C. and Wisner, A. (eds) *Age et contraintes de travail* (NEB, Paris), 236–290.

Teiger, C. 1989. Le vieillissement différential dans et par le travail: un vieux problème dans un contexte récent, *Le Travail Humain* **52** (1): 21–56.

Teiger, C. 1990. Travailleurs 'vieillissants' et formation: gageure ou enjeu? In H. David (ed.) Actes du Colloque 'Le vieillissement au travail, une question de jugement', Mars 1989, (IRAT Montréal). 40–54.

Teiger, C. and Laville, A. 1981. Conditions de travail, santé, emploi—De quelques problèmes posés par l'approche ergonomique. In *Conceptions, mesures et actions en santé publique* (Ed. de l'NSERM, Paris) **104**: 309–326.

Teiger, C., Laville, A. and Lortie, M. 1981. Travailleurs de nuit permanents, rythmes circadiens et mortalité, *Le Travail Humain*, **44** (1): 71–92.

Teiger, C. and Marquié, J. C. 1991. About aging at work: some further reviewing of current work and prospective considerations. In Y. Queinnec and F. Daniellou (eds) *Designing for everyone*, Vol. 2 (Taylor & Francis London), 1613–1615.

Teiger, C. and Villatte, R. 1983. Conditions de travail et vieillissement différentiel, *Travail et Emploi*, **16**: 27–36.

Tétreault, P. 1992. Vers l'an 2000: L'impact du vieillissement de la population sur la santé et la sécurité du travail. Rapport pour la Commission de la Santé et de la Sécurité du travail, Montréal.

Vézina M., Vinet, A. and Buisson, C. 1989. Le vieillissement prématuré associé à la rémunération au rendement dans l'industrie du vêtement, le *Travail Humain*, **52** (3).

Volkoff, S. 1990. Les salarié-e-s âgé-e-s et leurs postes de travail: ce que disent les statistiques franaises. In M. David (ed.) Actes du colloque 'Le vieillissement au travail, une question de jugement', Mars 1989 (IRAT, Montréal), 63–71.

Volkoff, S., Laville, A. and Maillard, M. C. 1992. Ages et travail: contraintes, sélection et difficultés chez les 40–50 ans, *Travail et Emploi*, **54**: 20–33.

Welford, A. T. 1958. *Aging and human skill* (Oxford University Press, Oxford).

8

Work and aging: two prospective longitudinal French surveys among retired people and the active population

B. Cassou and F. Derriennic

Summary. Do working conditions enhance the development of age-related diseases or age-dependent diseases? To answer that question, we have undertaken two prospective longitudinal surveys. The purpose of the first study is to establish whether occupational risk factors have a long-term effect on health after retirement. It focuses on 993 retired subjects (men and women) aged at baseline 60 and over who were randomly selected from the file of an interprofessional supplementary pension fund in the Paris area. Information about occupational exposures and health status was collected by means of a questionnaire in 1982 and 1987. The results suggest that occupational risk factors and a high job mobility during working life might be risk factors for several impairments (cardiorespiratory and osteoarticular) and disability after retirement. The second survey is, for the moment, a cross-sectional study and was begun in 1990. It concerns 21 378 men and women wage earners who were born in 1938, 1943, 1948 or 1953 and who live in seven French regions. The subjects were randomly selected from lists of wage earners who have been followed by 380 occupational physicians. Information was collected by a personal interview with the worker and a clinical examination. The first results have shown relationships between age, low-back pain, heavy physical work and the ability to choose how to perform work. Methodological problems of these two surveys are discussed.

Keywords: aging, work, epidemiology, prospective study, osteoarticular impairment, disability.

Introduction

Studies of relations between working conditions and biological aging usually distinguish between two problems: the effect of biological aging on the individual's ability to perform his or her job, and the effect of working conditions on the aging process (Laville 1985). Epidemiological studies have principally been concerned with the latter problem.

One of the main difficulties of these studies is the lack of a clear definition of aging. Fundamentalists (Hayflick 1985) have found that it was very difficult to formulate a biological definition of the aging of a living organism. This complicates the choice of parameters to be measured in order to assess the processes involved in aging on the macroscopic scale required for epidemiological studies.

So, in studies of the relations between work and aging, biological aging has generally been approached from three points of view:

(1) reduced probability of survival. Studies have concentrated on lifespan, describing the effects of aging in terms of its effect on survival. The effect of many factors (including occupation) on individual longevity are evaluated. In a French longitudinal study, life expectancy for men at age 35 varied from 34.3 years for manual workers to 43.2 years for teachers (Desplanques 1985).

(2) Decreased functional abilities. In this case, the effects of aging are described in terms of the degree of decrement of subjects of various ages from that of healthy young workers. The French study of the car industry (Clement *et al.* 1968) is representative of this approach. He found that the increase in systolic blood pressure, decrease in hand grip strength, and decline in memory were more marked with age in auto workers in the metal-stamping, steel-working, and machine workshops than in controls (elementary schoolteachers and well-to-do subjects living in the Paris area).

(3) Increased vulnerability to disease. In the third case, aging is approached in terms of wear and tear; that is, the increased incidence of illness and infirmity that occurs with time. Aging is viewed as the accumulation of imprints that life stamps on an individual. The long-term effects of working conditions are therefore believed to be detectable only after many years of exposure (Collins and Redmond 1986).

But it is very difficult to distinguish between those effects of aging that are a result of disorders and diseases related to age in the statistical sense and not dependent on age in the causal sense, and those effects that are specific to aging (Brody and Schneider 1986). In this conception, the question of the impact of work on aging can become: do working conditions enhance the development of age-related diseases or age-dependent diseases? To answer that question, we have undertaken two prospective longitudinal surveys: the first study among retired subjects in the Paris area (IPSIE study) and the second among an active population in seven french regions (ESTEV study).

ISPIE study

Methodology

The ISPIE study focuses on 993 retired subjects who were randomly selected from the files of an interprofessionnal supplementary pension fund. A total of 327 men and 300 women (63 per cent of those selected) answered a questionnaire during home interviews. They were aged from 60 to 84 years. The average age was 69 years at the beginning of the survey. Blue-collar workers were the largest occupational group (37 per cent). Clerks and salesmen represented 23 per cent and white-collar workers 27 per cent. The interviews were carried out in subjects' homes by specially trained investigators in 1982–1983 (T1) and 1987–1988 (T2). On average, the first interview was carried out 6 years after retirement.

Information about occupational exposures was obtained during the interview using closed questions concerning eight harmful working conditions. A subject was considered to have been exposed if he reported exposure that lasted for at least 10 years.

Disorders of five major functions (musculoskeletal, cardiorespiratory, audiovisual, digestive, mental) were defined as impairments by grouping symptoms. Physical disability was measured in terms of difficulty with, or the need for help with, seven basic activities of daily life.

Several factors like age, sex, socio-occupational status, duration of work activity, living alone, smoking habits and so on were considered to be confounding factors and have been taken into account in the statistical analysis. Statistical calculations were done with SAS software. Statistical significance was measured by the χ^2 test for qualitative variables or by variance analysis for quantitative variables. Qualitative adjustment methods were derived from the Mantel-Haenzel procedures. Tests with a threshold of 5 per cent or less were considered significant.

Results

The first result of this study concerns the relationships between the evolution of the number of impairments according to job mobility. Job mobility was defined by a dichotomic variable using the number of companies and the number of branches of economic activity in the subject's work history. Those data were provided by the pension fund, which has retraced the occupational history of each subject at the time of retirement. A subject was considered to have high mobility when both the number of companies and the number of

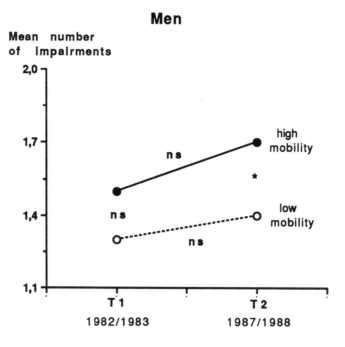

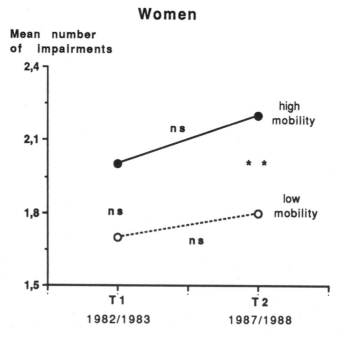

Figure 8.1. Number of impairments at T1 and T2 with relation to job mobility.
* p < 0.08; ** p < 0.05; ns: not significant

branches were higher than the average values observed in the sample. The rate of high mobility was 21 per cent for men 65/317) and 27 per cent for women (83/303).

In cross-sectional analysis (Iwatsubo et al. 1991), subjects with high mobility have statistically significant greater mean number of impairments than those with low mobility at T1 (for men 1.7 versus 1.3 and for women 2.1 versus 1.7). A total of 219 males and 245 females were seen at both interviews T1 and T2. For these subjects, the relationship between mobility and mean number of impairments is only statistically significant at T2, especially for women (figure 8.1). A slight increase in number of impairments was observed on average during the follow-up period in both sexes, but the difference was not statistically significant.

Adjustments in these relationships by potential confounding factors (age at the time of interview, previous health conditions, socio-economic categories, physical or chemical occupational exposures during working life) did not seem to modify the observed results. Health state was evaluated qualitatively by the variable 'two or more impairments' and multiple logistic regression models were used for the adjustments.

These results suggest that in our sample, the health level after retirement is statistically linked to job mobility during working life. There is no link concerning the evolution of health.

The second result concerns the frequency of osteoarticular pain according to past heavy physical work factors. Osteoarticular pain was defined as pain in the joints or spine for at least 6 months before the day of interview. In our cohort, the prevalence of pain was high at T1. Among men, 52 per cent had pain and 28 per cent restricted joint movements; these figures were higher for women: 71 per cent and 46 per cent (Derriennic et al. 1993).

Among men, the frequency of osteoarticular pain was higher among the subjects who reported more than 10 years of heavy physical work exposures (49 per cent for subjects without exposure, 58 per cent for subjects with exposure at T1). For women, the level of pain was higher than for men (figure 8.2). The relationship between pain and heavy physical work in women was significant at T1 and T2.

For men, there was a significant rise in the frequency of osteoarticular pain only for subjects exposed. For women, the rise of the prevalence was significant only for the subjects who said they had not been exposed. For men and women, these relationships didn't seem to have been affected by the possible confounding analysed factors. When the crude χ^2 test was significant, a separate adjustment was made for each variable liable to act as a confounding factor—namely, age, living alone, manual worker or not, cardiorespiratory impairment, obesity and smoking habits.

This second result suggests that the effects of heavy physical work are still felt long after retirement among men. For women, the results show a catching-up process among unexposed women as if the subjects exposed to heavy physical work presented premature osteoarticular disorders.

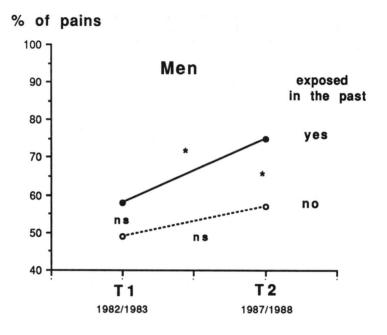

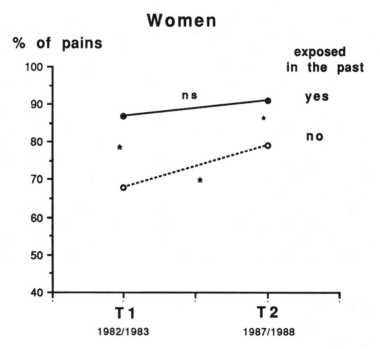

Figure 8.2. Frequency of osteoarticular pains with relation to past heavy physical work factors exposure and status between interview at T1 and T2.
* p < 0.05; ns: not significant

Table 8.1. Percentage of exposure to occupational risk factors with relation to physical disability after retirement.

Risk factor	Physical Disability			
		Yes		
	No $n = 439$ (%)	Without help $n = 130$ (%)	With help $n = 58$ (%)	p
Noise	13	14	26	*
Heat	6	5	16	*
Bad weather	10	11	12	ns
Dust	11	13	29	***
Toxic products	7	9	14	ns
Carrying heavy loads	12	16	28	**
Awkward postures	17	23	29	*
Vibrations	5	9	12	ns

* $p < 0.05$; ** $p < 0.01$; *** $p < 0.001$; ns: not significant

The third result concerns the relationships between occupational risk factors and physical disability after retirement. In a cross-sectional analysis (Cassou et al. 1992) nearly 30 per cent of the subjects reported difficulties in at least one daily life activity and 10 per cent required help for at least one. Disability was measured by the difficulties reported by subjects and by their dependence on assistance for seven basic activities of daily life: walking up and down stairs, walking on flat ground, dressing, going to the lavatory, eating, shaving and trimming toenails.

Whatever the occupational risk factor considered, the percentage of exposed subjects increased with the category of disability (table 8.1). The proportion of exposed subjects was higher among the group of subjects requiring help in basic daily life activities. Significant relationships were shown between physical disability and stressors such as noise, heat, exposure to dust, carrying heavy loads, and awkward postures.

In a logistic regression model, exposure to carrying heavy loads seemed to be specifically linked to the presence of physical disability independently of the presence of impairments or occupational status. This result suggests that some occupational risk factors might be risk factors for disability after retirement.

ESTEV study

Methodology

The second survey, called ESTEV, in French 'enquête santé, travail et vieillissement' (health, work and aging survey), is led by a multidisciplinary network

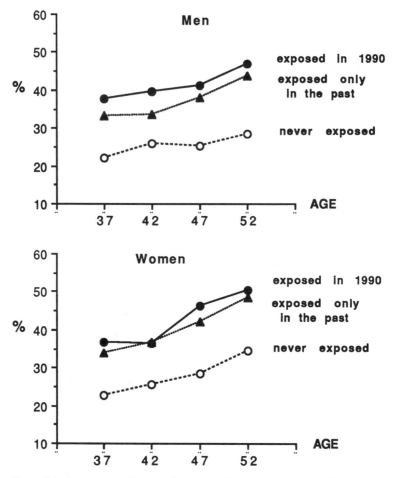

Figure 8.3. Percentage of low-back pains with relation to carrying heavy loads.

group including epidemiologists, ergonomists, gerontologists, statisticians and occupational physicians (Derriennic et al. 1992).

For the moment, this is a cross-sectional study in which the first stage was begun in 1990. It concerns 21 378 male and female wage earners who were born in 1938, 1943, 1954 or 1958 and who live in seven French regions.

The subjects were randomly selected from lists of wage earners who have been followed by 380 occupational physicians. The subjects were seen during their annual medical examination by the occupational physicians. They will be followed up and seen again in 1995.

The information was collected in a personal interview by means of closed questions, with the help of the occupational physician for the reconstitution of the working conditions at the time of the survey and during the past job. Health was measured by a list of symptoms and by the Nottingham Health

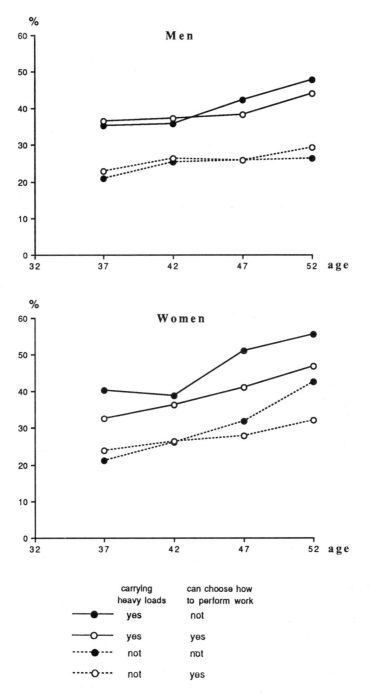

Figure 8.4. Percentage of low-back pains with relation to carrying heavy loads and 'can choose how to perform work'.

Profile, which is a subjective health scale. Various tests were performed by the occupational physician.

Results

We have chosen to present only the relationships between low-back pain, carrying heavy loads, job decision latitude and age.

The percentage of men who suffered low-back pains for at least 6 months prior to interview was higher in older age groups whatever the exposure category (figure 8.3). The percentage was higher among the group of men who were exposed to carrying heavy loads at the time of the survey, in all age groups. The percentage for men exposed in the past was a little lower. The prevalence of low-back pain in women is the same as for men, and we found the same relationships between low-back pain, age and carrying heavy loads. These relationships remained after adjustment with the other occupational constraints and with domestic and leisure physical activities for men and for women.

If we distinguish the subjects according to whether they carried heavy loads and the fact that the subjects can or cannot choose how to perform their work, we can see three kinds of relationship (figure 8.4). For men, two relationships were statistically significant: the frequency of low-back pain increased with age; and the frequency of low-back pain was higher in the group exposed to carrying heavy loads in all age groups. The third relationship, between low-back pain and the ability to choose how to perform work, was not significant. Moreover, there was no significant interaction between carrying heavy loads and age, between age and the ability to peform work, or between carrying heavy loads and the ability to perform work.

For women, we did not find the same relationships. First, there were three significant relationships between low-back pain and age, low-back pain and carrying heavy loads, and low-back pain and the ability to choose how to perform work. Moreover, there was a significant interaction between age and this ability. The percentage of low-back pain was much higher when the subject could not choose how to perform work at both 47 and 52 years of age, whatever the physical exposure.

These different results according to sex show the complexity of relationships between work and aging. Physical and non-physical work conditions could interact with aging processes, in particular in lumbar localization, for women.

It will be interesting, after the second stage in 1995, to check these cross-sectional relationships by undertaking a longitudinal analysis and to determine the true association with age.

Discussion

The interpretation of these results is not easy. In addition to the difficulties common to all fields of epidemiological research concerned with occupational risk, notably the measure of the risk factors in the past of the subjects, there are specific obstacles connected with the definition and evaluation of aging, the design of the survey and statistical analysis.

Although the incidence of disease increases with age, aging and disease are not synonymous. But the effects of aging and disease are difficult to separate. So the concept of 'age-related disease' or 'age-dependent disease' seems more convenient than aging, as there are no biological markers of aging. Moreover, with the aging of the population, we are mostly dealing with chronic conditions and most disability among elderly people is associated with age-dependent diseases.

The effect of work on aging can be measured in two ways: cross-sectional and longitudinal designs. The findings of a cross-sectional study cannot, however, be interpreted as demonstrating age changes. Only average differences between age groups are identified. On the other hand, while longitudinal studies identify age changes, they have a number of limitations of their own. For example, drop-outs owing to loss of contact or to subjects' refusal to continue participation raise a problem in the interpretation of results.

The time interval between each interview must be compatible with the evolution of the parameters which have to be statistically detectable on groups of populations. Moreover, an aging effect is only present if the dependent variable is a function of age regardless of the subjects' birth year or the period of observation. If cross-sectional differences include age and cohort effects, longitudinal changes include age and period effects.

Lastly, logistic and multilinear models are necessary in order to take simultaneous account of all the variables—the dependent variables, explanatory variables and the confounding factors. But statistical problems such as over-adjustment or lack of power can mask the relationship between work and aging. Analyses of cumulative exposures may often suffice to indicate whether or not there is an association between exposure and disease. If one wants to explore more deeply the exposure–disease relationship, however, for example to determine if a factor appears to operate at an early or late stage in the aging process, it is important to take into account the time pattern of the studied exposure and of the relevant confounders. The effects of working conditions may be a function of a factor's intensity, the period over which it acts and individual vulnerability at the time. But analytical techniques are very complex and you need a relevant work history. Time-related confounders are also particularly important in occupational cohort studies because of the 'health worker effect'. This effect declines with the length of follow-up (Pearce 1992).

In these two studies, the analytical model used is the one illustrated in figure 8.5. O symbolizes a biomarker of aging or a particular health characteristic liable to reflect some aspect of aging. Even when the significance of the parameter O is not certain, the ability to distinguish clearly between pattern 1 and 2 provides information on the rate of evolution of health or the aging criterion considered, and not solely on the level of this criterion at a given time.

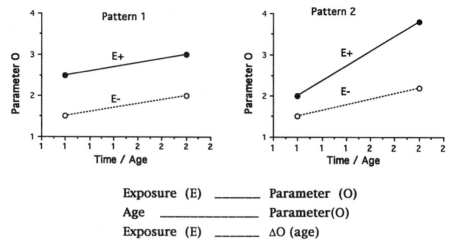

Figure 8.5. General patterns to longitudinal studies.

From this point of view, the work–aging relationship can be analysed as an interaction, in the statistical sense, between duration (that is, the time elapsed between at least two assessments) and an occupational characteristic. But, there are two implications of this model:

(1) that aging processes are intimately involved in the development of age-dependent diseases;
(2) that the risk for the expression of specific age-dependent disease will depend on working factors that may increase susceptibility. Obviously, other environmental or genetic factors can also modify the risk.

Conclusion

In conclusion, until recently, only very few researchers in the occupational epidemiological field were interested in the relationship between work and aging. Our results should encourage the organization of more ambitious studies with a longer longitudinal follow-up and more precise measurements of occupational exposure and work histories. On the other hand, there is a general agreement that prevention of unsuccessful aging should begin before retirement. Consequently, it is important not to forget the possible long-term effects of occupational risk factors. These effects could explain in part the inequalities in health found in later life.

References

Brody, J. A. and Schneider, E. G. 1986. Diseases and disorders of aging: an hypothesis, *Journal of Chronic Diseases*, **39**: 871–876.

Cassou, B., Derriennic, F., Iwatsubo, Y. and Amphoux, M. 1992. Physical disability after retirement and occupational risk factors during working life: a cross-sectional epidemiological study in the Paris area, *J. Epidemiol. Community Health*, **46**: 506–511.

Clement, F., Cendron, H. and Housset, P. 1968. Le vieillissement differentiel d'une population ouvrière de la région parisienne, *Bulletin de l'INSERM*, **23**: 889–920.

Collins, J. F. and Redmond, C. K. 1986. The use of retirees to evaluate occupational hazards, *J. of Occupational Medicine*, **28**: 595–602.

Derriennic, F., Iwatsubo, Y., Monfort, C. and Cassou, B. 1993. Evolution of osteoarticular disorders as a function of past heavy work factors: longitudinal analysis of 627 retired subjects living in the Paris area, *Br. J. of Industrial Medicine*, **50**: 851–860.

Derriennic, F., Touranchet, A. and Volkoff, S. 1992. Enquête ESTEV: un instrument d'étude des relations entre âge, santé et travail, *Arch. des Maladies Professionnelles*, **53**: 79–89.

Desplanques, G. 1985. La mortalité: un phénomène social, *Revue Française de Santé Publique*, **29**: 33–45.

Hayflick, L. 1985. Theories of biological aging, *Exp. Gerontology*, **20**: 145–159.

Iwatsubo, Y., Derriennic, F. and Cassou, B. 1991. Relation between job mobility during working life and health state after retirement: a cross-sectional study of 627 subjects living in the Paris area, *Br. J. of Industrial Medicine*, **48**: 721–728.

Laville, A. 1985. Travail et vieillissement: le point de vue de l'ergonomie, *Revue Française de Santé Publique*, **29**: 14–18.

Pearce, N. 1992. Methodological problems of time-related variables in occupational cohort studies. *Revue Epidemiologie et Santé Publique*, **40**: S43–S54.

9

Job demands and work stress in relation to aging

W. J. A. Goedhard

Summary. This study deals with the possible relationship between aging, the development of age-related diseases and the influence of working conditions such as job stress. When looking at the response in heart rate to an orthostatic manoeuvre, it can be observed that this response decreases with advancing age. This is owing to a diminished responsiveness of the baroreceptor reflex, in particular of the functioning of the aging myocardial cell. This decrease was also observed in this study. When the perception of job stress was studied, a small increase in stress was found with age. A more striking observation was the finding of a substantial decrease in baroreflex response with increased job stress. The contributions of job stress and age on the decrease of baroreflex function are of the same order of magnitude. Based on these findings, a model is proposed that may explain the relation between aging and the development of disease. The rate of aging is supposed to be the critical variable. This variable is influenced by environmental stimuli. Job stress may have a considerable deleterious effect on the rate of aging and therefore on the development of age-related diseases.

Keywords: work stress; baroreflex; aging; model.

Introduction

In the twentieth century a substantial increase in life expectancy was observed in the developed countries. This increase in calendar years has so far not been followed by a proportional number of healthy years for most people. On the contrary, the quality in the third phase of life, that is the period over 65 years of age, leaves much to be desired. Many older people suffer from chronic diseases or are disabled and have become dependent on care facilities. Many of these chronic diseases are age-related (Butler 1988). The development of these diseases has something to do with the aging process, although the exact nature of the relation between aging and disease is not clear. Gerontologists, among others, are diligently studying the question of how to maintain good health and preserve good quality of life with advancing age. Compression of morbidity (Fries 1980) is an important goal of present and future health care. Prevention may become an important instrument aimed at the delay or postponement of age-related diseases (Berg and Cassels 1990). Appropriate preventive measures should be initiated as early as possible. The many chronic and debilitating diseases of old age probably have their beginning some decades before clinical signs and symptoms begin. For example, the development of skeletal or cardiovascular diseases may be strongly enhanced by exposures earlier in life, for example at work. Therefore, it may be important to observe occupational exposures, since the active working life of most people precedes the period of increased morbidity, disablement and mortality.

One may speculate how and why various diseases show a relationship with age. It is likely that with advancing age the organism becomes more vulnerable because cellular or systemic functions begin to fail. It is not known what causes the aging process, but the effects on functions and structures can be observed and measured. The rate of aging may vary considerably from one individual to another. Environmental stimuli can probably influence the rate of aging. An increased rate of aging then may be associated with an increased vulnerability (Fabris 1992).

Work stress can be considered such an environmental stimulus. It has been demonstrated that adverse work stress conditions give rise to hypertension (Theorell et al. 1988) and increased cardiovascular morbidity (Johnson et al. 1989). Hypertension is one of the important chronic conditions at a later age, since it is a major risk factor for the development of cardiovascular diseases. It should be realized that the majority of cases of cardiovascular disease, and subsequent mortality, occur after the age of 65 (WHO 1987).

Theoretically, systemic arterial blood pressure should remain constant within narrow limits over the life-span because the organism is provided with a number of control systems (Guyton 1980). Yet, it is usually observed that blood pressure tends to increase with age. One of the reasons could be the decrease of function of the various control systems with advancing age. Baroreflex function can be considered a suitable model of aging. In earlier studies it was shown that baroreflex function shows an average decrease of about 1 per cent per year (Goedhard et al. 1985).

In this study the following aspects of work stress in relation to aging were studied:

(1) Does the perception of work stress change with age?
(2) Is baroreflex function influenced by stress factors?
(3) What might be a suitable model to explain the relation between job stress, aging and disease?

Methods and materials

Psycho-social work stress was examined in a group of 162 overtly healthy workers with an age range of 20–64 years. Use was made of a questionnaire containing 34 items, published by Frommer et al. (1986). It was translated into Dutch and gave satisfactory results in an earlier test examination (unpublished results). Information is obtained on the following six stress factors: sf1 = boredom, sf2 = lack of support, sf3 = quantitative overload, sf4 = qualitative overload, sf5 = unsatisfactory prospects or salary, sf6 = unsatisfactory physical working conditions; range of scores: 1 (eustress) to 5 (distress).

The baroreflex function was studied by means of the Finapres method. Mean arterial pressure and heart rate were recorded continuously on a four-channel Gould recorder. This was done at rest in supine position as well as during and after an orthostatic manoeuvre. In standing position BP was recorded for at least 1 minute. This procedure was followed in a subgroup of 25 randomly chosen healthy workers.

Results

Stress and age

Stress factors were measured in the group of 162 workers. The results, which are published elsewhere in more detail (Goedhard 1993), showed that the highest stress is experienced from sf3 (quantitative overload): mean score 2.1; sd = 0.6 (figure 9.1).

No statistically significant changes were observed between the scores of stress factors 1, 4, 5 and 6, respectively, and age; for stress factors sf2 and sf3 a small, but statistically significant, increase with age was observed; with linear regression the following results were obtained:

$$sf2 = 1.41 + 0.009 \times age \ (r = 0.20; \ p < 0.01)$$

$$sf3 = 1.61 + 0.013 \times age \ (r = 0.24; \ p < 0.01)$$

Blood pressure, baroreflex function and age

Mean arterial BP is derived from the recording preceding the orthostasis; baroreflex function is assessed by measuring the rise in heart rate (dF) on orthostasis or by calculating baroreflex sensitivity from the ratio of dF and dP (i.e., the change in pressure). In earlier studies it was found that dF is strongly negatively associated with age (Goedhard et al. 1985).

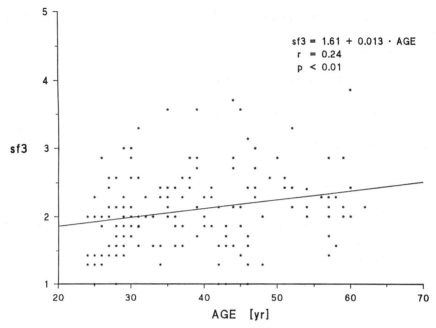

Figure 9.1. Quantitative overload (stress factor 3 = sf3) increases slightly but significantly with age (for explanation: see text).

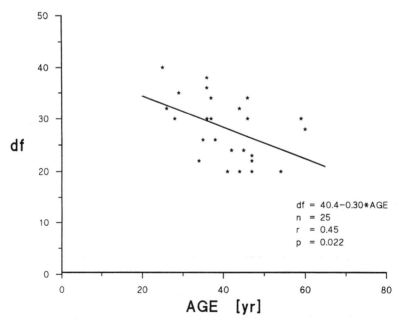

Figure 9.2. The response in heart rate (df, beats per minute) on an orthostatic manoeuvre decreases with advancing age.

On orthostasis, the increase in HR (dF) in the studied group appeared to be age-related (see figure 9.2).

The regression equation

$$dF = 40 - 0.3 \times age (r = 0.45; p < 0.05)(BPM)$$

indicates that this variable decreases by about 0.9 per cent per year (dF at 20 years = 100 per cent). This confirms the results of earlier observations (Goedhard et al. 1985).

Work stress and baroreflex function

The relations between blood pressure and stress factors were not statistically significant. However, a significant inverse relation was observed between some stress factors, for example sf3, and age (figure 9.3).

The regression equation

$$dF = 45.6 - 7.87 \times sf3 \ (r = 0.50; p < 0.02) \ (BPM)$$

demonstrates that the correlation is of the same order of magnitude as the relation of dF and age.

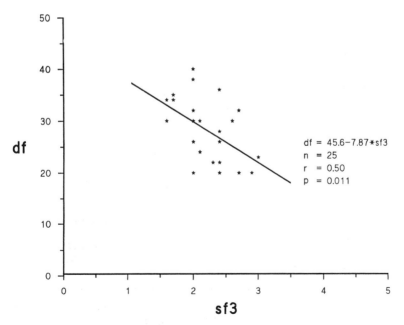

Figure 9.3. The response in heart rate on orthostasis in relation to perceived stress. The response is lower at higher stress levels: for sf3 = 1.00 (eustress) df = 38 bpm; for sf3 = 3.00 (moderate distress) df = 22 bpm.

By means of multiple regression the following best fit was obtained:

$$dF = 54 - 0.19 \times age - 3.2 \times sf1 - 5.4 \times sf3 \; (BPM)$$

$$(r = 0.65; \; p < 0.01; \; \text{explained variance: 42 per cent})$$

Discussion

Perception of job stress obviously varies in different age groups, but the differences observed in this study are not impressive. Besides, it should be realized that cohort effects might also be present. Anecdotal evidence during the past decade suggests that increased work pressure is experienced by many in comparison with the 1960s and 1970s. That might account for the perception of increased quantitative workload.

The relation of age and baroreflex function is quite another matter. The phenomenon is clearly based on aging mechanisms, probably situated in the myocardial cells (Lakatta 1986). It is rather surprising, then, that a similar relationship between perceived job stress and baroreflex function is observed. This study shows that aging as well as job stress exert a negative influence on baroreflex function. The contributions of the two variables are of the same order of magnitude. This would imply that too high job stress increases the rate of aging.

In order to obtain more insight into the relation between aging and disease, the rate of aging seems to be a variable of crucial importance (Hazzard 1985). If the above-mentioned mechanism is true, one has to assume a relationship between the perception of job stress and the functioning of the myocardial cell. The response of the latter seems to deteriorate as a consequence of increased job stress.

A model

When studying the relation between aging and chronic age-related diseases, considerable interindividual differences can be observed. With advancing age the differences in functioning of people seem to be increasing. This may be a result of a difference in the rate of aging. This variable is probably dependent on genetic factors as well as environmental factors. Description of a model about the relation between aging and disease was presented elsewhere (Goedhard 1993). In this model the rate of aging is presented as a variable to be controlled in analogy to physiological control systems. So far, only semi-quantitative data are available to prove that the model is correct. The findings of this study support this model. Stress is considered to be an adverse environmental factor that influences the rate of aging. In figure 9.4 a model is depicted that contains aging as well as stress as determinants of the rate of aging.

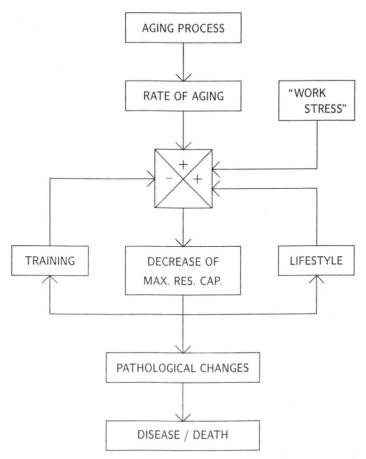

Figure 9.4. A model to describe the relation between aging and disease with job stress as an environmental stressor that might increase the rate of aging. With advancing age a decrease in maximum reserve capacity is observed in many functions. Training (physical exercise) or an adverse lifestyle (smoking, alcohol) may respectively decrease or increase the rate of aging.

Early retirement owing to ill health and disablement is often observed in older workers. In a great number of cases high job stress is the leading cause of premature retirement. Increased absenteeism and adverse lifestyles, as means of stress reduction, are usually observed in the years preceding the development of disabling diseases. Sensible preventive measures should be developed and implemented to reduce job stress and to slow the rate of aging.

References

Berg, R. L. and Cassells, J. S. (eds) 1990. *The Second Fifty Years. Promoting Health and Preventing Disability* (National Academic Press, Washington).

Butler, R. N. 1988. Health and aging: the new gerontology. In J. J. F. Schroots, J. E. Birren and A. Svanborg (eds) *Health and aging* (Springer, New York), 143–153.

Fabris, N. 1992. The aging process in humans; theories on aging. In W. J. A. Goedhard (ed.) *Aging and work* (Pasmans, The Hague), 9–31.

Fries, J. F. 1980. Aging, natural death, and the compression of morbidity, *New Engl. J. of Med.*, **303**: 130–135.

Frommer, S., Edye, B. V., Mandryk, J. A., Berry, G. and Ferguson, D. A. 1986. *Systolic blood pressure in relation to occupation and perceived work stress*. Scand. J. Work Environm. Health **12**: 476–485.

Goedhard, W. J. A., Wesseling, K. H. and Settels, J. J. 1985. Baroreflex pressure control responding to orthostatic changes with age. In J. F. Orlebeke, G. Mulder and L. J. P. van Doornen (eds) *Psychophysiology of Cardiovascular Control* (Plenum, New York), 191–202.

Goedhard, W. J. A. 1993. The relation between psychosocial stress and aging in a working population. In J. Ilmarinen (ed.) *Aging and Work*. Proceedings 4 (Institute of Occupational Health, Helsinki), 25–32.

Guyton, A. C. 1980. *Arterial Pressure and Hypertension* (Saunders, Philadelphia).

Hazzard, W. R. 1985. Aging and atherosclerosis. In M. C. Geokas (ed.) *The Aging Process. Clinics in Geriatric Medicine*, Vol. 1 (Saunders, Philadelphia), 251–284.

Johnson, J. V., Hall, E. M. and Theorell, T. 1989. Combined effects of job strain and social isolation on cardiovascular disease morbidity and mortality in a random sample of the Swedish male working population. *Scand. J. Work Environ. Health* **15**: 271–279.

Lakatta, E. G. 1986. An integrated approach toward understanding myocardial aging. In M. Bergener, M. Ermini and H. B. Stähelin (eds) *Dimensions in Aging* (Academic Press, London), 105–132.

Theorell, T., Perski, A. and Akerstadt T. 1988. Changes in job strain in relation to changes in physiological state: a longitudinal study, *Scand. J. Work Environ. Health*, **14**: 189–196.

WHO 1987. *Prevention of Cardiovascular Disease among the Elderly* (WHO, Geneva).

10

The aging shiftworker: adjustment or selection? A review of the combined effects of aging and shiftwork

B. C. H. de Zwart and T. F. Meijman

Summary. Compared with the total work-force in industrialized West-European countries, the mean age of shiftworkers is also expected to increase over the next few decades, as a result of economic, demographical and social changes. This process has given rise to a growing interest in the interaction between age and shiftwork. It is hypothesized that, with increasing age, people are less able to adjust to shiftwork and that health complaints become manifest. These theories will be studied in this paper by reviewing the results of the international scientific literature. First, a theoretical model is presented. Then the effects of age and shiftwork on sleep disturbances and health complaints will be discussed. Next, the influence of various selection mechanisms on shiftwork populations are addressed. Finally, the possibilities of preventive measurements and the need for further research will be stated.

Keywords: age, shiftwork, sleep disturbance, health complaints, selection effects.

Introduction

Shiftwork is and will continue to be a common way of organizing work in industry and services. There are several reasons for this trend: an increase in continuous technological process operations, the economic benefits of 24-hour production and, finally, an ongoing increase in social demands which have to be met round the clock—for example, in the fields of transport, health, security, information and retail distribution (Corlett et al. 1988).

The number of workers in The Netherlands engaged in shift- or unusual-hours-work is close to 10 per cent of the total work-force (CBS 1990). Over the next few decades, the mean age of the work population, including shiftworkers, is expected to increase. This will be the result of demographic, economic and social changes. Present low birth rates and the aging 'baby boom' generation will contribute to a greying society. Therefore, the proportion of workers aged between 50 and 65 years is expected to rise from 14 per cent in 1991 to 21 per cent in 2020 (CBS 1991). This trend can also be observed in other European countries (ILO 1992). Over the past decades, however, there has been an accelerating trend towards early retirement. This process has been encouraged by public policy in response to rising unemployment. Organizations and governments are now faced with the task of reversing this trend and controlling the rising costs of early retirement and the social security system.

The change in the economic value of the greying work-force leads to a growing interest in the impact of the aging process on the balance between work capacities and demands. A negative disturbance of this balance may result in health problems in the short and long term. Work capacities normally decline with increasing age. This is characterized by an ongoing decline of physiological variables (Dehn and Bruce 1972, Åstrand et al. 1973, Babcock et al. 1992; Åstrand 1960, Robinson et al. 1973, Nygård et al 1991). An age-related decline of the maximal work capacities results in a decrease of the reserve capacities, even when work demands remain at a stable level (Ilmarinen et al. 1991). Work in shift systems is associated with heavy work demands, as a result of continuous, mostly irregular, changes of circadian rhythms caused by a dephased sleep–wake rhythm (Thierry and Meijman 1993). Moreover, jobs in shiftwork are characterized by more unfavourable stressors than jobs in daywork (Rutenfranz and Knauth 1986). One would therefore expect that, especially among the aging shiftworker, the balance between work demands and work capacities will be increasingly, and dangerously, disturbed.

In this review paper, the findings of the combined effects of age and shiftwork, as reported in the international scientific literature, will be discussed. Shiftwork and aging; adjustment or selection?

Theoretical framework

Nowadays, a wide variety of work schedules can be found in work organizations (Tepas and Monk 1987). For example, Jansen (1987) identified 925 different rosters or schedules in a large-scale survey of Dutch industries. Most of these different shift combinations can be classified into one of the three following basic shift systems: (1) discontinuous systems, with an interruption in work at the end of the day and at the weekend—usually two crews; (2) semi-continuous systems, with an interruption in work at the weekend—usually three crews; (3) continuous systems, involving 7-day workplace operations, 24 hours a day—usually four crews or more.

Over the past 30 years, a majority of shift system studies—particularly those carried out in Europe—revealed the impact of shiftwork on human well-being. It is possible to categorize these studies under one of the following 'shiftwork' headings: circadian rhythms, sleep disturbance, consequences for health and well-being, impaired performance and adjustment. During the past decade there appears to be a growing interest in the interindividual differences in response and tolerance to shiftwork (Kerkhof 1985; Härmä 1993, Monk and Folkard 1985, Thierry and Meijman 1993). This review focuses mainly on the differences in reactions to shiftwork on a single factor, age. A number of studies revealed that age, as well as shiftwork experience, plays a significant role in adjusting to shiftwork (Tepas et al. 1993, Frese and Okonek 1984, Keran and Duchon 1990). According to the 'shiftworker phases of sequential development' model of Haider et al. (1981, see also Rutenfranz et al. 1985, Tepas et al. 1993), the shiftworker passes through several phases during his shiftwork career (figure 10.1). Within each phase, different time-contingent variables determine the individual tolerance to shiftwork. To delay intolerance and drop-out, different phase-specific coping mechanisms must be developed during each phase.

With his entry into a shift system, adaptation to the concomitant problems of working unusual hours directly becomes one of the major concerns for the shiftworker. During this 'adaptation phase' the shiftworker is faced with changes in sleeping and eating behaviour, family life, participation in social activities, work strain and so on. After 5 years, the shiftworker enters the 'sensitization phase', in which the development of work career, financial safety, satisfaction with shiftwork and family and social life are the most important factors for shiftwork tolerance. Within the 'accumulation phase', after about 20 years of work on shifts, the accumulation of environmental hazards and the hazards of coping strategies becomes manifest, strengthened by the effects of the biological aging process. Risk factors, sleep quality and attitudes to shiftwork then appear to have the strongest influence on health state and tolerance. Finally, following the accumulation phase, a proportion of aging shiftworkers may enter the 'manifestation phase', before or sometimes even after the retiring age. This phase is characterized by a strong rise in the incidence of disorders and diseases caused by the individual's inability to cope with shiftwork any longer. The age at which shiftworkers enter this phase show large interindividual differences as a result of individual differences in tolerance to shiftwork (Härmä 1993). Tepas et al. (1993) proposed that: 'at some point, the worker reaches a tolerance limit, where disturbance is probable, coping with shiftwork becomes impossible, and drop-out is likely'. As has been suggested by several authors (e.g. Frese and Okonek 1984, Koller et al.

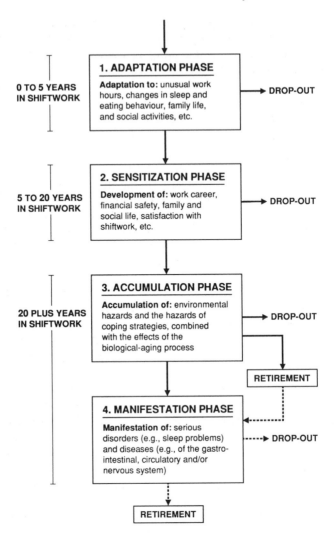

Figure 10.1. Shiftworker four phases of sequential development (after Tepas et al. 1993, adapted from Haider et al. 1981). During his or her career, a shiftworker may pass through four different phases: the adaptation phase, the sensitization phase, the accumulation phase, and, finally, a select group will reach the manifestation phase before or after retiring age. In this phase, serious disorders and diseases become manifest. Within each phase, drop-outs can be found as a result of the individual inability to cope with time-contingent variables.

1978, Frese and Semmer 1986), individual reasons for drop-out differ widely. Hence, drop-outs can be found in all phases of the model, as a result of differences in the individual ability to cope with time-contingent variables.

Analogous to the general aging process, the career of a shiftworker may be considered as a dynamic process, with a complex network of interacting factors. To develop a future policy which improves the quality of working conditions for the increasing number of older shiftworkers, it is of primary importance to gain a clear understanding of the structure of this process as well as the incidence of disorders and diseases.

Effects on health

Sleep disturbances

Although there is still a debate in the literature on the function of sleep, restoration of the human mind and body is supposed to be one of its manifestations (Carskadon and Roth 1991). This daily process of restoration is unfavourably affected by sleep disturbances. In a number of studies, disturbed sleep/wakefulness is reported as one of the main problems accompanying shiftwork (e.g. Rutenfranz et al. 1981, Torsvall et al. 1989, Tepas and Mahan 1989, Åkerstedt 1990, 1991).

Three factors are behind sleep disturbances. One, unusual working times interfere with normal sleep patterns. Two, day-time sleep is strongly influenced by circadian and environmental factors. Three, phase-shifted sleep affects human circadian rhythms (Rutenfranz et al. 1977, Åkerstedt 1991). The sleep deprivation model of Tepas and Mahan (1989) proposes that working one night shift exposes the worker to the effects of acute partial sleep loss. When consecutive night shifts are worked, the effects of acute partial sleep loss, plus the carry-over, must be considered. According to Åkerstedt et al. (1990), Tepas and Carvalhais (1990), Dinges and Kribbs (1991), and De Zwart et al. (1993), it is reasonable to assume that this chronic sleep deprivation or sleep disturbance of night shiftworkers may have hazardous consequences on physical performance capacity, productivity, safety, and health in the short and long term. During the night shift period, a day sleep is generally associated with difficulties in remaining asleep and a feeling of not having been sufficiently restored (Åkerstedt and Gillberg 1981, Åkerstedt 1990). Sleep length during the day after a night shift is normally reduced to 1 to 4 hours. This reduction seems to affect mainly stage two and REM sleep (Walsh et al. 1981, Åkerstedt 1984, 1990). It was found by Åkerstedt and Gillberg (1981) that this shortened daytime sleep is the result of a circadian rhythmicity of sleep and, to a lesser extent, of environmental factors.

Aging is considered to modify normal sleep patterns dramatically (Webb 1978, Ehlers and Kupfer 1989). Besides a reduced slow-wave activity of the EEG, an increase in the number and duration of arousals from sleep is reported (Miles and Dement 1980). The results of several studies also confirm a stronger disturbance of sleep and wakefulness of the aging shiftworker compared with his younger and non-shiftworker colleagues (e.g. Lavie et al., 1989, Härmä et al. 1992). In the extensive and well-known study of Foret et al. (1981), for example, sleep quality of 700 shiftworkers in a large oil refinery was investigated. In each group, shiftworkers showed the poorest sleep qualities–compared with non- and former shiftworkers. Increasing age and length of service in shiftwork were also found to affect sleep quality adversely. While for the younger workers (< 29 years) a poor sleep index was associated with difficulties in falling asleep, for the older ones (> 40 years) it was associated with night-time or early morning awakenings.

Recently, a laboratory study investigating young and middle-aged men who were isolated from environmental time cues for 15 days, also indicated aging as an aggravating factor for sleep disturbance (Moline et al. 1992). For the first 4-day interval, after a shift of daily routines by 6 hours backwards, middle-aged men reported larger increases in waking times during the sleep period and earlier termination of sleep compared with the younger subjects.

Åkerstedt and Torsvall (1981) also found reduced sleep quality and sleep length with increasing age and years in shiftwork. Age was found to be the main predictor for sleep length and well-being in the younger groups. For the older subjects, however, years of shiftwork experience was found to be the main predictor. It was suggested that at the age of 45 years, mostly corresponding to 20–25 years experience in shiftwork, deterioration of sleep length and well-being is the result of a process of accumulation. Within the younger groups, this effect may be modified by continuous habituation and selection.

Recent work by Tepas et al. (1993), also showed the main effects of shift, age and gender on sleep length. The longest average sleep length was reported by the youngest age group (18–29 years), while the shortest length was reported by the 40–49 age group. Significant interactions were found for age by shift, shift by gender and age by shift by gender.

Torsvall et al. (1981) studied two groups of locomotive engineers (age 25–35 and 50–60). During day sleep the older group showed relatively more stage shifts, awakenings, stage one sleep, higher diuresis and noradrenaline excretions. As frequently reported in other studies, however, no differences were found in sleep length between both groups.

Keran and Duchon (1990) reported a significant negative relationship between sleep length and age. During the afternoon shift, older workers tended to wake up earlier in the morning compared with the younger workers. During the nightshift period this older group started to sleep later in the morning. An age-and-shiftwork interaction was also found for sleepiness on the shifts.

According to Keran and Duchon, these problems of the older workers during the morning hours might be the result of their classification as 'morning types'. Several studies have reported that morning type is highly correlated with advancing age (Åkerstedt and Torsvall 1981, Kerkhof 1985) and difficulties in adapting to nightwork (Östberg 1973, Torsvall and Åkerstedt 1980, Breithaupt et al. 1980, Moog 1987). In addition to the influence of age on diurnal type, an interaction between age and personality (introvert/extrovert) was also found, resulting in disturbance of the sleep–wakefulness pattern (Tune 1969).

Health complaints

With the adverse combination of increasing age and long-term exposure to shiftwork and its concomitant effects, a rise in health risk may be expected. For many years, research studies have been designed in order to investigate the health risks of shiftwork. However, only a few of them focused on the interaction between age and shiftwork.

A retrospective cohort study over a period of 10 years, in which the sickness records of shift- and dayworkers with similar work were analysed, was carried out by Angersbach et al. (1980). No differences in sickness rates were observed between the groups, which was the same finding as reported in a similar study by Walker and de la Mare (1971). However, shiftworkers reported more gastrointestinal and skeletal diseases; absenteeism owing to gastrointestinal disease was twice as high as compared with dayworkers. Significantly more shiftworkers than dayworkers (46 per cent versus 37 per cent) had to visit their occupational health services because of gastrointestinal diseases. After the fifth year, the shiftworkers fell ill more often for the first time than the dayworkers.

Differences in cardiovascular diseases between both groups were not observed. Knutsson et al. (1986), however, reported an increase in the relative risk of ischaemic heart disease (IHD) with increasing years of exposure to shiftwork, as compared with dayworkers. A significant risk was associated with an exposure of 11–20 years, and age was found to be the main predictor, followed by shiftwork experience (see also Åkerstedt et al. 1984, 1986).

An age-and-shiftwork interaction on health state was also observed in a large, questionnaire-based study within a population of shiftworking and dayworking police officers (Ottman et al. 1989). Age and shiftwork effects influenced prevalence rates of autonomous and musculoskeletal symptoms (increase with age) and respiratory infections and disturbance of appetite and indigestion (decrease with age). In spite of the fact that all dayworkers had previous shift experience, shiftworkers complained more often than dayworkers in all age groups.

Keran and Duchon (1990) also found in older shiftworkers an increase of problems of digestion, breathing, fatigue, feeling sweaty/trembly, leg cramps and tight stomach. Koller et al. (1983) used a number of indicators to define a health score for shift- and dayworkers—for example, absence owing to sickness, morbidity, distribution and severity of diseases, and subjective complaints. These health scores were found to deteriorate with age, and in different patterns and proportions for the shift- and dayworkers. In all age groups, shiftworkers revealed worse health scores as compared with day workers. Whereas dayworkers reported a stable health score during their first years, followed by a constant decline from the age of 34, those of shiftworkers were characterized by a steep decrease during the initial years at work, a flattening phase in the middle age group—until 40 years of age—and, finally, a second steep decrease from the age of 41 onwards. Health scores were used also by Meers et al. (1978) as an indicator of a shiftworker's well-being. In a newly set-up plant, subjective scores were obtained from those who were about to start work in shifts. After 6 months a deterioration in health score was observed, which was even more pronounced 4 years later.

The results from the studies mentioned above show some variation. It is assumed that this can be ascribed primarily to the influences of selection processes within shiftwork and the type of research design (Angersbach et al. 1980).

Selection effects

It is assumed that those who leave shiftwork are distinct from the group of shiftwork 'survivors'. In addition to age and years of exposure, several other time-contingent variables seem to modify these selection mechanism (Frese and Okonek 1984, Frese and Semmer 1986). According to the 'four-phases' model (figure 10.1), drop-outs can be found in all phases. Hence, throughout the aging process the individual shiftworker is subject to several selection processes, which can be classified into a number of categories.

First, the decision to apply for shiftwork at an early age can be considered as a 'pre-selection'. The attractive financial benefits of working unusual hours appears to be one of the main reason for this choice (Koller et al. 1978, Frese and Okonek 1984, Tepas et al. 1993). Social and family reasons, on the other hand, seem to prevent workers from applying for this type of work schedule (Koller et al. 1978). According to a recent study by Knutsson and Åkerstedt (1992), it is also conceivable that pre-selection is based on sleep behaviour. In contrast with those who apply for daywork, shiftworkers showed less rigid sleep patterns.

Second, during the first years of shiftwork an 'adaptation-selection' can be noticed. If employees demonstrate a physical inability to adjust to unusual work/sleep hours or suffer from other adjusting problems—for example, family life, social activities—this could be a reason to drop-out from shiftwork at an early age.

Third, a 'career-selection' might appear. To improve their work career, some shiftworkers switch to day-time jobs with better prospects. Studies have often revealed that drop-outs have higher skills compared with those who pursue shiftwork (Frese and Okonek 1984, Frese and Semmer 1986). Therefore, it is assumed that a higher educational level, especially among younger shift-workers, increases the chances of switching to a day-time job, often without deterioration of financial income or work career.

Finally, a 'health-selection' can be observed, owing mainly to the manifestation of serious disorders and diseases with advancing age and exposure to shiftwork. In a number of studies, a strong decline in health complaints around the age of 50 was ascribed to this 'healthy-worker' selection effect (Ottman et al, 1989, Knutsson et al. 1986, Oginska et al. 1993).

A questionnaire-based study by Koller et al. (1978) indicated that health problems (28 per cent) were the main reason for dropping out from shiftwork, followed by family (15 per cent) and career (13 per cent). One would therefore expect to find the deterioration of health more pronounced within the drop-out group, as compared with the group of 'survivors'. This has often been the main hypothesis of research studies. Aanonsen's (1959) frequently quoted study, for example, revealed higher frequencies of observed disorders in the group who had stopped working shifts.

In another study, workers with a considerable prior exposure to shiftwork—drop-outs—reported exceptionally high rates of sick leave and cardiovascular diseases compared with those who continued working shifts and dayworkers (Koller 1983). For drop-outs, and shiftworkers as well, a different distribution of frequencies of diseases with age was observed. Frese and Semmer (1986) also carried out a study with former shiftworkers who stopped working because of health problems or other problems. Those who had dropped out because of health problems were more unhealthy than other shiftworkers and non-shiftworkers. However, no effects for age were observed. The results of Koller et al. (1978) indicate pronounced subjectively perceived health problems in drop-outs, even several years after leaving shiftwork. It was concluded that, as a result of shiftwork, 'long-lasting psychosomatic, pseudoneurotic or sensitization reactions may develop in certain individuals'.

These results support the idea of various selection processes among shiftwork populations (Knutsson and Åkerstedt 1992). These processes might result in an underestimation of the actual impact of shiftwork on the health and well-being of the aging worker, as reported in the scientific shiftwork literature.

Discussion

From the results reported in the literature, it seems clear that several time-contingent factors play an important role in the adjustment and selection to shiftwork throughout the shiftworker's career—for example, sleep disturbance, family life, social activities, work career, financial benefits, health state and so on (Frese and Okonek 1984). The strength of each factor on these processes is strongly related to age and, even more, to years of experience in shiftwork. In general, those who work in shifts can be divided into five groups:

(1) Those who drop out from shiftwork within a few years, owing to problems of adaptation.

(2) The group of—mostly young—workers who switch to a day-time job with better prospects to improve their work career.

(3) Shiftworkers who still continue in spite of adaptation and health problems. For this group, the attractive financial benefits of working

unusual hours and less chance of a career in daywork are the main
reasons for continuing. At a certain stage in their career, coping with
shiftwork becomes impossible, and drop-out is likely.

(4) The fourth group consists of those workers who adapted to shiftwork,
and finally have reached the manifestation phase where drop-out as a
result of serious health problems is expected.

(5) Finally, a small group of shiftworkers continues work up until retirement age without significant deterioration in their health state during
their active shiftwork career.

The aging shiftworker at risk can be placed mainly in the third and fourth
groups.

Koller et al. (1978) reported that more than one in two shiftworkers (51.5
per cent) were unsatisfied with their work. In this group, a shift to daywork
was preferred. It might therefore be hypothesized that, for many aging shiftworkers, work demands are often incompatible with their individual work
capacities. Again shiftworkers use different coping mechanisms to withstand
this situation by, for example, rearranging their family and social life and sleep
habits (Koller 1983, Härmä 1993). However, at some stage, coping strategies
fail to resist the accumulation of negative shiftwork consequences, and health
problems finally break through.

From the results of the studies mentioned in this paper, it may be concluded
that only a small percentage of shiftworkers are able to cope adequately with
shiftwork up to old age, especially with night shifts, without the manifestation
of health problems. Selection mechanisms seem to have a major role in the
formation of shiftwork populations. Research studies on aging 'shiftworker-
survivor' populations might therefore give an underestimation of the impact of
shiftwork on health and well-being. With increasing age, the shiftworker is
subject to several time-contingent variables, which determine individual shiftwork tolerance. However, the relation between shiftwork, aging and health
may not be considered as a simple cause-and-effect relationship. We must
assume, rather, that a complex network of interacting and time-contingent
factors is determining the aging shiftworker's well-being (Haider et al. 1981).
Finally, it seems clear that years in shiftwork appears to be a stronger risk
factor for complaints of health and well-being than chronological age.

For improving the quality of work circumstances and schedules for the
aging shiftworker the development of an age-conscious shiftwork policy must
be considered. Several preventive, age-related measures have already been proposed: for example, additional free weeks included in the work schedule for
rehabilitative measures (Koller 1983), selection limits for shiftwork at the age
of 50 (Thijs-Evensen 1958, Rutenfranz et al. 1977), frequent medical checks for
shiftworkers after the age of 40, as well as making nightwork after 40 years of
age voluntary (Härmä 1993).

Nevertheless, further extensive research is necessary to gain a clear understanding of the process and mechanisms underlying problems with the aging

shiftworker. Special attention should be paid to the interaction of age and occupation—for example, tasks, environmental stress, workload, shiftwork system and so on. Instead of cross-sectional analyses, longitudinal studies are recommended in order to control for several selection processes.

References

Aanonsen, A. 1959. Medical problems of shift-work, *Industrial Medical Surgery*, **28**: 422–427.

Åkerstedt, T. 1984. Work schedules and sleep, *Experientia*, **40**: 417–422.

Åkerstedt, T. 1990. Psychological and psychophysiological effects of shiftwork, *Scandinavian Journal of Work, Environment and Health*, **16**: 67–73.

Åkerstedt, T. 1991. Sleepiness at work: effects of irregular work hours. In T. H. Monk (ed.) *Sleep, Sleepiness and Performance* (John Wiley, Chichester), 129–154.

Åkerstedt, T., Alfredsson, L. and Theorell, T. 1986. An aggregate study of irregular work hours and cardiovascular disease. In M. Haider, M. Koller and R. Cervinka (eds) *Night and Shiftwork: Longterm Effects and their Prevention* (Peter Lang, Frankfurt a/M), 419–425.

Åkerstedt, T. and Gillberg, M. 1981. Sleep disturbances and shift work. In A. Reinberg (ed.) *Night and Shiftwork* (Pergamon Press, Oxford), 127–137.

Åkerstedt, T., Knutsson, A., Alfredsson, L. and Theorell, T. 1984. Shift work and cardiovascular disease, *Scandinavian Journal of Work, Environment and Health*, **10**: 409–414.

Åkerstedt, T. and Torsvall, L. 1981. Shift-dependent well-being and individual differences, *Ergonomics*, **24**(4): 265–273.

Åkerstedt, T., Torsvall, L. Kecklund, G. and Knutsson, A. 1990. The shift cycle and clinical studies of insomnia. In G. Costa, G. Cesena, K. Kogi and A. Wedderburn (eds) *Shiftwork: Health, Sleep and Performance* (Peter Lang, Frankfurt a/M), 421–426.

Angersbach, D., Knauth, P., Loskant, H., Karvonen, M. J., Undeutsch, K. and Rutenfranz, J. 1980. A retrospective cohort study comparing complaints and disease in day and shift workers, *International Archives of Occupational and Environmental Health*, **45**: 127–140.

Åstrand, I. 1960. Aerobic work capacity in men and women with special reference to age, *Acta Physiologica Scandinavia*, **49**: Suppl. 169.

Åstrand, I., Åstrand, P-O., Hallbäck, I. and Kilbom, A. 1973. Reduction in maximal oxygen uptake with age, *Journal of Applied Physiology*, **35**: 649–654.

Babcock, M. A., Paterson, D. H. and Cunningham, D. A. 1992. Influence of ageing on aerobic parameters determined from a ramp test, *European Journal of Applied Physiology*, **65**: 138–143.

Breithaupt, H., Hildebrandt, G., Döhre, D., Josch, R., Sieber, U. and Werner, M. 1980. Tolerance to shift of sleep, as related to the individual's circadian phase position. In W. P. Colquhoun and J. Rutenfranz (eds) *Studies of Shiftwork* (Taylor & Francis, London), 177–184.

Carskadon, M. A. and Roth, T. 1991. Sleep restriction. In T. H. Monk (ed.) *Sleep, Sleepiness and Performance* (John Wiley, Chichester), 155–168

Centraal Buro voor de Statistiek 1990. *Werknemers in Loondienst*. Voorburg.

Centraal Buro voor de Statistiek 1991. *Bevolkingsprognose*. Voorburg.

Corlett, E. N., Queinnec, Y. and Paoli, P. 1988. *Adapting Shiftwork Arrangements* (The European foundation for the improvement of living and working conditions Dublin).

De Zwart, B. C. H., Bras, V. M., Van Dormolen, M., Frings-Dresen, M. H. W. and

Meijman, T. F. 1993. After-effects of night work on physical performance capacity and sleep quality in relation to age, *International. Archives of Occupational and Environmental Health,* **65**: 259–262.

Dehn, M. M. and bruce, R. A. 1972. Longitudinal variations in maximal oxygen intake with age and activity, *Journal of Applied Physiology,* **33**: 805–807.

Dinges, D. F. and Kribbs, N. B. 1991. Performing while sleepy: effects of experimentally-induced sleepiness. In T. H. Monk (ed.) *Sleep, Sleepiness and Performance* (John Wiley, Chichester), 97–128.

Ehlers, C. L. and Kupfer, D. J. 1989. Effects of age on delta and REM sleep parameters, *Electroencephalogram Clinical Neurophysiologie,* **72**(2): 118–125.

Foret, J., Bensimon, G., Benoit, O. and Vieux, N. 1981. Quality of sleep as a function of age and shift work. In A. Reinberg (ed.) *Night and Shiftwork* (Pergamon Press, Oxford), 149–154.

Frese, M. and Okonek, K. 1984. Reasons to leave shiftwork and psychological and psychosomatic complaints of formers shiftworkers, *Journal of Applied Psychology,* **69**(3): 509–514.

Frese, M. and Semmer, N. 1986. Shiftwork, stress and psychosomatic complaints: a comparison between workers in different shiftwork schedules, non-shiftworkers, and former shiftworkers, *Ergonomics,* **29**(1): 99–114.

Haider, M., Kundi, M. and Koller, M. 1981. Methodological issues and problems in shift work research. In L. Johnson, D. I. Tepas, W. P. Colquhoun and M. J. Colligan (eds) *The Twenty-four Hour Workday* (NIOSH, US Department of Health and Human Services, Cincinnati), 197–220.

Härmä, M. 1993. Individual differences in tolerance to shiftwork: a review, *Ergonomics,* **36**(NOS 1–3): 101–109.

Härmä, M., Hakola, T. and Laitinen, J. 1992. Relation of age to circadian adjustment to night work, *Scandinavian Journal of Work, Environment and Health,* **18**(suppl. 2): 116–118.

Ilmarinen, J., Tuomi, K., Eskelinen, L., Nygård, C-H., Huuhtanen, P. and Klockars, M. 1991. Summary and recommendations of a project involving cross-sectional and follow-up studies on the aging worker in Finnish municipal occupations (1981–1985), *Scandinavian Journal of Work and Environmental Health,* **17** (Suppl. 1): 135–141.

International Labour Office (ILO) 1992. *The ILO and the Elderly* (ILO Geneva).

Jansen, B. 1987. *Daywork and Shiftwork Compared* (in Dutch). (Swets and Zeitlinger, Amsterdam/Lisse).

Keran, C. M. and Duchon, J. C. 1990. Age difference in the adjustment to shiftwork. *Proceedings of the Human Factors Society 34th Annual Meeting,* **1**: 182–185.

Kerkhof, G. 1985. Inter-individual differences in the human circadian system: a review, *Biological Psychology,* **20**: 83–112.

Knutsson, A. and Åkerstedt, T. 1992. The healthy-worker effect: self-selection among Swedish shift workers, *Work and Stress,* **6**(2): 163–167.

Knutsson, A., Åkerstedt, T., Johnsson, B. and Orth-Gomer, K. 1986. Increased risk of ischemic heart disease in shift workers, *Lancet,* **ii**: 89–92.

Koller, M. 1983. Health risks related to shift work, *International Archives of Occupational and Environmental Health,* **53**: 59–75.

Koller, M., Kundi, M. and Cervinka, R. 1978. Field studies of shift work at an Austrian oil refinery. I: Health and psychosocial wellbeing of workers who drop out of shiftwork, *Ergonomics,* **21**(10): 835–847.

Lavie, P., Chillag, N., Epstein, R., Tzischinsky, O., Givon, R., Fuchs, S. and Shahal, B. 1989. Sleep disturbances in shift-workers: a marker for maladaption syndrome, *Work and Stress,* **3**(1): 33–40.

Meers, A., Maasen, A. and Verhaegen, P. 1978. Subjective health after six months and after four years of shiftwork, *Ergonomics,* **21**(10): 857–859.

Miles, L. E. and Dement, W. C. 1980. Sleep and aging, *Sleep*, **3**: 119–229.

Moline, M. L., Pollak, C. P., Monk, T. H., Lester, L. S., Wagner, D. R., Zendell, S. M. and Graeber, R. C. 1992. Age-related differences in recovery from simulated jet lag, *Sleep*, **15**(1): 28–40.

Monk, T. H. and Folkard, S. 1985. Individual differences in shiftwork adjustment. In S. Folkard and T. H. Monk (eds) *Hours of Work* (John Wiley, Chichester), 227–238.

Moog, R. 1987. Optimization of shift work: physiological contributions, *Ergonomics*, **30**(9): 1249–1259.

Nygård, C-H., Luopajärvi, T. and Ilmarinen, J. 1991. Musculoskeletal capacity and its changes among aging municipal employees in different work categories. *Scandinavian Journal of Work and Environmental Health*, **17** (Suppl. 1): 110–117.

Oginska, H., Pokorski, J. and Oginski, A. 1993. Gender, ageing, and shiftwork intolerance, *Ergonomics*, **36**(NOS 1–3): 161–168.

Östberg, O. 1973. Interindividual differences in circadian fatigue patterns of shift workers, *British Journal of Industrial Medicine*, **30** 341–351.

Ottman, W., Karvonen, M. J., Schmidt, K.-H., Knauth, P. and Rutenfranz, J. 1989. Subjective health status of day and shift-working policemen, *Ergonomics*, **32**(7): 847–854.

Robinson, S., Bill, D. B., Ross, J. C., Robinson, R. D., Wagner, J. A. and Tzankhoff, S. D. 1973. Training and physiological aging in man, *Federation Proceedings*, **32**: 1628–1634.

Rutenfranz, J., Colquhoun, W. P., Knauth, P. and Ghata, J. N. 1977. Biomedical and psychosocial aspects of shift work, *Scandinavian Journal of Work, Environment and Health*, **3**: 165–182.

Rutenfranz, J., Haider, M. and Koller, M. 1985. Occupational health measures for nightworkers and shiftworkers. In S. Folkard and T. H. Monk (eds) *Hours of Work* (John Wiley, Chichester), 199–210.

Rutenfranz, J. and Knauth, P. 1986. Introductory remarks on shiftwork and combined effects. In M. Haider, M. Koller and R. Cervinka (eds) *Night and Shiftwork: Long-term Effects and their Prevention* (Peter Lang, Frankfurt a/M), 319–326.

Rutenfranz, J., Knauth, P. and Angersbach, D. 1981. Shift work research issues. In L. Johnson, D. I. Tepas, W. P. Colquhoun and M. J. Colligan (eds) *The Twenty-four Hour Workday* (NIOSH, US Department of Health and Human Services, Cincinnati), 221–261.

Tepas, D. I. and Carvalhais, A. B. 1990. Sleep patterns of Shiftworkers, *Occupational Medicine: State of the Art Reviews*, **5**(2): 199–208.

Tepas, D. I., Duchon, J. C. and Gersten, A. H. 1993. Shiftwork and the older worker. *Experimental Aging Research*, **19**(4): 295–320.

Tepas D. I., and Mahan, R. P. 1989. The many meanings of sleep. *Work and Stress*, **3**: 93–102.

Tepas, D. I. and Monk, T. H. 1987. Work schedules. In G. Salvendy (ed.) *Handbook of Human Factors* (John Wiley, New York) 819–843.

Thierry, H. K. and Meijman T. F. 1993. Time and behaviour at work. In M. D. Dunnette and L. M. Hough (eds) *Handbook of Industrial and Organizational Psychology*, Vol. 4 (Consulting Psychologists Press, Palo Alto), 341–414.

Thijs-Evensen, E. 1958. Shift work and health, *Industrial Medical and Surgery*, **27**: 493–497.

Torsvall, L. and Åkerstedt, T. 1980. A diurnal type scale: construction, consistency and validation in shift work, *Scandinavian Journal of Work, Environment and Health*, **6**: 283–290.

Torsvall, L., Åkerstedt, T., Gillander, K. and Knutsson, A. 1989. Sleep on the night shift: 24-hours EEG monitoring of spontaneous sleep/wake behaviour. *Psychophysiology*, **23**(3): 352–358.

Torsvall, L., Åkerstedt, T. and Gillberg, M. 1981. Age, sleep and irregular workhours. A field study with electroencephalographic recordings, catecholamine excretion and self-ratings, *Scandinavian Journal of Work, Environment and Health*, **7**: 196–203.

Tune, G. S. 1969. The influence of age and temperament on the adult human sleep–wakefulness pattern, *British Journal of Psychology*, **60**(4): 431–441.

Walker, J. and Mare, G. de la 1971. Absence from work in relation to length and distribution of shift hours, *British Journal of Industrial Medicine*, **28**: 36–44.

Walsh, J. K., Tepas, D. I. and Moss, P. D. 1981. The EEG sleep of night and rotating shift workers. In L. Johnson, D. I. Tepas, W. P. Colquhoun and M. J. Colligan (eds) *The Twenty-four Hour Workday* (NIOSH, US Department of Health and Human Services, Cincinnati), 451–466.

Webb, W. B. 1978. Sleep, biological rhythms and aging. In H. V. Samis and S. Capobianco (eds) *Aging and Biological Rhythms* (Plenum, New York), 309–323.

Conclusions

In this part I, literature surveys and original investigations are brought together concerning the interrelated effect of work and the aging process on workers' health. The chapters stress the importance of prospective longitudinal research to assess the effects of work on the worker's health, although the majority of research has studied the same relationship with cross-sectional designs. Such designs are not adequate since they only deal with workers who have more or less 'survived'. This is the reason why their complaints on work and health will be grossly underestimated, because of the so-called healthy worker effect. Moreover, since complaints and diseases can be registered only at the time of measurement or recalled by questionnaire and/or interview, they may be biased.

It is a fact that, when people get older, their physical capacities gradually decrease. The decrease is on average about 10 per cent per decade, although there are large interindividual differences. It is still not clear if this decrease is solely a result of the aging process or whether it is also caused by the general change in lifestyle such as less physical activity during free time and/or the occurrence of diseases.

The increase in prevalence rates of diseases with age that is demonstrated in follow-up studies in both sexes does not only apply to physically heavy jobs but also to jobs that are mentally stressful and jobs that include both aspects. In addition to the physical and mental load of the job that is related to disease rates, environmental (heat and cold) and organizational factors (responsibility, decision latitude) seem to have an important impact on the worker's health, sooner or later.

Two solutions for the lower physical capacity in the aging worker are suggested:

(1) The decline in physical capacity will be prevented or at least be delayed by adequate physical training organized by the company, such as employee fitness programmes.

(2) Adjusting work demands in order to maintain the same work reserve (the difference between maximal work capacity and daily physical strain) of the aging worker by job redesign, diminishing heavy demanding tasks with compensation of skilful tasks and/or other ergonomic measures.

Shiftwork is an extra factor that can harm the workers' health. There are, however, several factors that play a role in the selection of and adjustment to shiftwork, such as family life, social activities, career planning, sleep disturbances and health. Only long-term longitudinal studies can disentangle the impact on health of factors that apparently are causative for the drop-out rate on the short-term (1–3 years) and long-term (20+ years) base.

This section raises more questions on the interaction between aging, work and health than it answers. When the reader realizes how complicated the matter is and that only carefully designed, longitudinal studies over a relatively long period of time with a multidisciplinary approach can give small but significant answers about the true relationship between age, work and health, the authors of this section will be satisfied.

Part II
Aging and Mental Work Capacity

Introduction

Age has been studied for a long time, in both the applied and academic fields. In the applied field, papers were presented in order to emphasize the importance of adapted work performance, work conditions and communication devices for old people; in the academic field, experiments were designed in order to investigate age-related differences in task performance. It has become more and more clear, however, that the interpretation of these studies gains value when both approaches are combined. This part II is aimed at discussing the relevance of age-related differences in task performance for everyday work, as well as age-related adaptations in work conditions for both employees and employers. The strength of the present section lies in the overall view emerging from the papers, although each paper might emphasize only one aspect of age-related differences in performance.

Zeef, Snel and Cremer, in chapter 11, start by observing a consequence of age-related slowing in information processing in an everyday situation, namely traffic participation. They present an aging study, in which they investigate individuals' ability to track an object that moves invisibly with a constant velocity. Their results are explained in a information-processing rate model.

In chapter 12, Kok, Lorist, Cremer and Snel focus on the nature and extent of decline in fluid abilities of the cognitive information-processing system by using a processing resource model. The assumption is that with advancing age there is a decline in resources, especially under stressful conditions such as lack of sleep and auditory noise. They also assume that caffeine intake will have a beneficial effect on the amount of processing resources to be allocated. The authors test their ideas by regressing performances of young and old subjects, which were obtained in various reaction-time tasks in Brinley plots.

Rudinger and colleagues report in chapter 13 on an interesting study concerning the difficulties elderly people face when they have to use modern, everyday technology such as ticket machines and communication apparatuses.

They show that, with simple measures, the improvement in quality of usage in terms of errors and time loss is striking. They propose to design a kind of technical grammar with universal instructions of use to prevent problems of usage.

Rabbitt and Carmichael set up another simple but promising framework for undertaking human-factors design for communication systems for elderly people (chapter 14). They hypothesize that healthy, normal aging is accompanied by an accumulation of mild but progressive loss of efficiency in all sensory systems. They also assume that in aging adults, severe loss in one sensory modality is usually accompanied by slight losses in others. It is emphasized that human-factors design needs a good methodology to examine all sensory demands relevant for communication systems. Such a methodology should also be able to predict how well elderly people will accept new communication systems.

Chapter 15 presents Baracat and Marquié's attempt to point out the difficulties likely to be encountered by old employees facing the computerization of the workplace. The authors analyse how these difficulties may become manifest in a real-life training situation where new computer techniques are being learned.

Finally, in the last chapter 16 of this part II, Molinié and Volkoff describe the generation effect employers encounter when computers are applied in the workplace. They also observe this effect in the areas of shift- or nightwork and in time constraints imposed on workers at the work-site.

11

Judgement of the position of an invisibly moving object in young and old adults

R. Cremer, E. Zeef and J. Snel

Summary. The present study focuses on the question of whether representations of dynamic spatial schemata lead to differences between old and young adults in the accuracy of judgement of the position of invisibly moving objects. Young and old subjects had to estimate the stop position of a moving bar as accurately as possible. The moving bar followed its path invisibly after 25 per cent of its trajectory. After hearing a tone, subjects had to estimate its stop position. Only the accuracy of estimation was stressed, whereas the time needed to make a judgement was not taken into account. By examining performance on six blocks of trials it was tested whether the (in)accuracy of estimations was structural or was reduced after many trials. In addition, the presentation of four different movement velocities (within blocks) enabled us to test the proportional effect of processing limitations. The results indicated that both young and old subjects underestimated the velocity of the object's movement, but that older subjects did so to a far greater extent than did the young ones. Positions further away from the starting-point were estimated less accurately. This effect was more outspoken with higher velocities. It was concluded that the accuracy of estimation may have suffered from age-related cognitive slowing. The results suggest that especially the elderly underestimate velocities of moving objects. In particular, for elderly persons this may mean that insufficient visual information on moving objects may increase the risk of having accidents, for example at the work-site or in traffic situations.

Keywords: judgement accuracy, perceptual judgement, age-related inaccuracy, estimation of velocity.

Introduction

In everyday life humans frequently have to make decisions on the position and speed of moving objects in the environment. The relevance of this ability is demonstrated, for example, in work or traffic situations when one has to judge positions of both stationary and moving objects. This cognitive skill requires adequate control in processing spatial representations of the outside world. Elderly people may find that it takes more effort to control movement in their environment, and to locate objects in space, particularly in relation to their own movement.

For reasons of elderly individuals' and others' safety, for various forms of work participation at the work-site or in traffic situations the accurate judgement of the (forthcoming) position of objects and subjects is required. When in such situations they have to make such judgements as accurately and as quickly as possible. When judging movement we rely heavily on the availability of feedback. In many practical situations this feedback may be interrupted for brief periods by contextual items such as other subjects, machines, walls and so on. In such cases we have to base the predictions of ongoing movements on incomplete information. When feedback is unavailable for a while, a degree of error in position judgement may occur and only the timely reappearance of a target object allows corrections.

Since the accuracy of our judgements of movement in several divergent types of situations is of utmost importance for optimum functioning and since there are differences in people's ability to do so, it is interesting to compare the performance of young and old subjects. It is predicted that old subjects are slower on the timing of position. Based on theories on aging and information processing and supported by experimental evidence (Korteling 1990, Cremer et al. 1990), it is expected not only that old subjects are slower in the timing of position but also are less accurate in velocity judgement.

Judgement of movement is a complicated skill; it is a compound process that requires continuous revision (Salthouse and Prill 1983). In order to understand its nature and to hypothesize about the factors contributing to superior performance, Gerhard's (1959) model describes the way a trajectory interception task is performed. He suggests that different information-processing mechanisms may contribute to variations in proficiency associated with alternative conceptualizations of the skill.

He explains the ability to track an object that moves with a constant velocity as follows: 'the human mind does not follow the physical definition and dimension of the concept of velocity: does not experience velocity as cm/sec; but rather the reciprocal value sec/cm is valid. Consequently the velocity of movement is not estimated according to the distance traveled in a certain unit of time but rather in accordance with the temporal duration required to travel a distance unit' (Gerhard 1959:).

During the visible running of a moving object, an internal representation of the observed movement (velocity) is formed, which is used as the basis of its subsequent positions. As a consequence of the age-related slowing of information processing per time unit, a smaller amount of task-relevant information can be processed than in a non-slowed system. Analogous to this velocity estimation, task-processing limitations must inevitably lead to less effective processing, that is underestimation of the object's position. The second presumption is that age-related processing limitation is structural. Judgement will increase with the distance travelled and be a multiple of an error constant per position unit. An accumulation of judgement error over all traversed positions will account for the final total inaccuracy score. Similarly, expressed in a regression equation, the slope represents the inaccuracy as a function of traversed positions and reflects the error constant per time unit. The intercept reflects the structural, age-dependent error of the perceptual system or in other words represents the processing efficiency during visible movement.

The assumption is that accuracy can hardly be improved by practice since the quality of processing is determined by structural limitations of the processing system, that is the nervous system.

Based on these assumptions the prediction is that over repeated measurements in experimental blocks of trials the intercept of the error function over estimated positions remains unchanged regardless of age. Higher velocities require that more information needs to be processed per time unit. This inevitably must affect the slope parameter. The error proportion can be expressed in a kind of spatial Weber fraction, which is higher for old than young adults. It is spatial in the sense that the change in slope with age is independent of the speed of movement or, put differently, for slopes there will be no interaction between age and velocity. In a study of predicting situations in which part of the trajectory of a moving light was obscured, Welford (1973) reported on a systematic relationship between the variability of subjects' performance and the time the light was visible.

Since the speed of cognitive processing declines with age, this age-related slowing should manifest itself primarily under conditions of time pressure (Sharps and Collin 1987) and should consequently be reduced when speed demands are absent. Indeed, Sharps' results of a mental-rotation task indicate that old subjects perform as well as young participants when the speed factor is eliminated. When speed was a factor, the accuracy scores of elderly subjects were significantly lower than those of young subjects. Differences in the speed of performance on mental-rotation tasks between young and old subjects have been shown in many other studies (e.g. Berg et al. 1982). A general conclusion drawn from experimental evidence is that elderly people usually tend to be slower, but show equal or higher accuracy scores on such tasks than younger people (Welford 1976). When speed of response is included in the instructions of a tracking task (Cremer et al. 1990), age-related motor slowing may be a determinant of response latency. In the present manipulation this speed factor was excluded.

It should be noted that in the present study 'speed demands' may be related unintentionally to the judgement of movement, a skill that is closely related to timing. Stated in other words: to perform, it is required to have an internal clock which matches the flow of real time. The same holds true for complex cognitive tasks in general. When more processing units are needed than in elementary cognitive tasks, a proportional decline in performance with age has been demonstrated (Cerella 1990) as a function of task complexity.

The present experiment aims to determine whether aging affects the estimation of the position of an invisibly moving object, when no feedback on performance is given and without taking choice reaction time into account—that is, without any time pressure on responding. The skill manipulated in the present study is the visualization of the movement of a small vertical bar that is blocked from vision after 25 per cent of its trajectory. As exposure time is limited, like in Korteling's platoon-car-following task, subjects cannot take extra time to analyse the stimulus. Hence, the fact that the moving object is visible only for a small part of its trajectory may constitute a kind of speed factor in itself.

On account of the slowing theories of aging, the prediction is that old compared with young individuals will underestimate the velocity of the invisible moving object to a greater extent, and that this estimation error will increase with higher velocities. For example, it is a well-known observation that younger subjects are much more familiar with computer tasks such as speed games than older subjects. For this reason we expect to find a training effect in older subjects that will be more pronounced, the more trials that are given.

Design and method

Subjects

The subjects are 20 students (age 19–31 years, , $\bar{x} = 22.9$); five men and 15 women, and 20 older individuals (age 62–79 years, $\bar{x} = 69.4$); 13 men and seven women, all with higher education. The participants filled out a health questionnaire. None of them reported major health problems. Subjects had normal or corrected-to-normal visual acuity. In order to measure fluid aspects of intelligence (Salthouse 1978) and to check for the comparability of both groups, the Digit Symbol Substitution test was administered. The data indicated (old: $\bar{x} = 54.1$, sd $= 8.8$; young $\bar{x} = 64.95$, sd $= 9.3$; $t = 3.78$, df $= 38$, $p < 0.001$) lower scores for the old group.

Apparatus

Stimuli were presented on an IBM PC-AT monochrome computer screen, measuring 32 cm diagonally (29.1°). The PC's standard keyboard was used by subjects to indicate and register their responses. Beeps were generated by a Hewlett-Packard 3310B function generator at a level of about 30dB, frequency 2000 Hz, duration about 500 ms and presented over Sennheiser headphones (HD414 SL).

Procedure

While performing the task, the subject is in a stationary position and monitors a bar moving to the right at a constant speed. The moving bar is only visible for the first 25 per cent of its trajectory. After the initial visible part of the trajectory, the bar proceeds invisibly along the remaining 75 per cent of the trajectory. Along a range of evenly spaced markers covering the invisible part of the trajectory, the bar is stopped at the signal of a beep. After the beep the subject is requested to estimate the stop position by pressing a letter on the keyboard corresponding to its estimated distance reached along the hidden trajectory. The subject is urged to be accurate rather than fast in his or her response. Subjects sat in front of the PC screen at a distance of 60 cm.

Task

On the screen a rectangular frame was shown on the monitor screen, in which the task information was presented. In this continuously visible, rectangular frame a thin, vertical bar was projected to the left of the frame (figure. 11.1).

When the spacebar was struck, a trial started. A trial consists of the movement of the bar to the right in the visible section up to the marker letter 'a' at 25 per cent (5.34 cm) of the full trajectory (21.37 cm), followed by suggested movement at the same speed in the invisible section. This invisible part was marked with equally spaced (0.64 cm) vertical stripes, indicated by the letters 'a' to 'z'. Subjects wore headphones. When the invisible bar arrived at the exact position of a letter marker, a beep sounded. Subjects were asked to indicate the position (that is, the letter of the alphabet) at which they imagined the bar had arrived by striking the appropriate letter key on the keyboard. There were no time restrictions on responding.

The next trial started when the spacebar was struck. In the experiment the first two and the last two stop positions (letters a, b and y, z) were never used as stop positions in order to widen the range of under- and overestimation. In order to prevent the use of strategies, four different constant velocities (3.3, 4, 4.7 and 5.3 cm/sec) were presented randomly in balanced blocks. The approximate duration between two letter positions of the four velocities were: 121, 138, 162 and 194 ms, respectively. The experimental session consisted of two

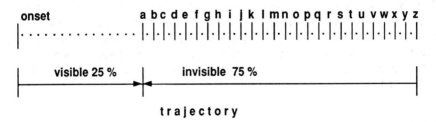

Figure 11.1. Layout of the tracking task. A bar moving with a constant velocity is hidden from view after a quarter of its trajectory at the letter position 'a'. The subject is instructed to indicate the stop position (type the letter) as indicated by a beep.

practice blocks with 44 trials each (two highest velocities with each 22 letter positions) and six experimental blocks of 88 trials each (four velocities with each 22 letter positions). After the third experimental block subjects took a short break. On purpose, no feedback was given on performance in order to prevent changes in strategy or motivation. Subjects were told they could respond at their own pace and were encouraged to visualize 'mentally' the movement of the bar once it had disappeared in order to predict as accurately as possible its stop position. They were told that the bar would move at four different, but constant velocities and that response time (after the beep) was not important at all. The total experiment took up to 2 hours per subject.

Results

Inaccuracy of performance, that is visualization of the distance travelled by the invisibly moving object, is expressed as the discrepancy between actual and estimated position of the stopped bar. For each subject the inaccuracy (number of bar positions misjudged) was regressed against the baseline (true bar positions). Means of the regression parameters (see table 11.1) were calculated over blocks (ignoring velocities) and over velocities (ignoring blocks). Individual regression functions with R^2 of 0.80 and higher were accepted for further analyses. Data of two elderly subjects had to be discarded because of this criterion. For each subject it was found that with greater distance to estimate, the inaccuracy accumulated linearly. The statistical difference of inter-

Table 11.1. Linear regression weights of young and old subjects, calculated per block and velocity. In all individual functions at least 80 per cent of the variance was accounted for.

Averaged over experiment level: total trials 528	Intercept		slope (sd)	
	young	old	young	old
4*22 = 88 per block	$n = 20$	$n = 18$	$n = 20$	$n = 18$
block 1	−1.91	−0.76	0.37 (0.11)	0.59 (0.09)
block 2	−1.51	−0.89	0.39 (0.13)	0.62 (0.10)
block 3	−1.83	−0.92	0.38 (0.13)	0.62 (0.10)
block 4	−1.88	−1.15	0.42 (0.12)	0.62 (0.11)
block 5	−2.17	−1.14	0.39 (0.12)	0.61 (0.10)
block 6	−2.04	−1.27	0.39 (0.12)	0.62 (0.11)
6*22 = 132 per velocity	$n = 20$	$n = 18$	$n = 20$	$n = 18$
velocity 1	−1.46	−1.01	0.33 (0.12)	0.55 (0.11)
velocity 2	−1.76	−0.93	0.37 (0.12)	0.59 (0.10)
velocity 3	−2.31	−1.24	0.44 (0.11)	0.67 (0.09)
velocity 4	−2.09	−0.90	0.42 (0.10)	0.63 (0.09)

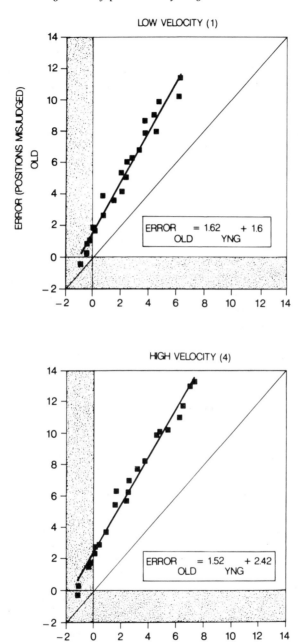

Figure 11.2. Means of positional (24 positions) misjudgements of elderly adults regressed against those of young adults in the lowest (left) and highest (right) velocity

cept and slope values between groups and conditions (blocks and velocities) was tested by analysis of variance.

Slope differences

Between blocks no significant difference was found, but with higher velocities the slopes increased (or the accuracy decreased $F(3,4) = 9.141$, $p < 0.001$. Larger slopes for old than for young adults were found over velocities $F(1,4) = 170.139$, $p < 0.001$. No interaction effects of blocks or velocities with age group were found. The within age group variation of the slopes is extremely small and consistent within blocks and velocities (SDs ranging between 0.09 and 0.13).

Intercept differences

No significant differences were found between blocks or velocities and no interaction effects of blocks or of velocities with age were found. Over velocities, an inaccuracy constant was larger for old than young subjects $F(1,4) = 12.168$, $p < 0.001$.

For further analysis, the data was pooled over blocks as this factor did not vary systematically. For each of the four velocities mean accuracies between young and old groups were regressed in an XY space. The mean accuracy on trajectory positions of young adults is denoted on the X axis and that of old adults on the Y axis. The points in the XY space represented combined means of age groups on the trajectory positions. For each relation a linear regression function was calculated (no better fit was obtained by curvilinear functions). The functions accounted for at least 95 per cent of the variance and were significant ($p < 0.0001$).

Regression comparisons for each of the four velocities were:

velocity one: OLD.err = 1.62*YNG.err + 1.59;
velocity two: OLD.err = 1.56*YNG.err + 2.0;
velocity three: OLD.err = 1.48*YNG.err + 2.4;
velocity four: OLD.err = 1.51*YNG.err + 2.42.

These functions indicate that, for example, for velocity one, old adults constantly underestimate 1.59 positions more than young adults and that a multiplication factor of 1.62 shows the proportional inaccuracy rate for old subjects relative to the young subjects over all positions which had to be estimated.

The difference between the function for the lowest and highest velocities (BMDP 4R, $F(2,40) = 5.541$, $p = 0.008$) indicated that, in particular, the intercepts of the regression function differed beyond chance (figure 11.2). Stated in other words, scores do not simply shift position on the same function, but the entire function (with almost similar slopes for the two velocities) is shifted.

Discussion

Skills required for a specific task are dependent on the task's specific characteristics or demands. When participating in such a task or situation, frequent updates of spatial schemata of movement of objects should be made in relation to changes of one's own position. For example, in the industrial setting, information on movement is often interrupted for short or long intervals by objects that block vision or because the observer has to switch or divide visual attention to other stationary or moving objects in the environment. In the present study only a few of the multitude of the involved variables were manipulated.

In a tracking task, the displacement of a bar moving invisibly from left to right across a video screen had to be estimated. It was argued that processing efficiency per time unit is limited in old as compared with young adults, that practice would not change accuracy and that increased velocities would impair effective processing even further. The fact that the moving object was only visible for 25 per cent of its trajectory might have created a kind of time pressure; in other words, subjects had a limited amount of time to analyse visually the speed of movement.

As suggested by Sharps and Collin (1987), age-related slowing should manifest itself primarily under conditions where high demands are made on subjects and age-related differences in performance may disappear by giving subjects sufficient time to process information adequately. Applying these considerations to the results of this study, we think it is legitimate to conclude that the larger estimation error made by older subjects was because they were less able than young subjects to complete the process of visual analysis in the time available.

The predictions were confirmed by the data. Age-related effects became apparent from regression functions expressing inaccuracy over trajectory positions. In old subjects the slopes and intercepts of judgement error were larger than in young adults. The slope reflecting a constant proportion of lower accuracy over trajectory positions was larger in elderly than in younger subjects. The larger intercept in the old subjects compared with the younger ones demostrated a larger 'onset' error, caused by perceptual misjudgement of position during the visible projection. Practice on the task did not change the intercept or the slopes, lending support to the assumption that the deficit is structural. Higher velocities increased the slope values for subjects in both age groups; this finding was explained by processing limitations, which become more manifest with higher processing demands.

Although we support the view that age-related decrements in information processing capacity are caused by structural deficits in the nervous system, other interpretations are possible. For example, the greater inaccuracy in the old than in the young can be explained by changes in attention or concentration. Lapses of attention owing to variation in concentration or temporary gaps in processing continuity could have interfered with and delayed the tracking process. However, such interference would be irregular over blocks and would have shown a low fit of individual regression functions.

Another explanation for the age-related inaccuracy could be that 'timing' is an important ability, especially in tracking tasks. However, various studies have reported that older adults are not worse than young adults in time estimation, which might be expected if their internal clock runs slower (Salthouse 1985, p. 415). So, besides 'inspection time', the proposed structural deficit (a degraded neural system) in the continuity of information processing remains a likely candidate for the age-related deficit.

Finally, it should be noted that, contrary to the present task, visual tracking in real-life situations is a process that is usually performed under conditions of continuous feedback and, as a consequence, is very accurate. The initial representation of perceived movement is adjusted continually, depending on visual relations with other contextual images. On the other hand, the severity of the effects is perhaps mitigated by the observer's stationary position. The combined and changing movement of both the observer and moving objects at the same time might tax the processing capacity of the subject to the limit.

The present study suggests plausibly that elderly people are move vulnerable to trespassing processing limits than younger ones. For the sake of the safety at the workplace, programmes should take note of these findings in ergonomic work design.

Acknowledgements

This research was conducted at the Faculty of Psychology, Department of Psychonomics of the University of Amsterdam. We thank Bertie Maritz for collecting the data as part of her master's degree.

References

Berg, C., Hertzog, C. and Hunt, E. 1982. Age differences in the speed of mental rotation, *Developmental Psychology*, **18**(1): 95–107.

Birren, J. E. 1974. Translations in gerontology- from lab to life: psychophysiology and speed of response, *American Psychologist*, **29**: 808–815.

Cerella, J. 1990. Aging and information-processing rate. In J. E. Birren and K. W. Schaie (eds) *Handbook of the Psychology of Aging*, 3rd ed (Academic Press, San Diego).

Cremer, R. and Zeef, E. J. 1987. What kind of noise increases with age? *Journal of Gerontology*, **42**: 515–518.

Cremer, R., Snel, J. and Brouwer, W. 1990. Age related differences in timing of position and velocity identification, *Accident Analysis and Prevention*, **20**: 467–474.

Gerhard, D. J. 1959. The judgment of velocity and prediction of motion, *Ergonomics*, **2**: 287–304.

Korteling, J. E. 1990. Perception-response speed and driving capabilities of brain-damaged and older drivers, *Human Factors*, **32**(1): 95–108.

Salthouse, T. A. 1978. The role of memory in the age decline in digit symbol sustitution performance, *J. Gerontology*, **33**: 232–238.

Salthouse, T. A. 1985A. *A Theory of Cognitive Aging* (Elsevier, Amsterdam; North-Holland).

Salthouse, T. A. and Prill, K. 1983. Analysis of a perceptual skill, *Journal of Experimental Psychology: Human Perception and Performance*, **9**(4): 607–621.

Sharps, M. J. and Collin, E. S. 1987. Speed and accuracy of mental image rotation in young and elderly adults, *Journal of Gerontology*, **42**(3): 342–344.

Welford, A. T. 1973. Stress and performance, *Ergonomics*, **16**: 567–580.

Welford, A. T. 1976. Thirty years of psychological research on age and work, *Journal of Occupational Psychology*, **49**: 129–138.

12

Age-related differences in mental work capacity: effects of task complexity and stressors on performance

A. Kok, M. M. Lorist, R. Cremer and J. Snel

Summary. The present study focuses on the question of whether there is a decline in the capacity of the information-processing system with advancing age. Attention is first paid to general theoretical issues like the distinction between fluid and crystallized abilities, the reduction of processing capacity, and the effect of task complexity and stressful conditions on task performance in elderly subjects. Then, results of a series of experiments are reported that were carried out to determine to what extent the age-related decline in performance in more demanding cognitive tasks is also affected by stressors such as noise, sleep loss and caffeine. The results clearly demonstrated that, with respect to speed measures, the complexity effect, that is the greater slowing of old subjects in more complex than in simple RT tasks, is a robust phenomenon. It was further concluded that the combined effects of task complexity and stress conditions yielded a complex pattern of performance measures, which could not be explained simply on the basis of a simple resource-depletion model of aging.

Keywords: aging and mental work capacity, processing resources, general slowing, complexity function, stressors, noise, sleep deprivation, caffeine.

Introduction

Satisfactory job performance depends both on the characteristics of the individual employee and the conditions of the work environment. An important characteristic of employees is their information-processing capacity. In a broad sense, processing capacity or mental 'resources' have been conceived as the composite of both structural and energetic components of information processing (Salthouse 1985, Gopher 1986, Wickens 1986). Structural processing components have traditionally been identified with specific information-processing operations (Sternberg 1969, Sanders 1980) or a limited capacity working memory system (Baddeley 1981). In addition, energetic components have been associated with some form of mental energy, effort or attention (Kahneman 1973).

The present study focuses on the question of whether there is a decline in the capacity of the information-processing system with advancing age. It has been argued that processing capacity or mental resources cannot explained sufficiently by a unitary non-specific process (Sanders 1983, Wickens 1986). Nevertheless, the notion of a non-specific pool of processing capacity may provide a parsimonious interpretation of the decline in the efficiency of performance in older subjects. As will be explained in greater detail in the following paragraphs, one of the most compelling reasons for accepting the view of a non-specific reduction of a single pool of processing capacity with advancing age is that the absolute magnitude of age-related differences in performance increases with greater task complexity (Salthouse 1985).

A second class of variables that may influence job performance in elderly subjects is stressful environmental conditions. By using the terms 'stress' or 'stressor' we do not primarily refer to extreme biological, subjective or environmental conditions as proposed by Selye (1936), but rather to the more usual transactions between subjects and their environments. Especially in the context of studies of human performance, the term stress has been identified frequently with the non-specific effects of environmental conditions such as sleep loss, noise or subjective and biochemical factors that are known to affect the quality of human performance (Broadbent 1963, 1971, Hockey and Hamilton 1983). In a broader sense the term stress has also been used as a label for a whole range of conditions that involve some form of 'energy mobilization' (Cannon 1936, Duffy 1962, Hockey et al. 1986). Systematic studies on the effects of stressful conditions on performance in elderly subjects are still relatively scarce. This is to be regretted, since it is likely that elderly workers especially are more vulnerable to the noxious effects of stressful environmental conditions on the quality of their job performance.

In the present paper attention will first be paid to general theoretical issues like the distinction between fluid and crystallized abilities, the reduction of processing capacity and the effect of task complexity and stressful conditions on task performance in elderly subjects. Then, we will report the results of a series of experiments that have recently been carried out in our laboratory to determine to what extent the age-related decline in performance in more demanding cognitive tasks is also affected by stressors like noise, sleep loss and caffeine.

Fluid and crystallized abilities

With advancing age there is a decline in physical and sensory functions, and also in the level of mental functioning (cf. Birren and Schaie 1990 for a comprehensive recent survey). A central but still unresolved issue in aging research is the question of whether all mental abilities are equally affected by age, or that different cognitive subsystems are affected at different times to different degrees (Salthouse 1985, Rabbit 1992). There is substantial evidence to suggest that the decline in mental functions is most conspicuous for 'fluid' abilities, in contrast to 'crystallized' abilities which are much more resistant with age (Horn 1982). Fluid abilities refer to cognitive functions such as speed of decision making, short-term memory and attention, which mainly play a role during the acquisition and transformation of information. Crystallized abilities are associated with verbal, acquired or specialized skills. These functions depend strongly on experience, and are mainly activated when previously acquired information which is stored in long-term memory has to be retrieved or used. Since performance is positively related to experience, crystallized abilities may even improve with age owing to gradual accumulation of experience and knowledge. The dependence of fluid abilities on attention is also substantiated by psychometric studies: the decline in fluid abilities with increasing age disappears if attentional factors are separated out. Similarly, the increase in crystallized abilities with advancing age becomes even greater if one controls for attention effects (Stankov 1988). It is generally assumed that the decline in fluid abilities is closely related to biological factors such as the neural degeneration of cortical structures and brainstem nuclei that occur late in life (Petit 1982, Selkoe 1992).

The function of fluid abilities in real-life settings

The impact of the decline in fluid cognitive abilities with advancing age in our society is substantial, especially if we consider the increased life expectancy of elderly persons, and the many changes of modern life. Consider, for example, the skills required for word-processing as compared with normal typing. In addition, modern electronic appliances and various number codes such as credit card and bank numbers have become customary even in simple financial transactions. These new technologies of our 'information society' have introduced extra complications for both employed middle-aged and retired old persons (Bromley 1988). Another factor is the increase in traffic density and the complexity of the road system against the background of a much higher number of traffic accidents in the elderly population, and the timing and pacing of events in urban environments, which may be too fast for many older pedestrians. Finally, according to several surveys, older workers receive lower ratings for speed of work and learning abilities, especially in technical jobs, although their overall efficiency is the same. Old subjects are also found much less frequently in jobs involving time stress (Davies and Sparrow 1985). Clearly, such developments of modern society and work conditions present greater hazards for elderly persons, who need more time to scan the environment and to prepare appropriate actions.

It should be noticed, however, that although the elderly as a group show a decline in performance, there are many old individuals who show performance levels that are similar to or even better than the average of that many years younger (Welford 1984). These differences may sometimes arise from specific crystallized abilities, specialized skills or intellectual abilities. In other cases, however, some old subjects (the 'young-olds') seem to be preserved better functionally, or biologically, than other old subjects (the 'old-olds'), and therefore show only little or no decline in their innate, fluid abilities. In view of the large variability in the levels of mental functioning of subjects of the same chronological age, Birren has introduced non-chronological age scales, such as biological and functional age, and contrasted them to chronological age (Birren and Cunningham 1985).

The decline in fluid cognitive abilities is probably not restricted to the retired elderly population, but may also affect middle-aged individuals who are still actively involved in various professions. Although little research has been conducted in this area, it may be hypothesized that a considerable number of middle-aged individuals, especially in the category of the 'old-olds', suffer from the adverse effects of early aging. This may present problems for those middle-aged persons who have to work under strenuous working conditions, such as, for example, bus drivers, air-traffic controllers, managers or teachers. It has been proposed that early functional aging depends on intrinsic factors like genetic and constitutional characteristics, but also on extrinsic factors such as stress, diet and health hazards. For instance, a recent study carried out by Houx and associates (Houx et al. 1991, Houx 1992) has shown that old subjects who had suffered mild head injury, general anaesthesia or excessive alcohol consumption earlier in life had lower performance levels in cognitively demanding tasks than old subjects in whose life these events had not occurred. Interestingly, these effects were seen most clearly in cognitive tasks that depended on typically fluid cognitive functions requiring rapid decision making, memory scanning and selective attention. The division between fluid and crystallized abilities should not be seen as too absolute. With extensive training fluid abilities can sometimes become crystallized. For instance, learning to find one's way in a new supermarket may initially be a very demanding task, since it requires permanent search in working memory and concurrent visual search for locations of objects in an unfamiliar environment. In the long run, however, this task will probably become much less demanding, given that the locations of various products in the supermarket remain constant. In addition, many cognitive and psychomotor tasks in which people engage in daily life are amalgamations of task components that rely on both crystallized and fluid abilities. A suitable example is the skill of driving a car, which depends on the ability of the driver to make rapid decisions under time pressure, and his ability to manoeuvre the car in unfamiliar or complex traffic situations. These situations usually require a high amount of attentional control. Driving skill also depends on general driving experience, however, and specific psychomotor abilities which seem to be carried out with little or no attentional control.

Difficulties in the assessment of the decline in fluid abilities

The last example may serve to illustrate that the use of complex real-life tasks, that is tasks in which people engage on a day-to-day basis, may not always be a good starting-point for aging research. At least not for the type of research that is concerned primarily with assessment of the nature and rate of decline of more basic, fluid cognitive processes (see also Salthouse 1985, p. 115–116 for a similar view). In contrast, in cognitive tasks that are developed in the laboratory, it is much easier to disentangle and measure the relative contribution of fluid and crystallized components of task performance than it is in natural task settings. The assessment of age-related differences in task performance in more natural occupational settings also faces other methodological problems (Welford 1958, Davies and Sparrow 1985, Sparrow and Davies 1988). First, in many working environments, selection processes like 'survival of the fittest' may cause voluntary or involuntary transfer or even dismissal of older employees. This could result in an underestimation of differences in production records of elderly and young subjects, since the older people that remain in the job are not representative of their age group. Second, with advancing age fluid abilities often decrease in importance, as occupational specialization and narrowing of interests will gradually restrict exposure to new sources of information. Finally, a third methodological problem is that age-related decrements in fluid abilities can probably be partly compensated by experience-based specific skills and expertise of workers. This could result in almost identical average performance levels of young and old workers (see figure 12.1).

Note, however, that the last supposition lacks empirical foundation, and that insufficient research has been conducted to answer the question of how experience and skills may be used to compensate for decline in fluid cognitive abilities in complex, real-life task conditions (Salthouse 1984, 1990). A careful

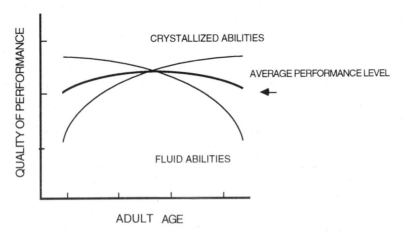

Figure 12.1. Schematic illustration of the contribution of fluid and crystallized task components to changes in the overall quality of task performance with advancing age.

decomposition of job behaviour in terms of the demands on fluid and crys-tallized abilities and the various ways they may interact, is therefore an impor-tant requirement for understanding the complex relationship between age and effectiveness of job behaviour. In the following paragraphs our focus will be mainly on the nature and extent of decline in fluid abilities that are measured by using basic laboratory tasks that are relatively free of cultural and experien-tial influences.

Age and task complexity

An issue of much recent debate is if the age-related decline in cognitive per-formance is global, that is affects all cognitive functions to an equal degree, or specific. Proponents of the view that cognitive aging is a specific process have usually argued that there are large individual differences both between and within elderly subjects in the pattern of their cognitive performances, and in the rates of decline of their cognitive functions. With increasing age the varia-bility between and within subjects increases. Performance across different cog-nitive skills and abilities becomes increasingly uneven and domain specific (Rabbit 1992).

Salthouse (1992) has argued recently that aging research should focus on discovery of both general and specific age-related influences. The point is that before proposing highly specific interpretations of age-related decrements in performance, influences of more general, simple or 'low-level' factors should be considered that could describe these effects in more parsimonious terms. The fluid–crystallized dimension that was described in the preceding section may serve as a good example of a general factor that could explain, at least in part, why variability between cognitive task components increases with age. Crys-tallized abilities, that is highly domain-specific skills, show little change with advancing age, while fluid abilities such as speed of information processing and attention-demanding processes are much more vulnerable to the effects of aging.

There is substantial evidence that other even more general, or basic, age-related factors than the fluid–crystallized dimension could also play an impor-tant role. One of the most robust phenomena in cognitive aging seems to be a general slowness that affects performance in a variety of qualitatively different tasks, but is manifested most clearly in complex,[1] mentally demanding task situations. This slowing with age has also been referred to as the 'generalized slowing deficit' or 'complexity function' (Salthouse 1985, Cerella 1990, Myerson et al. 1990). A complexity function is usually computed by per-forming a meta-analysis across a wide range of reaction tasks that show a large variation in response latencies. The results of these analyses are then depicted in regression functions or 'Brinley plots' (Brinley 1965, Birren 1965, Cerella et al. 1980, Cerella 1985, 1990, Salthouse 1988). In these plots the average RTs of a group of young adults is plotted on the *x*-axis and the average RTs of a group of old adults is plotted on the *y*-axis. Typically, the relationship between performance of old and young subjects is linear, and can be fitted with the following regression line:

$$\text{Old}_{i,j} = a + b. \text{Young}_{i,j} + E_{i,j}$$

in which a (the intercept) represents an additive slowing factor that is attributable to deficits in peripheral sensorimotor processes, and b (the slope) represents a proportional slowing factor representing the slowing of central processes as a function of task complexity. The implication of the significant linear relation is that the speed of performance of old subjects can be predicted with fairly high precision on basis of the speed of young adults, while no information about the specific tasks is required. Most of these meta-analyses have been obtained in typical laboratory tasks, such as choice reaction and memory search tasks, that require rapid decision making. As indicated earlier, these tasks specifically engage fluid cognitive abilities. Interestingly, Myerson et al. (1992) recently demonstrated in a meta-analysis of semantic priming and word-recognition tasks that the complexity function also holds for lexical processes.

The generalized slowing effect does not necessarily preclude the existence of more specific or 'localized' age-related deficits in information-processing abilities. It has been argued in various recent studies that meta-analyses are a suitable method to describe cross-age comparisons in age-homogeneous groups, but may overlook deviations in the general pattern that result from variations within tasks (Salthouse 1992, see also Fisk and Rogers 1991, Cerella 1991 and Fisk et al. 1992, for a lively debate on this issue). Another shortcoming of most meta-analyses is that they failed to control an important source of variance, namely accuracy of responding. Salthouse (1988, 1991), however, has recently made the observation that the complexity effect is not necessarily restricted to speed measures, but can be applied to any performance measure (e.g. percentage errors) that reflects the demands on limited processing resources. Thus, it may be proposed that in elderly subjects error rates could provide a second suitable index, in addition to speed of responding, to monitor processing capacity.

Several authors have proposed an alternative, more 'strategic' explanation for the slowing-with-age phenomenon. For instance, Rabbitt (1965) and Welford (1984) have suggested that the slower responding of old subjects might be attributable to a greater emphasis on accuracy. However, the few studies that have specifically addressed this issue have not provided clear support for the latter suggestion. Salthouse (1979), Salthouse and Somberg (1982a) and Strayer et al. (1987) were the first investigators that computed speed–accuracy trade-off of functions in young and old subjects. It appeared from their data that old subjects remained slower than young subjects even when comparisons were made between age groups at the same levels of accuracy. They therefore concluded that a speed–accuracy trade-off cannot account for all age differences in speed. The same suggestion was made more recently by Cerella (1990), who described the results of a meta-analysis assembled from a great number of RT tasks that reported both RTs and accuracies from groups of young and old subjects (Plude et al. 1984). This analysis also did not provide clear evidence of a shift towards more accurate responding in elderly subjects with slower responses. Instead, elderly subjects showed a slight decrease in accuracy with increased response latencies.

To conclude, it has not yet been demonstrated convincingly that the increased response latencies in more complex cognitive tasks of old subjects can be attributed to a decreased tolerance of error, or greater caution. To solve this problem there is still a need for more studies, using experimental designs, in which the response strategies of old and young subjects are varied systematically in tasks of different complexities (Salthouse and Somberg 1982b, Bashore et al. 1989).

The role of processing resources in aging

It has frequently been argued in the aging literature that a general slowing factor is of great theoretical importance, since it could indicate that age-related slowing in performance is quantitative or general, rather than qualitative or specific. However, a crucial question that remains is why and how the general factor operates. Salthouse (1985, 1988) suggested that more complex tasks make greater demands on an underlying commodity, denoted as limited processing resources (see figure 12.2). He further proposed that processing resources may be conceptualized in three different ways, namely as limitations in (a) time, or rate of processing; (b) space or working memory capacity; and (c) energy or attentional capacity.

In a similar vein, Gopher (1986) and Wickens (1986) have argued that resources represent influences of both energy and structure. The amount of processing power at the disposal of the system, or the capacity limitation, is the function of the joint influence of structural and energetic processes. A suitable metaphor to illustrate the latter conceptualization of processing resources is the engine of a car. The cylinders are the structural resource and the fuel is the energetic resource, and their joint operation determines the total capacity or power of the engine.

At the brain level, it is also possible to distinguish between structural and energetic components of processing resources. Structural processes in the brain have usually been associated with typical 'computational' functions of the cerebral cortex; for instance, feature analysis in the visual areas of the brain. In addition, it is assumed that the neurotransmitter, brainstem and limbic systems play an important role in energizing cortical computations, and providing sufficient activity necesssary for mental functions like consciousness, memory and attention (Beatty 1986, Foote 1987). Note that the human brain also undergoes a number of anatomical and chemical changes late in life that

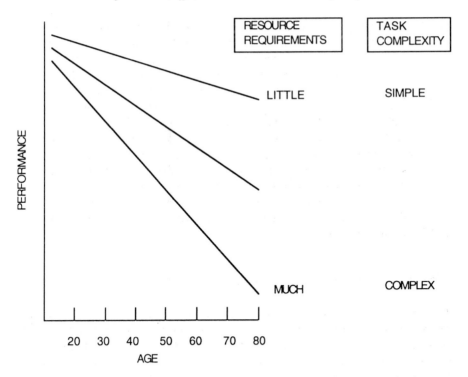

Figure 12.2. Hypothetical illustration of the changes in task performance with advancing age as a function of resource requirements and task complexity (adapted from Salthouse 1985).

may potentially impair its processing capacity (Scheibel and Scheibel 1975, Scheibel 1980, Petit 1982, Giaquinto 1988). The weight of the brain declines with age, resulting in an average loss of 20 per cent above 70 years (Selkoe 1992). In addition, the same amount of nerve cells appears to vanish in lower structures of the brain, like the brainstem and limbic system. The neural degeneration of brainstem structures also affects neurotransmitter systems such as the afferent ascending cholinergic and noradrenergic innervations of the cortex. As indicated previously, these systems play an important role in the regulation of intensive, energetic aspects of imformation processing.

Finally, the electroencephalogram, which is assumed to reflect the state of arousal of the human brain, also undergoes changes with advancing age in frequency and topographical characteristics. The same holds for the amplitude and latency of components of the event-related potentials of the brain, which reflect more specific cognitive characteristics of cerebral processes (Dustman and Snyder 1981, Pfefferbaum et al. 1984, Donchin et al. 1986, see also Kok and Zeef 1991).

It is further important to emphasize that the notion of task complexity, which plays a crucial role in the present paper, only makes sense in single capacity models. In a single capacity view, performance failures such as a decrease in speed and accuracy of performance are a consequence of capacity overloading (cf. Kahneman 1973, Shiffrin and Schneider 1977). It has been argued frequently in the past that a single capacity notion of processing resources may be too simplistic. For instance, Sanders (1983) and Wickens (1986) have proposed that processing resources are composed of several independent energetic sources associated with different brain mechanisms. As soon as processing resources are considered to be multidimensional, performance deficits can no longer be attributed to general task demands or capacity overloading, since the performance deficit may be attributed to overloading of one type of resource but not another (Sanders 1983, 1986, Wickens 1986).

The position taken in the present paper is that the multidimensional capacity view does not rule out the possibility that more general information-processing mechanisms could also play an important role. More specifically, it is proposed that a single capacity notion of processing resources can be useful to understand why slowing in performance in the elderly is larger in complex than in simple tasks. According to the capacity view, old subjects show greater decrements in the quality of performance in more demanding tasks than young subjects, because old age is accompanied by a smaller quantity of a non-specific pool of processing resources. As argued earlier, the limited-capacity notion may provide a parsimonious interpretation of performance decrements with advancing age, as far as fluid mental functions are concerned.

The effect of stressful conditions on processing resources in elderly subjects

It is generally assumed that the amount of processing capacity invested in a task, also denoted as mental effort, is mainly determined by intrinsic demands such as the difficulty or complexity of the task (Kahneman 1973, Mulder 1986). Mental effort is involved if a task requires the use of central, attention-demanding control operations. This is the case when a task:

(a) is carried out in the early stages of acquisition of a skill;
(b) involves difficult internal transformation of information like search of items in short-term memory; or
(c) has to be carried out under time pressure.

A second factor that may influence the allocation of processing resources in elderly persons, in addition to task demand, is stressful conditions such as lack of sleep, fatigue and noise. These conditions ususally interfere with an optimal level of performance in human operators (Broadbent 1963, 1971, Hockey and Hamilton 1983). For instance, noise can lead to impairment of accuracy of responding to reaction tasks and less efficient functioning of short-term memory (Loeb 1986, Cremer 1993). Additionally, factors such as sleep deprivation and shifts in circadian rhythms that occur frequently in shiftwork are generally associated with a reduced alertness, and may therefore be detrimental for efficient mental functioning (Johnson 1982, Hockey and Hamilton 1983).

Drugs are another type of environmental stressor that could have profound effects on the performance of elderly subjects. Beta-blockers and psychotropic drugs including sedatives, stimulants and anti-depressant medications, are known to be taken in greater quantities by old than by young persons. Furthermore, older persons are likely to be more sensitive to the effects of drugs, and to suffer from adverse drug interactions, leading to confusional states, memory loss and a drop in psychomotor performance. These effects may occur not only because of changes in the intrinsic sensitivity to the drug, but also because old people are more likely to be on more than one medication than young persons (Wheathley 1983, Burns and Phillipson 1986). The same holds for 'non-medical' substances such as caffeine, nicotine and alcohol. There is convincing evidence that alcohol may reduce processing capacity substantially in tasks that require attentional processing and application of fluent mental abilities (Craik 1977, Fisk and Schneider 1982). Since processing capacity decreases with age, it is likely that alcohol intoxication may have profound adverse effects on higher cognitive functioning such as memory, attention and complex psychomotor performance in elderly people. In contrast, caffeine, a drug that is presumed to increase mental alertness and concentration, could have positive effects on mental functioning in the elderly, because there is evidence to suggest that the level of cortical arousal is lower in old than in young subjects (Welford 1984, Woodruff 1982, 1985, Kok and Zeef 1991, van der Stelt and Snel 1993).

The precise ways in which drugs, substances and stressors interfere with the quality of human information processing still remain unknown, but it is usually assumed that these conditions affect mainly the energetic, or intensive, components of processing resources. As indicated earlier, energetic factors such as arousal or mental effort also contribute to the amount of processing power that is at the disposal of the information-processing system (Sanders 1983, Gopher 1986). With advancing age the loss of 'gain' in these energizing systems could gradually affect the quality of cognitive task performance, especially during stressful or suboptimal environmental conditions (Parasuraman and Nestor 1986).

Regression analysis of performance as a function of task complexity and stressors in young and old subjects

It is surprising that, although many studies have been conducted on the effects of stressful conditions on the quality of performance in young subjects, very little is known about the ways in which these conditions can affect the mental functioning and efficiency of performance of old persons. A hypothesis that was tested recently by Cremer (1993) and Sumulders (1993) in our laboratory was that old subjects would be more handicapped than young subjects in their cognitive performance by stressors such as noise and sleep loss.

This hypothesis was based on the following rationale:

(a) old subjects presumably suffer from a lack of processing resources;
(b) mental work performance under noise or sleep loss is generally considered to be more demanding;
(c) these stressful conditions will therefore be likely to aggravate the decline in performance in elderly subjects, especially under highly demanding task conditions.

A similar but reversed reasoning was followed in a third study on the effect of caffeine. Caffeine is an agent that is assumed to enhance concentration and the level of alertness, mental conditions that are positively related to processing resources (Lorist et al. 1994, van der Stelt and Snel 1993, Snel 1993). Accordingly, it was predicted that the age-related decline in performance in more demanding tasks would be less conspicuous after the administration of caffeine as compared with the placebo condition.

To answer these questions we computed Brinley plots on the performance data of young and old subjects that were obtained in various RT tasks and conditions. The primary purpose for using these tasks was to isolate the specific effects of age on stages of information processing, such as encoding, decision, preparation and execution of the motor response (see also Simon and Pouraghabagher (1978) and Salthouse and Somberg (1982b)). In addition, in two experiments attentional functions such as focused and divided attention abilities (visual search and memory search) were measured. The results of these analyses are described elsewhere in greater detail (Smulders 1993, Cremer 1993, Lorist et al. 1994). Apart from specific age and stressor effects, it appeared from these experiments that the effects of age on information-processing components were also interpretable in more general terms. In most of the conditions, older subjects were slower and less accurate than young subjects. Closer inspection of the data further suggested that, in general, the age-related drop in the quality of performance was larger in the more difficult than in the easy RT task conditions. To test this impression, regression analyses were carried out on the performance data over all task conditions that were used in these experiments. In comparison with traditional meta-analyses, two important innovations were applied in our approach. Recall that Brinley plots represent the global slowing in speed of reactions with an increase in task complexity, which is larger for older than for younger people. This effect, reflected in the slope of the linear regression function, is taken as a manifestation of the reduction in the amount of available resources, or processing capacity, with advancing age. As stated earlier, practically all meta-analyses reported in literature have been carried out on speed (RT) measures.

However, since in some of the conditions of our studies old subjects also showed a greater drop in accuracy in the more demanding tasks than young subjects, the error scores were also subjected to regression analyses. A second innovation of our analyses was that we computed separate regression functions for the normal and stressful conditions. Our assumption was that if task complexity and a stressor affect a common central factor such as a limited capacity system, this should become manifest in a change in the slopes of the regression functions of both the speed and accuracy measures.

Finally, the present analyses also differed in two other important aspects from preceding meta-analyses. First, each regression function was computed on RT tasks that had been performed by the same groups of subjects receiving the same instructions in the same laboratory, which precluded that differences in response tendencies or strategies between subjects or tasks would affect the data. Second, since each regression function was derived from RT tasks in which the same type of visual display and response panel was used, it was unlikely that these tasks differed in peripheral sensory and motor requirements.

Noise

Figure 12.3 illustrates the regression functions that were computed on reactions times and error percentages of the young male subjects ($n = 24$, 20–30 years) and old male subjects ($n = 24$, 65–75 years) in the noise experiment. Noise consisted of a mixture of realistic sounds recorded from an airport terminal, a swimming pool, a metal factory and road traffic, which was presented binaurally by headphones. The intensity level was 85 dB (A). The experimental conditions were: intact and degraded stimuli (digits), compatible and incompatible S–R associations, single versus double button press conditions, high and low display load, and high and low memory load conditions.

The analysis confirmed prior findings that an increase in task complexity was associated with a proportional (around 40 per cent) slowing of old subjects, if we leave the intercept effects out of consideration. The predicted variance r^2 was 0.94 and 0.90 for the quiet and noise conditions, respectively, suggesting that a linear function tends to represent the data well. It is also interesting to observe in figure 12.3 (upper panels) that the two data points that were associated with the longest RTs in young subjects, deviated somewhat from the general trend. Although we did not test these deviations statistically, they could be taken as an indication that the conditions that elicited these long RTs (the double responding and high memory load condition, respectively) also produced more specific age effects.

It further appeared that under noise, accuracy of performance in old subjects also decreased proportionally with increasing task complexity, as compared with young subjects (see figure 12.3: lower-right panel). Thus, old subjects became progressively less accurate than young subjects in their performance on more complex or demanding tasks when these tasks had to be

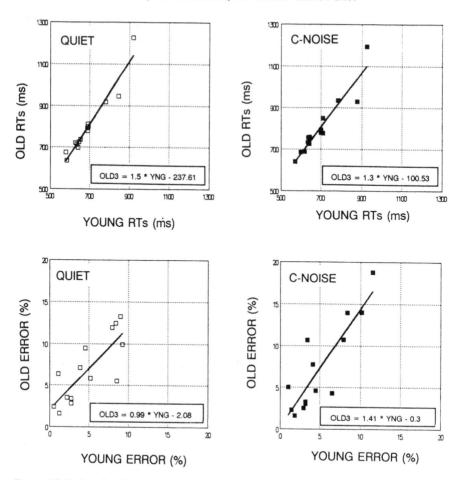

Figure 12.3. Speed and accuracy of performance of older adults as a function of the speed and accuracy of performance in young subjects. Left: quiet condition; right: noise condition. Data points represent averages of 14 visual RT conditions. Conditions were: intact and degraded stimuli, compatible and incompatible S–R (stimulus–response) associations, single versus double button press conditions, high and low display load, and high and low memory load conditions. The last two conditions comprised both target and non-target stimuli.

performed under noise, but not in the quiet condition. The r^2 values of error scores were much lower than those for RTs, namely 0.65 and 0.75 for the quiet and noisy conditions, respectively, indicating that the linear fit of these measures was much less than that of the speed measures. An additional correlation analysis was carried out to verify if the higher number of errors under noise (mostly false alarms) could be attributed to a shift in the relationship between speed and accuracy of responding. To estimate these changes in performance more directly, difference scores were calculated by subtracting RT and error scores in the quiet condition from those in the noisy condition, for each individual subject. Note that if subjects show a general tendency to become faster and less accurate under noise, we may expect a negative correlation between the difference scores of speed and accuracy. The results showed that in both age groups the correlations were significant and positive (young: 0.37; old: 0.29; $p < 0.001$), which indicated that there was no overall (i.e. across subjects and tasks) trade-off between speed and accuracy of responding in the noisy condition. Thus, both young and old subjects became slower and less accurate under noise (see also Cremer 1993 for a more detailed discussion of the latter results).

Sleep deprivation

The second study was carried out to investigate the effects of sleep deprivation on information processing (see, for details, Smulders 1993). Two groups of male young and old subjects (young, $n = 12$, 18–24 years; old, $n = 12$, 62–73 years) were kept awake for one night in the laboratory and tested in the afternoon (12.00 h) of the following day. Experimental conditions were: intact and degraded stimuli, compatible and incompatible stimulus (sleep–response) associations, time-certain and time–uncertain stimulus presentation conditions. Again, it was found that old subjects were about 40 per cent slower in their reactions across all RT tasks, if the intercept effects were not considered (figure 12.4: upper panels). Both young and old subjects were slower and less accurate in their performances after sleep deprivation than after normal sleep. In addition, r^2 values of errors were much lower than those of the speed scores, confirming the visual impression that the data points of accuracy of performance were more scattered than for the speed measures. In contrast with the noise study, no clear indications were found that old subjects suffered more from the stressor than young subjects. Young subjects even showed a higher amount of omission errors after sleep loss, which probably indicated that they were more prone to short lasting lapses of attention than old subjects (figure 12.4: middle-right panel). This last impression was strengthened by the finding that the number of omission responses increased with longer task duration: the frequency of attentional lapses usually increases with longer task duration (Johnson 1982). Furthermore, RT measures showed only small effects of sleep loss, suggesting that between attentional lapses subjects may work relatively normally.

Young subjects also showed significantly higher ratings of subjective effort during task performance after sleep loss than old subjects. Thus, young subjects not only suffered more from attentional lapses but also found mental work more strenuous after sleep loss than old subjects.

Caffeine

The third experiment was conducted to investigate the effects of caffeine in young and old subjects (young: eight males and seven females, age: 18–23

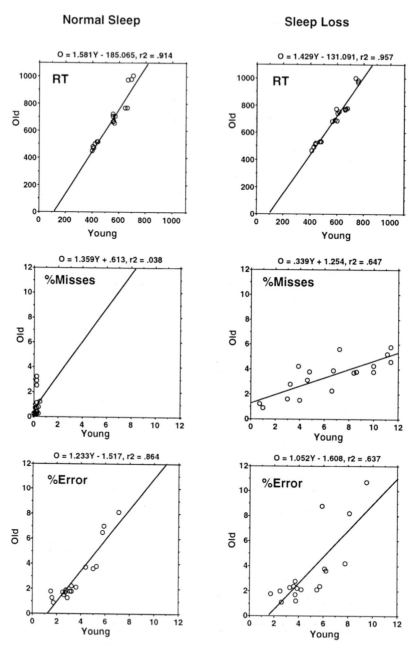

Figure 12.4. Speed and accuracy of performance of older adults as a function of the speed and accuracy of performance in young subjects. Left: normal sleep condition; right: sleep-deprivation condition. Data points represent averages of six task conditions (intact and degraded stimuli, compatible and incompatible S–R associations, time-certain and time-uncertain RT conditions) that were administered in three successive blocks of trials (total: 18 data points).

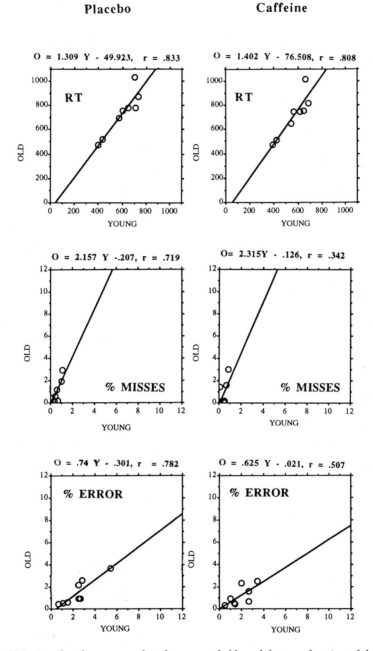

Figure 12.5. Speed and accuracy of performance of older adults as a function of the speed and accuracy of performance in young subjects. Left: placebo condition; right: caffeine condition. Data points represent averages of eight task conditions (intact and degraded stimuli, compatible and incompatible S–R associations, time-certain and time-uncertain RT conditions, high and low memory load conditions).

years; old: four males, 11 females, age: 60–72 years). Subjects received 200 mg + 50 mg (maintenance dose) caffeine or lactose as a placebo, dissolved in a cup of decaffeinated coffee. The treatment was double blind and subjects were made to believe that normal coffee was served in both conditions. Saliva samples were taken to verify that caffeine was still active during task conditions. Experimental conditions were: intact and degraded stimuli, compatible and incompatible S–R associations, time-certain and time-uncertain stimulus presentation conditions, high and low memory load conditions.

Results showed that all subjects reacted faster in the caffeine than in the placebo condition. No effects of caffeine were found on the accuracy of reactions. Old subjects made slightly fewer errors (false alarms) and more omissions (misses) than young subjects. Accuracy was relatively high in the caffeine study (>95 per cent), which means that the regression analyses of the error scores are not very reliable. Figure 12.5 further illustrates that age-related slowing was greater in more complex than in simple tasks: old subjects were about 30–40 per cent slower in their reactions than young subjects (slope effect). Owing to the smaller number of data points, r^2 values were slightly smaller for caffeine than for the noise and sleep loss conditions. It can also be clearly seen from figure 12.5 that the slopes of the RT functions were not affected by caffeine as compared with the placebo condition. Stated in more general terms: there were no clear indications that age-related differences in processing resources, as manifested in the slope of the regression functions, were attenuated by caffeine. Note that the latter conclusion is based on a global analysis of performance measures over all task conditions, and does not rule out the possibility that caffeine had specific effects on age-related differences in individual task conditions.

Conclusions

The results clearly demonstrated that, with respect to speed measures, the complexity effect, that is the greater slowing of old subjects in more complex than in simple RT tasks, is a robust phenomenon. With respect to the accuracy of response, the regression functions were much more variable and predicted less age-related variance.

With the exception of the effects of noise, the results did not confirm our initial assumption that variation in the state of subjects would either enhance or compensate for an age-related deficit in processing resources. The combined effects of task complexity and stress conditions yielded a complex pattern of performance measures, which could not be explained simply on the basis of a straightforward resource depletion model of aging. This pattern depended strongly on the specific type of stressor that was applied in the experiment. Noise clearly aggravated the drop in accuracy in more complex tasks in old subjects as compared with younger subjects. Alternatively, sleep deprivation produced a paradoxical effect on the accuracy of responding: old subjects showed a smaller drop in accuracy (the number of misses) after sleep deprivation than young subjects. One possible reason why sleep loss was less detrimental for old than for young subjects is that, with advancing age, the biological restorative function of sleep is altered. For instance, old people show a virtual absence of slow-wave sleep, particularly stage 4 sleep (Prinz 1976) along with increased night-time wakefulness (Prinz et al. 1990). Although these findings are still poorly understood, they could indicate that night-time sleep loss is less detrimental for day-time performance in old than in young subjects. Finally, there were no indications that caffeine compensated or reduced the age-related decline in speed of performance in more demanding tasks. Snel (1993) has suggested that beneficial effects of caffeine on mental performance are usually seen more clearly in low-alertness conditions. Additionally, there is evidence that the alleged lower cortical arousal or alertness of old subjects is more conspicuous in low-demand situations than in effortful tasks (Prinz et al. 1990). Thus the absence of appreciable effects of caffeine on the slope of the regression functions could have resulted from the fact that in the more demanding task conditions, old subjects had reached the same level of arousal as young subjects.

In summary, the present laboratory studies have presented evidence that older subjects show a decline in fluid cognitive abilities, which is most conspicuous when cognitive tasks become more demanding, or when complex tasks have to be carried out under conditions of loud environmental noise. If we extrapolate from laboratory to real-life working conditions, the present results could indicate that elderly workers will probably show the largest decline in their job performance in complex, mentally demanding tasks, which require rapid decision making and which have to be carried out in noisy environments. Alternatively, the perspectives from our studies for elderly workers are more optimistic when mentally demanding tasks have to be carried out after loss of sleep. The present findings may also serve to illustrate the utility of linear regression analysis in assessing the effects of environmental stressors on the resource-dependent performance of older subjects, regardless of the identity of specific information-processing components of the task. A change in a single regression parameter can be used as an indication of the global effect of state on age-related decline in resource-dependent performance across a great variety of tasks.

Note

1. Task complexity is usually defined as the number of processing operations required to perform the task. For instance, a difficult memory-search task is considered to be more complex than a simple reaction task because it is reasonable to assume that the former task consists of a greater number of processing operations or processing 'steps' than the latter task (Salthouse 1985, Myerson et al. 1990). Cerella et al. (1980)

have defined complexity in more operational terms as the processing time of a task minus the latency of purely sensory and motor components. To avoid terminological confusion we will use the same operational definition of complexity that has been used by Cerella in the present study. Furthermore, the terms task difficulty and task complexity will be used interchangeably to indicate any task operation that will increase (central) processing time.

References

Baddeley, A. 1981. The concept of working memory, *Cognition*, **10**: 17–23.

Bashore, T. R., Osman, A. and Heffley, E. F. 1989. Mental slowing in elderly persons: a cognitive psychophysiological analysis, *Psychology and Aging*, **4**: 235–244.

Beatty, J. 1986. Computation, control and energetics. In R. J. Hockey, A. W. K. Gaillard and M. G. H. Coles (eds) *Energetics and Human Information Processing* (Martinus Nijohff, Dordrecht), 43–52.

Birren, J. E. 1965. Age changes in speed of behavior: its central nature and physiological correlates. In A. T. Welford and J. E. Birren (eds) *Behavior, Aging and the Nervous System* (Charles C. Thomas, Springfield, IL), 191–216.

Birren, J. E. and Cunningham, W. J. 1985. Research on the psychology of aging: principles, contents and theory. In J. E. Birren and K. W. Schaie (eds) *Handbook of the Psychology of Aging* (2nd edn) (Van Nostrand Reinhold, New York), 3–34.

Birren, J. E. and Schaie, K. W. 1990. *Handbook of th Psychology of Aging* (3rd edn) (Academic Press, San Diego).

Brinley, J. F. 1965. Cognitive sets, speed and accuracy of performance in the elderly. In A. T. Welford and J. E. Birren (eds) *Behavior, Aging and the Nervous System* (Charles C. Thomas, Springfield, IL), 114–149.

Broadbent, D. E. 1963. Differences and interactions between stresses, *Quarterly Journal of Experimental Psychology*, **15**: 205–211.

Broadbent, D. E. 1971. *Decision and Stress* (Academic Press, London).

Bromley, D. B. 1988. *Human Ageing. An Introduction to Gerontology* (Penguin, London).

Burns, B. and Phillipson, C. 1986. Drugs, *Ageing and Society* (Croom Helm, London).

Cannon, W. B. 1936. *Bodily Changes in Pain, Hunger, Fear and Rage* (Appleton-Century-Crofts, New York).

Cerella, J. 1985. Information processing rates in the elderly, *Psychological Bulletin*, **98**: 67–83.

Cerella, J. 1990. Aging and information processing rate. In J. E. Birren and K. W. Schaie (eds) *Handbook of the Psychology of Aging* (3rd edn) (Academic Press, San Diego), 201–212.

Cerella, J. 1991. Age-effects may be global, not local. Comment on Fisk and Rogers (1991), *Journal of Experimental Psychology: General*, **120**: 215–223.

Cerella, J., Poon, L. W. and Williams, D. M. 1980. Age and the complexity hypothesis. In L. W. Poon (ed.) *Aging in the 1980s. Psychological Issues*, (American Psychological Association, New York), 332–340.

Craik, F. M. 1977. Similarities between the effects of aging and alcoholic intoxication on memory performance, construed within a 'levels of processing' framework. In I. M. Birnbaum and E. S. Parker (eds) *Alcohol and Human Memory* (Erlbaum, Hillsdale, New York), 9–21.

Craik, F. I. M. 1984. Age differences in remembering. In L. R. Squire and N. Butters (eds) *Neuropsychology of Memory* (Guilford, New York), 3–12.

Cremer, R. 1993. Cognitive aging, visual task demands and noise. Ph. D. thesis, University of Amsterdam, Department of Psychonomics.

Davies, D. R and Sparrow, P. R. 1985. Age and work behaviour. In N. Charness (ed.) *Aging and Human Performance. Studies in Human Performance* (John Wiley, Chicester), 293–332.

Donchin, E., Miller, G. A. and Farwell, L. A. 1986. The endogenous components of the event-related potential—a diagnostic tool? In D. F. Swaab, E. Fliers, M. Mirmiran, W. A. van Gool and F. van Haaren (eds) *Progress in Brain Research* (Elsevier, Amsterdam), 87–103.

Duffy, E. 1962. *Activation an Behavior* (Wiley, New York).

Dustman, R. E. and Snyder, E. W. 1981. Life-span changes in visually evoked potentials at central scalp, *Neurobiology of aging*, **2**: 303–308.

Fisk, A. D. and Schneider, W. 1982. Type of task practice and time-sharing activities predict performance deficits due to alcohol ingestion. In Proceedings of the Human Factors Society, Santa Monica, California.

Fisk, A. D. and Rogers, W. A. 1991. Towards understanding of age-related memory and visual search effects, *Journal of Experimental Psychology: General*, **120**: 131–149.

Fisk, A. D., Fisher, D. L. and Rogers, W. A. 1992. General slowing alone cannot explain age-related search effects: reply to Cerella, *J. of Experimental Psychology: General*, **121**: 73–78.

Foote S. L. 1987. Extrathalamic modulation of cortical function, *Annals of Neuroscience*, 10, 67–95.

Giaquinto, S. 1988. *Aging and the Nervous System* (John Wiley, Chicester).

Gopher, D. 1986. In defence of resources: on structures, energies, pools and the allocation of attention. In G. R. J. Hockey, A. W. K. Gaillard and M. G. H. Coles (eds) *Energetics and Human Information Processing*, (Martinus Nijhoff, Dordrecht), 353–371.

Hockey, R. J., Coles, M. G. H. and Gaillard, A. W. K. 1986. Energetical issues in research on human information processing. In R. J. Hockey, A. W. K. Gaillard and M. G. H. Coles (eds) *Energetics and Human Information Processing* (Martinus Nijhoff, Dordrecht), 3–21.

Hockey, R. J. and Hamilton, P. 1983. The cognitive patterning of stress states. In R. J. Hockey (ed.) *Stress and Fatigue in Human Performance* (John Wiley, London), 331–362.

Horn, J. L. 1982. The theory of fluid and crystallized intelligence in relation to concepts of cognitive psychology and aging in adulthood. In F. I. M. Craik and S. Trehub (eds) *Aging and Cognitive Processes. Advances in the Study of Communication and Affect*. vol. 8 (Plenum, New York), 238–278.

Houx, P. 1992. Cognitive aging and health-related factors. Ph. D. thesis, Maastricht, University of Limburg.

Houx, P. J, Vreeling, F. W. and Jolles, J. 1991. Rigorous health screening reduces age effect on memory scanning task, *Brain and Cognition*, **15**: 246–260.

Johnson, L. 1982. Sleep deprivation and performance. In W. B. Webb (ed.) *Biological Rhythms, Sleep and Performance. Studies in Human Performance* (John Wiley, Chicester), 111–141.

Kahneman, D. 1973. *Attention and Effort* (Prentice-Hall, Englewood Cliffs).

Kok. A. and Zeef, E. 1991. Arousal and effort: a review and theoretical synthesis of studies of age-related changes in event-related potentials. In C. H. M. Brunia, G. Mulder and M. N. Verbaten (eds) *Event-related Brain Research*, Supplement no. 42 to *Electroenceph. Clin. Neurophysiol* (Elsevier, Amsterdam), 324–341.

Loeb, M. 1986. *Noise and Human Efficiency. Studies in Human Performance* (John Wiley, New York).

Lorist, M. M., Snel, J. and Kok, A. 1994. The influence of caffeine and information processing stages in well rested and fatigued subjects, *Psychopharmacology*, **113**: 411–421.

Mulder, G. 1986. The concept and measurement of mental effort. In R. J. Hockey, A. W. K. Gaillard and M. G. H. Coles (eds), *Energetics and Human Information Processing* (Martinus Nijhoff, Dordrecht), 175–198.

Myerson, J., Hale, S., Wagstaff, D., Poon, L. W. and Smith, G. A. 1990. The information loss model: a mathematical theory of age-related cognitive slowing, *Psychological Review*, **97**: 475–487.

Myerson, J., Ferraro, F. R., Hale, S. and Lima, S. D. 1992. General slowing in semantic priming and word recognition, *Psychology and Aging*, **7**: 257–270.

Parasuraman, R. and Nestor, P. 1986. Energetics of attention and Alzheimer's disease. In R. J. Hockey, A. W. K. Gaillard and M. G. H. Coles (eds) *Energetics and Human Information Processing* (Martinus Nijhoff, Dordrecht), 395–407.

Petit, T. L. 1982. Neuroanatomical and clinical neurophysiological changes in aging and senile dementia. In F. I. M. Craik and S. Trehub (eds), *Aging and Cognitive Processes* (Plenum, New York), 1–21.

Pfefferbaum, A., Ford, J. M., Wenegrat, B. G., Roth, W. T. and Kopell, B. S. 1984. Clinical application of the P3 component the event-related potentials. 1. Normal aging, *Electroenceph. Clin. Neurophysiol*, **59**: 85–103.

Plude, D. J., Cerella, J. and Raskind, C. L. 1984. Speed—accuracy tradeoffs in cognitive aging research. Paper presented at the meeting of the Gerontological Society of America, San Antonio, Texas.

Prinz, P. N. 1976. EEG during sleep and waking states, *Experimental Aging Res.*, **1**: 135–163.

Prinz, P. N., Dustman, R. E. and Emmerson, R. 1990. Electrophysiology and aging. In J. E. Birren and K. W. Schaie (eds) *Handbook of the Psychology of Aging* (3rd edn) (Academic Press, New York), 135–149.

Rabbit, P. M. A. 1965. Age and the discrimination between complex stimuli. In A. T. Welford and J. E. Birren (eds) *Behaviour, Aging and the Nervous System* (Charles C. Thomas, Springfield, IL), 35–53.

Rabbit, P. M. A. 1992. Cognitive changes with age must influence human factors design. In H. Bouma and J. A. M. Graafmans (eds) *Gerontechnology* (IOS Press, Amsterdam), 113–139.

Salthouse, T. A. 1979. Adult age and the speed—accuracy tradeoff, *Ergonomics*, **22**: 811–821.

Salthouse, T. A 1984. Effects of age and skill in typing, *Journal of Experimental Psychology: General*, **113**: 345–371.

Salthouse, T. A. 1985. *A Theory of Cognitive Aging* (North-Holland, Amsterdam).

Salthouse, T. A. 1988. Resource-reduction interpretations of cognitive aging, *Developmental Review*, **8**: 238–272.

Salthouse, T. A. 1990 Cognitive competence and expertise in aging. In: J. E. Birren and K. W. Schaie (eds) *Handbook of the Psychology of Aging* (3rd edn) (Academic Press, San Diego), 310–319.

Salthouse, T. A. 1991. *Theoretical Perspectives of Cognitive Aging* (Lawrence Erlbaum, Hillsdale, NJ).

Salthouse, T. A. 1992. Shifting levels of analysis in the investigation of cognitive aging, *Human Development*, **35**: 321–342.

Salthouse, T. A. and Somberg, B. L. 1982a. Time–accuracy relationships in young and old adults, *Journal of Gerontology*, **37**: 349–353.

Salthouse, T. A. and Somberg, B. L. 1982b. Isolating the age-deficit in speeded performance, *Journal of Gerontology*, **37**: 59–63.

Sanders, A. F. 1980. Stage analysis of the reaction processes. In G. Stelmach and J. Requin (eds) *Tutorials on Motor Behaviour* (North Holland, Amsterdam), 331–354.

Sanders, A. F. 1983. Towards a model of stress and human performance, *Acta Psychologica*, **53**: 61–97.

Sanders, A. F. 1986. Energetical states underlying task performance In R. J. Hockey, A.

W. K. Gaillard and M. G. H. Coles (eds) *Energetics and Human Information Processing* (Martinus Nijhoff, Dordrecht), 139–154.

Simon, J. R. and Pouraghabagher, A. R. 1978. The effect of age on the stages of processing in a choice reaction task, *Journal of Gerontology*, **33**: 553–61.

Scheibel, A. B. 1980. Aging and senescense in selected motor systems of man. In D. G. Stein (ed.) *The Psychobiology of Aging: Problems and Perspectives* (Elsevier North Holland, Amsterdam), 273–282.

Scheibel, M. E. and Scheibel, A. B. 1975. Structural changes in the aging brain. In H. Brody, D. Harmon and J. M. Ordy (eds) *Aging*, vol. 1 (Raven Press, New York), 11–37.

Selkoe, D. J. 1992. Aging brain, aging mind, *Scientific American*, **267**: 97–103.

Selye, H. 1936. A syndrome produced by various noxious agents, *Nature*, **138**: 32.

Shiffrin, R. M. and Schneider, W. 1977. Controlled and automatic human processing: II. Perceptual learning, automatic attending and a general theory, *Psychological Review*, **84**: 127–190.

Smulders, F. 1993. Selective age effects on information processing: responses times and electrophysiology. Ph. D. thesis, University of Amsterdam, Department of Psychonomics.

Snel, J. 1993. Coffee and caffeine. In S. Garratini (ed.) *Caffeine, Coffee and Health* (Raven Press, New York), 255–290.

Sparrow, P. R. and Davies, D. R. 1988. Effects of age, tenure, and job complexity on technical performance, *Psychology and Aging*, **3**: 307–314.

Stankov, L. 1988. Aging, attention and intelligence, *Psychology and Aging*, **3**: 59–74.

Sternberg, S. 1969. The discovery of processing stages: extensions of Donders method, *Acta Psychol.*, **30**: 276–315.

Strayer, D. L., Wickens, C. D. and Braune, R. 1987. Adult age differences in the speed and capacity of information processing: 2. An electrophysiological approach, *Psychology and Aging*, **2**: 99–110.

van der Stelt, O. and Snel, J. 1993. Effects of caffeine on human information processing. In S. Garratini (ed.) *Caffeine, Coffee and Health* (Raven Press, New York), 291–316.

Welford, A. T. 1958. *Ageing and Human Skill* (Oxford University Press, London).

Welford, A. T. 1984. Between bodily changes and performance: some possible reasons for slowing with age, *Experimental Aging Research*, **10**: 73–87.

Wheathley, D. 1983. *Psychopharmacology of Old Age* (Oxford University Press, Oxford).

Wickens, C. 1986. Gain and energetics in information processing. In R. J. Hockey A. W. K. Gaillard and M. G. H. Coles (eds) *Energetics and Human Information Processing* (Martinus Nijhoff, Dordrecht), 373–384.

Woodruff, D. S. 1982. Advances in the psychophysiology of aging. In F. I. M. Craik and S. Trehub (eds) *Aging and Cognitive Processes* (Plenum, New York), 29–53.

Woodruff, D. S. 1985 Arousal, sleep and aging. In J. E. Birren and K. W. Schaie (eds) *Handbook of the Psychology of Aging* (2nd edn) (Van Nostrand Reinhold, New York), 261–295.

13

Aging and modern technology: how to cope with products and services

G. Rudinger, J. Espey, H. Neuf and E. Paus

Summary. Modern gadgetry, ranging from automatic dispensing machines to video telephony, is supposed to be able to make life easier. As a matter of fact, the elderly might profit especially from many of these developments—for example, the medical care call—if the sets are usuable. But, all too often, interfaces and manuals of those sets are so confusing that they cannot be operated. In two research projects (ALTEC, LUSI) we studied, in both an observational and experimental setting, how people cope with ticket machines, video-recorders, ISDN telephones and a video-conference system (a type of video telephony). These projects have been conducted to test how difficult it really is, particularly for the elderly, to operate such sets successfully. After having explored the degree of the difficulties and the reasons underlying them, it is now possible to develop improved interfaces which should allow users—especially elderly users—to use modern technology more easily. From 1989 to 1991, the research project ALTEC was conducted. ALTEC stands for 'Aging and Techonology', and was funded by the state of North Rhine Westphalia. Furthermore, we are conducting the EU Project LUSI together with English, Spanish and French partners. LUSI is an abbreviation for Likable and Usable Service Interfaces. The aim of LUSI is development of likable and usable service interfaces for modern telecommunications services in Europe.

Keywords: usability, field observations, laboratory experiments, intervention, technological grammar, everyday technology, telecommunication.

Introduction

The automation of a great number of day-to-day services with the help of public automats, for example in banking houses and public transport, the use of facsimiles for both private and business communication, multifunction telephones and card operated telephones, is ever on the increase. Likewise, the number of functions available and, along with them, their intricacies have risen considerably in the field of home technology and entertainment technology. There are, for example, the TV set, video, videotex, teletext and the answering machine. This change has become a burden for the user.

The elderly, in particular, seem to have difficulty in keeping up with the pace of automation of everyday life (Kruse 1992). More and more frequently they are confronted with an unfamiliar situation where the only way to gain access to a service is to apply the syntax of the appliance. Moreover, the number of counter clerks providing a service is decreasing. The ability to deal with modern technology has become an essential qualification for coping successfully and independently with everyday life. The importance of such a qualification, however, is not only restricted to the ability to meet with the more immediate requirements of technology. More important are the repercussions of failures experienced in a significant field, such as administering an account using an automatic bank teller or buying a ticket (Breakwell and Fife-Shaw 1988, Danowski and Sacks 1980). Such a failure may thus leave them with the erroneous impression that their general competence has diminished. A deficient 'person–environment' fit may be accounted for by both a deficiency in user-friendliness of many modern automats and a certain lack of adaptability on the part of the elderly (Gilly and Zeithaml 1985). This issue is intensified by the increasing speed with which technical innovations are offered. The increasingly large percentage of elderly people in all highly industrialized societies means that there is an increasing number of people affected by this problem, (Caplan, 1975) which may be regarded as a quantitative intensification of the issue.

A number of tests focusing on a connection between elementary cognitive abilities on the one hand and complex performance of intelligence and of day-to-day performance on the other hand have corroborated that the more complex performances do not decline to the same extent as the basic cognitive processes do (Willis 1987). Elderly people usually counteract the decline of basic capacities with certain compensatory strategies (Baltes and Baltes 1991). Compensatory strategies may comprise, for example, deliberately giving more time to actions, more frequent control of the results of actions and limiting oneself to a few actions and aims. That there is a way of compensating for deficiencies raises the question of whether it should not more generally be possible to enhance the capacity of the elderly to deal with technology by teaching them reasonable compensatory techniques or by improving their skills in the field of basic processes. Such an approach may be said to form part of a 'person-centred-approach'. A person-centred intervention in the area of basic processes may be thought of as a complement to a technology-centred approach if essential basic processes that affected a number of complex performances could be identified—the approach followed here.

The ALTEC-project

ALTEC investigated how elderly people cope with technical apparatuses that are used by the general public in daily life (Rudinger, G., Espey, J., Neuf, H. and Simon U. 1992). Let us assume there are two categories of machines used by the general public:

(1) Machines that are used in public (ticket machines, banking and so on)
(2) Machines that are used in people's homes (TV, video-recorder, washing machines and so on).

The decision was made to select:

(1) The ticket machine of the Bonn-Köln transport system as a prototype of the non-private machines (machines, which are used in public like ticket machines, banking machines and so on). The surface of this ticket machine corresponds to the valid DIN-norms (cf. Reinig and Arnold 1983).
(2) A TV set and video-recorder, made by Grundig as a prototype of the 'private' machines which are used at home.

The usability of a ticket machine

A field study was conducted with 1400 persons (about 200 individuals were older than 60, and can therefore be defined as elderly) in order to discover the full range of problems users have in dealing with ticket machines. Then, on the basis of the results of this field study we tested certain improvements of the user interface in order to facilitate usage.

Before describing the field study let us take a look at the psychological abilities involved in successful use of a ticket machine. The user, who has to know the exact name of the station, has to look for that name in the alphabetical list or on the map of short distances, and he or she has to read and to remember the character next to the station name. After that he or she has to press the corresponding key and to insert coins into a slot. Finally, he or she has to take any change and the ticket out of a box at the bottom of the machine.

Each of these operations requires power of concentration, attention, the ability to select the important information and sufficient working memory (Anderson, 1988). The co-ordination of all actions depends especially on an effective working memory. In particular, the working memory of the elderly could be affected more easily.

Results from the field study

In our study we found that more than 40 per cent of our 1400 subjects and more than 60 per cent of elderly people had difficulties in understanding information about the price system. During the test, 25 per cent of the elderly gave up using the machines after various ineffective trials. This result shows that the usability of a ticket machine could be a real threshold to mobility. The operation time for the elderly subjects was approximately 35 per cent above average (younger persons: 42 sec; elderly: 62 sec).

Laboratory Experiment

A sample of 30 younger (<60 years) and 30 older (>60 years) subjects was drawn from the population of Bonn and Rhein-Sieg-Kreis. In front of a real ticket machine, installed in our laboratory, people were asked to buy the cheapest ticket possible for different destinations (we chose stations at short, medium and long distances).

The first independent variable was age (old/young), the second variable was the surface of the ticket machine ('original display vs. modified display').

Most of the time needed could be attributed to the display, which seemed to be an inconsistently arranged display of information. To prevent further confusion we covered the map of short distances.

It was our aim to show that even slight and easily realizable alterations in the design of technical machines will exercise a powerful influence on the quality of operation and on operation time.

Results from the laboratory experiment

All main effects and interaction effects of the experiment concerning the ticket machine (respectively the video-recorder) presented in this text are significant at the 5 per cent level. They are as follows:

(1) The required operation time for elderly subjects was reduced by almost 40 per cent in the modified display condition.
(2) The quality of usage (number of tickets selected correctly) was increased from 78 per cent to 93 per cent in the elderly group. In the modified display condition, in particular, elderly individuals tend to buy a valid but too expensive ticket.

Conclusions

The graphical display of destinations confuses the user with supplementary information; it causes an information overflow. The graphical display can be substituted easily and effectively by an alphabetical list. This compact list contains all the information needed to buy a valid ticket.

The usability of a video-recorder

It was decided to select a video-recorder manufactured by Grundig. As our investigations have shown, this type of video-recorder is representative of the marketable devices. The most important functions of a video-recorder were analysed and selected. Then, certain tasks were formulated, such as 'Please record the broadcast on TV Channel 10'.

The statistical design was the same as in the experimental study on the ticket machine. The first independent variable was age, the second the comparison between two types of instructions for use. We call them the 'Plus version'and the 'Minus version'.

In comparison with the highly complicated orginal instructions for use, both versions were improved with regard to the formulation of sentences. Sentences were short and superfluous technical terms were excluded. Moreover, the Plus version differed from the Minus version by some formal characteristics:

(1) Bigger characters.
(2) All operation actions were marked separately.
(3) With the help of colours, which were used in the text as well as in the graphics of the remote control, the discovery of the correct key should be facilitated.

This design was devised in order to show that a mere formal alteration of the instructions for use will lead to an enormous increase in usage quality and an enormous decrease in operation time.

The task of operating a video-recorder successfully includes reading the text, identifying the next key to press, recognizing it on the graphics of the remote control and pressing the correct key on the real remote control. This procedure may have to be repeated many times.

The psychological abilities mentioned above, which are involved in operating a ticket machine, are also involved in this case. Because of the necessary co-ordination of many operation steps a sufficient working memory is highly important.

Results and conclusions

Some interesting results concerning the elderly were:

(1) The association between the type of version (Plus vs. Minus version), the number of successfully solved tasks and the operation time. Each person was confronted with the same sequence of 6 tasks. Both groups have profited from the use of the improved version. Elderly individuals in particular have had an enormous increase in the quality of usage. This result shows that it is not absolutely necessary to develop special improvements for the elderly.
(2) The same result was found for the decrease of operation time.
(3) Apart from the question of which version is more useful, it is interesting to analyse the association between the two variables operation time and quality of usage. Many elderly people try to compensate for their difficulties in understanding the instructions and in putting them into practice by taking more time. Surprisingly, there was no one who

needed a lot of operation time and reached a high level of usage quality.

Hence, we might cautiously conclude that a short operation time is a pre-condition for successful use, otherwise the user's power of concentration and/or his patience is probably overcharged.

Let us have a look at one additional aspect of operating modern technology: the use of shift keys. By the use of shift keys, certain other keys attain another function. This modern operation procedure does not correspond to the expectations of the elderly (with their grammar of technology that one key serves for one function only). We call it key-function unity.

When we take a look at the results from our experimental tasks, involving shift keys, we can see that the understanding of shift-key tasks differs enormously between the two age groups. Long operation times and low usage quality could often be explained by a misinterpretation of shift keys. But, using the improved version of the instructions for use, this difference was highly reduced.

The LUSI project

The project LUSI (1992–1995) is funded by the European Union and aims to develop likeable and usable service interfaces for telecommunication sets. With advanced telecommunication sets such as ISDN Telephone especially, video-telephony is expected to become widespread throughout the whole of Europe. This development is of great importance for the elderly as well. The disabled elderly in particular need future telecommunication technology in order to call medical help if needed. If the user is immobile, the ability to hear and see one's partner during videotelephony will serve as a kind of substitute for a personal meeting, if the personal meeting is troublesome for the partners.

To detect all possible shortcomings of the interfaces, 50 per cent of the subjects (12 individuals), who were invited to operate an ISDN Telephone and a PC-integrated video conference system, were above 60 years. In addition to the operation of the two sets, subjects were asked to fill in a questionnaire, which measures the trait 'General technical experience'. In this questionnaire people have to describe, with regard to 60 machines or computer programmes, if they have ever used them and how much control they feel they have over the machine.

Results and conclusions

The following results should not be regarded as complete, because the trials are still running and developing improved interface versions.

Nevertheless it is possible to make the following statements regarding the elderly. Quality of usage and the time of operation for elderly people is dramatically lower than for the younger subjects. The use of the video-conference system seemed to be impossible for the majority of elderly subjects. All possible comparisons between the two age groups showed a great difference, which was clearer than for the case of the ticket machine or the videorecorder.

The questionnaire, which was developed by our research group, showed a clear difference in the general experience with modern technology. The mean score of the younger subjects amounts to 247 points, as compared with a mean score for the elderly of only 93 points. There is no doubt that this lack of familiarity with modern technology has also influenced the capability of elderly people to operate the ISDN Telephone and the video conference system.

Conclusions

Practical statements from the ALTEC project

(1) There are enormous differences between the achievements of elderly and younger people regarding the use of technology, used by the general public in their daily life.

(2) It is possible to develop improvements which are effective and could easily be put into practice at the same time. The elderly especially could profit from these improvements.

(3) The approriate method of testing these improvements is the observation of naive users.

(4) Objective observations often are just applicable to register a critical spot of an interface. To obtain information about the exact reasons for a mistake, it is necessary to ask the user. The user should be confronted with every mistake on a videofilm and immediately afterwards he or she should be asked to give reasons for the mistakes.

(5) To ensure that changing an interface is really an improvement, the usability researcher ought to perform a statistical analysis within the frame of a real life scientific experiment. Improved interface versions are to be tested against certain alternatives and the former version. In many cases this might be the only way to reject subjective fixed ideas which in many cases have nothing to do with reality.

Conclusions regarding the design of user interfaces

The following guidelines have emerged from our projects:

(1) *Reduction of complexity:* information or functions which are superfluous or rarely used have to be excluded or covered.

(2) *Clear structure* of the task: the starting point of operating a machine and the sequence of all following steps have to be easily recognizable.

(3) The *consistency of information*: contradictions or inconsistencies in the arrangement and labelling and colouring of the keys are to be prevented.

(4) *Rapid and distinct* (optical and acoustic) feedback: it is necessary that the user receives rapid and perceptible feedback, which indicates 'success' or 'failure' after all operating actions.

General Conclusions

Day-to-day technology should meet the requirements of and, above all, enable the elderly to lead their lives independent of immediate help for as long as possible. There are two ways of improving the usability of technical appliances.

(1) Compiling a catalogue of criteria for designing user-friendly technology.

(2) Ensuring that operating instructions essential to the use of user surfaces of the most varied kinds of machines will be summarized in a 'universal directions for use' for technical appliances of daily life.

Both approaches abstract from the specific appliance and its applications. Thus, the results are valid irrespective of a fast innovation of products.

The first approach requires an analysis of user models of how to deal with technical apparatuses. According to Norman (1988), the correspondence of 'a design model' and a subjective 'user model' reflecting someone's preconceptions about the operational aspects and the function of the machine forms the basis for a successful use of technology. Typical ideas about the connection between the functioning and the operation are seized with the help of behaviour-oriented procedures and cognitive psychological methods. Knowledge of user models supplies us with basic knowledge for compiling a catalogue of criteria for the novel design of technical appliances.

The second approach tries to encompass the structural resemblances of different sets of machines as to their cognitive and motorial demands on the user. The most important of these common features, including typical operational sequences, might be condensed to a kind of 'technological grammar' ('universal directions for use'). This catalogue of criteria may well serve as an orientation for engineers and designers, that is for professions related to the design and use of technical appliances.

References

Baltes, P. B. and Baltes, M. M. (eds) 1991 *Successful Aging* (Cambridge University Press, New York).

Breakwell, G. M. and Fife-Schaw, C. 1988. Aging in the impact of new technology, *Social Behaviour*, **3**: 119–130.

Clark, M. C., Stern, E. J., Gaide, M. S. and Czaja, S. J. Adaptions of technology for the elderly. A human factors perspective, *Gerontologist*, **26**: 38.

Danowski, J. A. and Sacks, W. 1980. Computer communication and the elderly, *Experimental Aging Research*, **6**: 125–135.

Gilly, M. C. and Zeithaml, V. A. 1985. The elderly consumer and adaption of technologies, *Journal of Consumer Research*, **12**: 353–357.

Kruse, A. 1992. Altersfreundliche Umwelten: Der Beitrag der Technik. In P. B. Baltes and J. Mittelstraß (eds), *Zukunft des Alterns und gesellschaftliche Entwicklung* (de Gruyter, Berlin), S. 668 – 6240.

Norman, D. A. 1988. *The Psychology of Everyday Things* (Basic Books, New York).

Rudinger, G., Espey, J., Neuf, H. and Simon, U 1992. Abschlußericht Zum Projekt 'Alter und Technik' (ALTEC): Kognitive Verarbeitung moderner Technologie. (Universität, Institut für Psychologie, Bonn).

Willis, S. L. 1987. Cognitive training and everyday competence, *Annual Review of Gerontology and Geriatrics*. **7**: 159–188.

14

Designing communications and information-handling systems for elderly and disabled users

P. M. A. Rabbitt and A. Carmichael

Summary. Human-factors design for elderly users must recognize that three different descriptions of human aging are equally valid. The incidence of serious sensory and other disabilities increases markedly in old age, so that most disabled people are also elderly. For the overwhelming majority of elderly people, however, the problem is that mild disabilities and sensory losses potentiate with each other and with slowing of information-processing rate to reduce everyday efficiency. These points are illustrated by concrete examples of recent applied human-factors research on the accessibility of communications equipment to elderly users carried out for the European Union 'TIDE' and 'RACE' initiatives. These offer warnings on the validity and guidelines for use and interpretation of questionnaire surveys and a framework within which designers can improve accessibility of new communications technology to elderly and disabled users.

Keywords: human-factors design, sensory loss, design of communication systems, multiple minor disabilities, multi-modal environment, applied human-factors research, questionnaires.

Introduction

When we consider how to optimize the design of communications systems for older users we find that three quite different frameworks, used to describe the process of human aging, are equally useful. First, aging may be treated as a process, independent of any pathologies or disabilities, causing characteristic

changes in the efficiency of all systems that collectively determine total physiological and cognitive competence and which ergonomists must consider in
order to optimize design of equipment and systems for elderly users. These
include the sense organs and the entire central nervous system (CNS), the
respiratory and cardiovascular systems, and the skeleto-muscular system.
Within this framework of description the problem is to assess the scope, relative severity and rates of these changes in order to predict the problems that
elderly users may have with communications systems.

A second description of aging is the remorseless accumulation of more or
less severe pathologies. The impact of each may be slight, and they may
progress slowly. Independently and interactively, they will, however, increasingly circumscribe everyday competence. An extreme version of this view is
that there is a very marked increase in the incidence of handicaps and disabilities with increasing adult age. From this standpoint, good design for the
elderly tends to be regarded as an extension of design for the handicapped;
with the rider that, in addition to their multiplying disabilities, older people
may also suffer from changes in efficiency brought about by 'normal' aging.

It is important to emphasize a third and equally valid view of aging as the
culmination of a lifetime's acquisition of knowledge, sophisticated
information-handling procedures and expert skills which young adults have
not had time to learn. The negative changes associated with normal, 'non-
pathological' biological aging may erode some of these acquired skills but
others may be almost indefinitely retained at a very high level of competence
(see Rabbitt 1993). Possession of these skills means that older people can often
compensate for physiological limitations in ways that may not be possible for
the young disabled. Further, while the marked increase in the numbers of
elderly people in affluent societies has been widely publicized, it has not been
sufficiently recognized that this also implies marked increases in the absolute
numbers and in the proportions of elderly people who remain healthy and
active into their late 70s and early 80s and who experience only modest
reductions in ability to cope with daily life. For human-factors engineers this
also means that condescending or patronizing design solutions are certain to
be rejected by individuals whose expectations are at least as sophisticated as
those who design for them.

This paper is an attempt to set up a simple framework for undertaking
human-factors design for communications systems for the elderly, illustrated
from practical experience gained on the EU RACE project 'Tudor 1088' and
the TIDE projects 'Audetel' and 'Telecommunity'.

A first essential point is that even healthy, normal old age is accompanied
by an accumulation of mild but progressive losses of efficiency in all sense
organs. These have been documented comprehensively by Corso (1981). For
example, in vision these changes include marked reductions in the amount of
light that reaches the retina with a corresponding loss of contrast sensitivity;
this is accompanied by loss of retinal receptor cells which causes reduction of
visual acuity, restriction of the visual field, loss of efficiency of dark-adapted

vision and reduction of colour sensitivity. Changes in hearing include increases in detection thresholds for pure tones, especially for higher frequencies. This is accompanied by a relatively sharper increase in thresholds for speech perception. It has been suggested (by Bergman 1971: 164) that increases in speech perception thresholds are less related to loss of peripheral hearing ability *per se*, than to 'time-related processing abilities' which become less efficient in old age and so are increasingly challenged by the rapid flow of dense and complex information that forms the speech wave. A further, correlated change is a selective loss of ability to distinguish speech sounds against a background of noise. There are also changes in vestibular function leading to some loss of sensitivity to changes in bodily orientation, changes in tactile sensitivity, loss of sensitivity of the skin to changes in temperature, and loss of efficiency of the senses of smell and taste. For designers, the point to bear in mind is that the combined impacts of mild losses cannot be understood in terms of a simple checklist of their individual effects or, as it were, by 'profiles' of sensory decrements across modalities. Sensory losses in separate modalities, such as decrements in hearing and vision, or in taste and smell, often act in combination to produce greater restrictions in performance than we assume from the simple sum of their individual effects. For example, when we follow television programmes, sight and sound cues reinforce each other in complex ways. A spoken sentence is often only intelligible if we can also see what is happening, and an action or change in facial expression only makes sense if we also hear what has just been said. We live in a rapidly changing multi-modal environment and we can only keep pace with it effectively if we can simultaneously receive and integrate sensory information from diverse sources.

A second characteristic of sensory changes in old age is that severe loss in one sensory modality is usually accompanied by losses in others. For example, elderly blind people are, typically, also slightly deaf. They may have poorer memories and so become less efficient at remembering the spatial lay-outs of their environments. They have usually lost some degree of tactile sensitivity so that they find braille more difficult to read.

A third characteristic of sensory losses is that while most people reach the end of their lives having experienced only mild sensory losses which are easily, and satisfactorily, remedied by prostheses, it is also the case that the incidence and prevalence of severe sensory, and other, disabilities markedly increases with the age of the populations for which we must design our systems. A good example is that the vast majority of people registered in the UK as blind or deaf are aged over 65 years. Even more important than the sharp increase in incidence of single disabilities is the fact that the joint incidence of severe multiple disabilities markedly increases in old age. This occurs because particular conditions, such as peripheral neuritis caused by diabetes, will produce marked loss in more than one sensory modality, such as vision and hearing.

Multiple sensory losses can occur at any age. In old age, however, multiple sensory losses not only become more common but are also more difficult to cope with because they take place against a background of changes in the efficiency of the CNS that gradually reduce information-processing speed, learning rate and short-term memory efficiency. In effect, this amounts to a loss of back-up computing power that reduces the ability to compensate for degradations of sensory input. It is not sufficiently recognized that, at any age, even mild degradation of sensory input makes additional demands on information-processing capacity. As we shall see, loss of information-processing speed and degradation of sensory input interact to place a double demand on the aging cognitive system.

The initial approach to design: discovering what older people need and will use

Given these problems, human-factors designers need to develop new methodologies in order to assess whether older users will accept new communications systems. An obvious first step is to consider precisely what demands any prospective system will make on sensory capabilities. These demands can then be assessed in terms of what is known about the incidence and relative degrees of severity of changes in these relevant sensory modalities within the populations for which the system is intended. Thus, in the applied projects we shall refer to in this chapter, such as the 'AUDETEL' (AUdio DEscription of TELevision) and 'Tudor 1088' projects, it was necessary to obtain some picture of the age distribution, the demographics and the incidence of other disabilities in the populations for which communications services were intended. This initial step is essential, but insufficient. We also need to discover what potential users feel they need and so what they are likely to expect from the system we offer them. This will allow us to gauge how well the system we are designing can fulfil these particular needs and expectations. This is essential because, even if a system can potentially meet the demands of a particular elderly population, it does not follow that they will completely recognize its advantages and enthusiastically welcome it. An unfortunate characteristic of new technology is that it is almost never completely transparent to novice users. People need to invest time and effort to realize the potentialities of the systems we offer them. This means that we must try to attain a clear idea of precisely what problems potential users will have in mastering a new system, and whether this target population think that the advantages that it offers them are sufficient to merit the amount of time and effort they will have to invest in order to derive its full benefits.

It is an unfortunate axiom among designers that older people are too easily intimidated by new technology, or have grown 'too rigid' to use it. This is simply untrue. For the 'RACE' 'Tudor 1088', Rabbitt, Collins, Hazel and Bhatti undertook to assess the extent to which elderly and disabled European Union citizens saw advantages in the new communications technologies imminently available to them. To do this they obtained answers to questionnaires from 3229 disabled and 2790 able-bodied elderly individuals from four member states of the European Union (Portugal, The Netherlands, Sweden and the UK). The questionnaires were designed to elicit information about the extent to which people in different age groups, and with different kinds of physical and sensory disabilities, used the telephone systems available to them, the extent to which they felt the need for currently available extensions to the capabilities of these systems and the extent to which they would welcome new technology that superseded these systems.

A first discovery was that people with different types of disabilities had very different attitudes to new technology. The extent to which they welcomed new technology, were confident of their ability to make use of it and were willing to invest resources to acquire and master it, depended, of course, on the particular type of disability from which they suffered and also on how far they believed that the new devices would help them. It equally depended, however, on their current interest in, and reliance on, new technology. For example, individuals with speech difficulties, who had already become accustomed to using new communication systems to run their lives and maintain their social contacts, eagerly looked forward to further innovations that might further benefit them. Other groups of disabled, who could clearly articulate what advantages new technology might have but who had little or no experience of new communications systems, were less enthusiastic about potential new developments.

The questionnaires were also analysed to examine the common assumption that older people are not interested in new technology or are even 'technophobic' in the sense that they are uneasy and afraid of novel systems. Elderly people were polled on their potential interest in particular items of imminently available technology such as videophones and home banking systems. To interpret their attitudes in terms of their current knowledge of technology we also asked them to describe their experience with examples of recently devel-

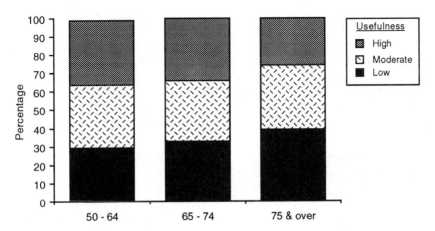

Figure 14.1. Perceptions of the usefulness of new technology.

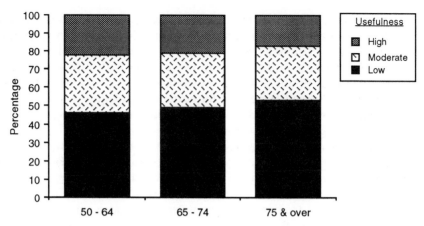

Figure 14.2. Low users of existing technology: usefulness of new technology.

oped pieces of electronic equipment such as compact-disc players, video-recorders, home computer systems and extended facilities on currently available telephone systems.

As figure 14.1 shows, when data are pooled over 2180 elderly respondents it does indeed seem that the degree to which they consider that new technology would be useful to them declines sharply with their average ages. However figures 14.2 and 14.3 show that the overall trend in figure 14.1 is a result of the sub-group, whose data are presented in figure 14.2, and who have had little experience with electronic equipment now commonly available, such as cassette players, videorecorders or compact-disc systems. We see from figure 14.3 that among individuals who already have some experience with such readily available equipment there is no effect of age. Indeed, a multiple regression analysis over the data set showed that when level of experience with current

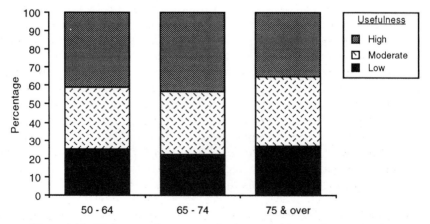

Figure 14.3. High users of existing technology: usefulness of new technology.

technology is taken into consideration ($t(1) = 12.35$, $p < 0.0001$), chronological age ($t(1) = 0.99$, $p = 0.3237$), considered on its own, does not significantly predict attitudes to new technology.

The message from this analysis is that although most people currently aged over 60 are less interested than are young adults in new technological developments, this state of affairs will not continue for long. People who are now in their 40s and 50s already have considerable interest in and competence with information-processing devices and communications systems such as computer communications networks and fax machines. We may expect them to maintain or increase this interest as they pass into their 60s and 70s, and begin to experience social or functional restrictions for which new devices offer solutions.

Notes on the methodology of evaluating new systems for use by older people

Having discovered what help people feel that they really need and are willing to invest resources of money to purchase, and of time and effort to learn how to obtain, the next step is to try and design equipment which optimally meets their hopes. The most obvious solution would seem to be simply to ask people what their problems are, and how these could be mitigated: in other words, to use structured interviews or questionnaires.

Caveats on questionnaires

Because questionnaires are so easy to design and administer, and because they can rapidly and cheaply give us so much useful information about individuals' interests and potential capabilities, with respect to hypothetical systems, they seem to offer a powerful and convenient methodology for human-factors designers. It is tempting to assume that older people are the best experts on the difficulties that they currently experience, or may anticipate in the future. If we ask the right questions, surely they can give us comprehensive and detailed advice on how best to help them?

It is unfortunate that elderly and disabled people have almost invariably been excluded from active participation in the early stages of design of equipment intended to meet their needs. Actual, concrete, 'hands on' involvement in design has been very rare. Typically, questionnaires are used. A large body of work shows, however, that in order to get people reliably to assess or predict their own competence, we must be careful to ask them very specific questions about the particular functions or skills we wish to evaluate, rather than general questions about their global competence in everyday life.

The realization that laboratory simulations cannot accurately probe their competence in complex everyday tasks has led psychologists to develop ingenious questionnaires which ask people to describe particular lapses of memory or attention that they have experienced during their daily lives (Reason 1979, Reason and Mycielska 1982) or to estimate the relative frequency with which they experience specific categories of cognitive lapses (Broadbent et al. 1982, Herrman and Neisser 1979). It soon became clear that while all these questionnaires have reasonable *reliability*, in the sense that individuals tend to give the same answers to them on separate occasions, their *validity* in predicting performance is questionable. There was no evidence that the general impressions of their everyday competence that people expressed in these questionnaires predicted their performance on specific laboratory tasks (Herrman 1982) or, indeed, on other, more global tests of everyday memory and cognitive competence (Rabbitt and Abson 1990).

Failures of correlation between self-reports and performance on laboratory tasks are unsurprising, and may, indeed, be grounds for increased confidence in questionnaires because, as we have seen, a main motivation for developing questionnaires has been that scores on simple laboratory tasks may only poorly reflect everyday competence! It is more disquieting that people's self-ratings on different questionnaires do not correlate very well with each other. Self-ratings also do not systematically change with age between 50 and 86 years, or with current performance on IQ tests (Rabbitt and Abson 1990). This last finding is particularly disturbing because, in spite of their notorious imperfections (see Gould 1981), IQ tests still remain the best measures of overall general competence that we have. Another disquieting point is that people's scores on some cognitive failure questionnaires correlate more strongly with their self-ratings on depression inventories than they do with their scores on other self-rating questionnaires which, to all appearances, probe the same areas of competence. This suggests that people's self-judgements of their own abilities may be influenced as much by their subjective feelings of inadequacy as by any objectively valid assessment of their abilities (Rabbitt and Abson 1990). Note that the correct conclusion is not that people's self-evaluations are unreliable, but rather that human competence is intensely domain-specific. A person may have remarkable ability on one task but perform poorly in other, apparently quite similar, situations. A different, and much more general logical problem with self-rating questionnaires is that people can never make absolute judgements about their own abilities. They can only make relative judgements, gauging the degree of success and effort with which they meet the demands of tasks that they know others can accomplish, or by comparing their own performance directly with that of colleagues, spouses or acquaintances. In other words, people's knowledge of their own abilities depends entirely on their degree of experience with the particular situations and tasks about which we question them. It also depends on the extent to which these tasks or situations can give them accurate feedback about their own levels of performance. Finally, even when accurate feedback is available, it depends on how accurately they can assess and remember such feedback as they get.

This last point turns out to be an especially important determinant of the reliability with which older people can evaluate their own competence. As we have said, although absolute changes in intellectual ability with age may be very slight, there are good objective grounds for supposing that they do occur and do reduce the ability to cope with everyday life in old age. Nevertheless, recent large-scale and brilliantly analysed studies (Hulsch 1993, Reason 1993) and earlier evidence (Rabbitt and Abson 1991) paradoxically show that while individuals aged from 50 to 60 years often give pessimistic accounts of their own everyday competence, those aged from 61 to 86 years make significantly fewer complaints of cognitive lapses and everyday memory problems.

Designers of communications systems, who must be concerned with the sensory capabilities of potential users, should be particularly aware that this paradoxical improvement of self-ratings with increasing age occurs even when people are asked to assess their own vision and hearing. Holland and Rabbitt (1993) compared objective measurements of acuity of vision and hearing obtained from a group of elderly motorists aged from 50 to 79 years with their subjective reports of perceived changes in efficiency. Objective tests showed the normal declines in vision and hearing which have been documented in healthy elderly populations, (Corso 1981). In contrast, younger people in their 50s, who had objectively better vision and hearing, reported noticing more change than did individuals in their 60s and 70s.

The clue to these paradoxical and unrealistic self-ratings of sensory ability and of everyday lapses is that people can only judge their own abilities in the context of very specific and very familiar tasks, or against the performance of their spouses, friends or colleagues. People in their 50s and early 60s are very much involved in life and have frequent opportunities to evaluate their own performance against everyday challenges and to compare themselves with the able young. Older people who have retired and begun to live much more circumscribed lives may indeed experience fewer lapses because fewer demands are made on them. They usually also have fewer opportunities to directly compare their performance with that of younger people. It is also likely that the lapses they experience will be relatively trivial, will occur in private and so will pass without comment or embarrassment and be readily forgotten. In short, the apparently overoptimistic account that elderly people give of their own sensory abilities and cognitive performance is likely to occur because they get increasingly less feedback from their restricted involvement in demanding situations.

It is also possible that older people may become less able to evaluate such feedback as they do receive. There is evidence that if this problem exists, it is probably not serious. A further study by Rabbitt and Abson (1991) found that even people in their late 70s could give reasonably accurate predictions of their future performance on unfamiliar laboratory tasks, and that their ability to do this did not depend on their current ages, but rather on their current levels of performance on IQ tests. A more recent study by Winder (1993) suggests that the crucial determinant of accurate self-prediction is neither age nor IQ test score, but rather the degree of specificity and familiarity of the particular task on which people are asked to asses their performance, and the degree of precision of the feedback that they can obtain from it. Volunteers aged from

50 to 84 years were asked to rate their own ability at solving cryptic crossword puzzles and were then given an objective test of puzzle-solving ability. All reported that they frequently attempted to solve the puzzles they came across in newspapers and some were avid crossword enthusiasts. They all, therefore, received frequent and unambiguous feedback of their current levels of performance. This was reflected by the fact that their self-ratings on a five-point scale correlated at 0.82 with their objective performance on crossword puzzle questions. Accuracy of self-ratings did not change with individual difference in age between 50 and 84 years or in IQ test performance. It does, therefore, seem that people of any age or level of general intellectual ability can be very accurate judges of their own performance on very tightly specified and familiar task domains. Thus, if appropriate precautions are taken, subjective self-ratings can contribute greatly to good system design. Some further problems with subjective ratings of system characteristics are illustrated, however, by system evaluations carried out by Carmichael and Rabbitt (1993) for the pilot phase of the 'AUDETEL' project funded by the current EU 'TIDE' (Technological Innovation for Disabled and Elderly people) initiative.

'AUDETEL' project

The aim of the 'AUDETEL' project is to develop and evaluate an audio-description service for blind and partially sighted television viewers. An initial engineering constraint was that the channel capacity available for audio description was restricted, so that the relative acceptability of various possible compromises on speech quality had to be assessed. In this case, sound quality was directly related to the maximum rate at which the digital information, which specifies the speech wave, could be transmitted (the 'bit rate'). An obvious approach was to play paired samples of speech transmitted at a variety of bit rates to elderly and partially sighted volunteers and ask them to make subjective judgements as to which of each pair was the clearest. They were also asked to compare pairs of speech samples presented in a second list to decide whether they were played with the same or with different sound quality. In addition, as an objective index of their competence, volunteers were also required to 'shadow' passages of speech played at different bit rates; that is, to repeat them continuously as they were played. Performance on this shadowing task was quantified in terms of the numbers of errors that they made. It might be argued that if volunteers could not reliably tell whether particular pairs of speech samples were played at the same or different bit rates, any difference between them would be unlikely to affect their intelligibility. We might further assume that if volunteers judged a pair of speech samples as being equally clear, they would also find them, objectively, equally intelligible.

Comparison of figures 14.4 and 14.5 shows that these assumptions were mistaken. Figure 14.4 shows that accuracy of subjective discriminations of

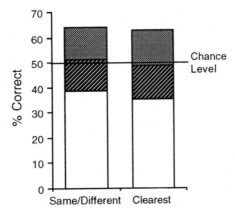

Figure 14.4. Subjective judgements.

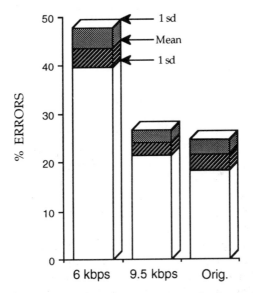

Figure 14.5. 'Shadowing' performance, by 'bit rate'.

clarity and of difference between speech samples was no better than chance, while figure 14.5 shows that, for the same samples, shadowing errors were significantly higher for the 6 kbps (kilo-bits per second) and 9 kbps rates than for the original, undistorted speech samples presented without bit-rate attenuation.

It seems that subjectively imperceptible differences in speech quality may, nevertheless, significantly affect their objective intelligibility.

These and other, similar results show that while it may be useful to obtain subjective judgements, we should always, in addition, make such objective tests as we can. The converse is also true, because alterations in speech quality that do not affect objective intelligibility may, nevertheless, prove so subjectively irritating to some listeners that they are unacceptable in practice.

The effects of multiple minor disabilities compound each other

As people grow old they typically experience multiple losses of sensory efficiency. Though each particular loss may be mild, their compounded effects may be unexpectedly serious. It is essential to bear this in mind when designing systems for the elderly disabled which attempt to mitigate losses of efficiency of one sensory modality by providing supplementary information on another. A good example is that because most blind and partially sighted people are elderly, they also usually suffer from some, albeit quite mild, hearing loss. Of course, this compound problem may be dramatic in cases when blindness and deafness are both secondary effects of a pathology, such as diabetes, which is increasingly common in later life.

Our experience on the 'AUDETEL' project also provides a practical example of the way that human-factors engineering for the elderly must typically be concerned with combinations of multiple minor disabilities. Because most users of audio description are likely to be elderly, it was important to assess the extent to which the relative intelligibility of speech at different bit rates might be affected by the low levels of hearing loss common in old age. Accordingly, we measured the hearing thresholds for pure tones of all volunteers who took part in experimental assessments of intelligibility using a straightforward 'hear word, write word' method. Data for individuals with little or no hearing loss (averaged across frequencies), and those with hearing losses of 30 to 77.5 dB were separated. These data are shown in figure 14.6.

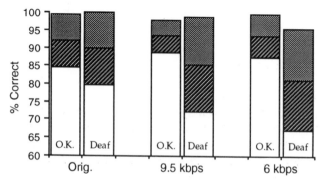

Figure 14.6. Intelligibility performance, by 'bit rate' and deafness level (OK and Deaf).

Figure 14.6 shows how intelligibility at different bit rates was affected by the level of hearing loss. Reducing sound quality from 'original samples' (frequency range restricted only) to 9.5 and 8 kbps (frequency range and bit rate restricted) significantly impairs intelligibility for those individuals with a 30 to 77.5 dB hearing loss, but not for those with better hearing. It should be noted that these levels of hearing loss, while appreciable, are by no means severe, and are very common in older populations. As a rough guide, people are only considered for hearing aids if they have hearing losses of more than 40 to 50 dB.

When we design systems to compensate for degradation of information on one sensory channel by providing supplementary information on another, we must bear in mind that the multiple minor sensory losses that people commonly experience as they grow old with require that we stringently assess the quality of the additional channel with reference to *norms for elderly rather than for young adults*. We must also bear in mind the further point that, as we shall see below, even slight sensory degradations imposed by age are compounded by a reduction of back-up information-processing capacity in aging brains. The effects of sensory loss in the elderly are made more acute by the reduction in information-processing resources that accompanies old age.

Some perceptual consequences of losses of information-processing capacity

It has been argued that the most obvious and pervasive change in mental abilities in old age is a marked slowing of the maximum rate at which the CNS can process information (Salthouse 1985, 1991). If sensory input is degraded by visual and hearing impairments, additional information-processing capacity is required to overcome this problem. A simple example is that when spoken or written words are presented in random noise, people take much longer to recognize them, and to repeat or read them correctly. We cannot follow speech degraded by our own deafness, or by a suboptimal channel, as rapidly as we can follow speech presented in clear. This makes the point that sensory losses, by themselves, make increased demands on central information-processing capacity. If available central information-processing capacity is further reduced by age, the effects of peripheral sensory loss will be compounded in complex ways. Obviously this will be particularly true in the case of listening to speech, when the rate at which information is presented cannot be controlled directly by the listener, but less true when, as when reading text, the reader can adjust her rate of information uptake to maintain accuracy. This makes it unsurprising that elderly people who suffer from mild hearing losses of 50 dB or less find that their reduced information-processing capacity makes it difficult for them both correctly to make out the words spoken to them and to carry out the rehearsal and elaborate encoding necessary to remember what they have heard (Rabbitt 1990).

Engineers evaluating speech communication systems must always realize that simple tests of channel intelligibility do not pick up the difficulties that elderly users may experience. It is obviously useless for individuals to be able accurately to repeat all the words that are played to them if they cannot remember, or precisely understand, the messages that they receive. For the elderly an additional constraint is that because their information-processing rates have declined, they cannot follow very rapid speech even if it is presented under optimal conditions. When speech quality is degraded, either by suboptimal communication channels or by their own hearing losses, their maximum rates of comprehension are even further reduce.

This has important practical implications for evaluating communication channels for elderly users. It is not enough to use conventional tests for intelligibility in which listeners take their own time to identify and repeat aloud brief samples of transmitted speech. A direct example of this comes from an unpublished study by Rabbitt and Fleming which tested what levels of pulse-modulated random noise could be tolerated by users of a telephone system. They found that young adults could achieve 100 per cent accuracy in repeating, one at a time, lists of random words played at a rate of one per second through a system in which noise remained 15 dB above the level of the speech signal. As the rate at which words were presented was increased, however, accuracy could only be maintained if noise levels were correspondingly reduced. Rabbitt and Fleming further found that this trade-off between speech rate and noise level both occurred earlier, and proceeded more sharply for older than for younger adults.

This trade-off between channel quality and maximum information-processing rate can be turned around to provide a useful methodology for assessing speech systems for elderly users. Although conventional intelligibility tests, in which words or phrases are presented one at a time with pauses for written or spoken repetitions can be useful, a better technique is to require listeners to 'shadow' passages of speech, as described above. This allows us to test the effects of varying speech rate on maintenance of shadowing accuracy, and to establish for audiences of different ages, and with different levels of sensory impairment, the trade-off plane between presentation speed, age, and hearing loss or channel degradation. A hypothetical example of such a plane is given in figure 14.7.

As we might expect, the trade-off between the quality of sensory information and the level of information-processing resources which are available to back up perceptual processing operates for vision as well as hearing. Dickinson and Rabbitt (1991) asked young adults to read text as fast and accurately as they could, either through lenses that artificially induced slight astigmatism, or with no lenses and so no distortion. Young adults could read the text equally quickly and accurately with or without distorting lenses. However, they made significantly more errors *recalling* material that they had flawlessly read aloud

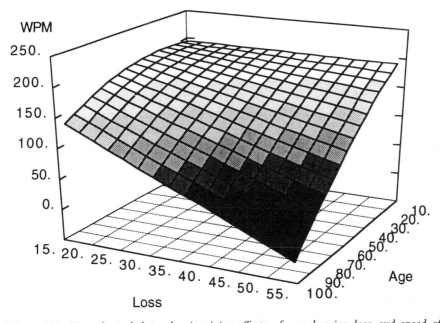

Figure 14.7. Hypothetical data showing joint effects of age, hearing loss and speed of presentation (wpm) on ability to 'shadow' accurately.

through distorting lenses than material they had read without lenses. Further analysis showed that the particular individuals who showed this effect of sensory degradation on memory had significantly lower unadjusted scores on IQ tests. People with high test scores recalled material equally well whether they had read it with or without lenses. It is known that individuals achieve higher IQ test scores partly because they process information faster. The faster information-processing rates of high IQ test scorers thus apparently allowed them sufficient spare channel capacity to both read distorted text and to understand, rehearse and remember it. Consistently with this finding that higher central information-processing rates allow people to overcome the effects of degradation of peripheral sensory input, Rabbitt (1991) also found that older adults with high IQ test scores show less effect of mild deafness on accuracy of recall of spoken text than do those whose lower information-processing speeds are reflected by their relatively lower IQ test scores. It seems that these 'knock-on effects' of sensory degradation and of reduced information-processing speed on the accuracy of higher cognitive judgements are quite general across all sensory modalities, and must be borne carefully in mind as a factor determining the efficiency with which any projected communications system can be used.

Although restrictions of information-processing speed operate both for hearing and vision, we must bear in mind that, in practice, there is a crucial difference between these two sensory modalities. The ear is a passive sense organ which can select between different frequencies or directions of input but cannot control the rate of information uptake. The eye is not only selective, in the sense that it can focus on different parts of the visual environment but, by varying scanning rate, it can also control the rate at which it accepts information. We cannot adjust our speed of listening to improve accuracy of speech perception, but, when reading or searching the visual world, we can decide whether to scan slowly and accurately, or rapidly and risk errors.

This turns out to be critical for choice of appropriate text for use with computer and communications monitors. A concrete example of this was provided by experiments to assess the relative legibility of typefaces on monitor displays for the RACE project Tudor 1088. We compared how four typefaces, Oxford, Monaco, Geneva and Zapf Chancery affected the speed and accuracy with which people aged from 50 to 86 years could discriminate words and carry out visual search for some letters among others. We found that the least legible typeface, Zapf Chancery, produced the slowest scanning speeds but greatest accuracy, while more legible typefaces, such as Geneva, produced relatively fast scanning speeds but significantly higher error rates.

This shows that humans actively adjust their performance to maximize speed and accuracy. It also shows that they often have difficulty in maintaining any consistent balance between search rate and errors across different types of material. This is because, in order to optimize speed–accuracy trade-off, people have to know when they make errors. In some tasks, such as those in which signals occur one at a time and a new response has to be made to each before the next appears, for example choice reaction time (CRT) tasks, people notice, and can correct, nearly all of the errors that they make (see Rabbitt 1968). In visual search, however, where most errors consist of failures to detect target signals embedded among others, humans are aware of very few of these lapses (Rabbitt et al. 1978).

If people get immediate and unambiguous feedback from their errors, it is easy for them to use this information to adjust their speed of information uptake to maintain accuracy. It is much more difficult to strike the correct speed–accuracy balance when feedback is not available. This is why in serial CRT tasks the effects of old age, of differences in competence at IQ tests and of ingestion of drugs such as alcohol express themselves as differences in speed, with errors remaining relatively constant. In contrast, in visual search, age, low IQ test scores and alcohol tend to increase error rates while leaving speed relatively unaffected (Rabbitt and Maylor 1991). Thus it is important for human factors engineers to consider whether the systems they propose provide immediate and unambiguous feedback about errors, and so allow subjects to monitor and adjust their own performance continuously, or whether they provide no, or poor, feedback, so that joint control of speed and accuracy becomes problematic. When designing systems for the elderly and disabled it is important to bear in mind that even when excellent feedback from errors is available, older people may become increasingly less able to sensitively monitor and utilize it to adjust their performance (Rabbitt 1990).

Further limitations imposed by reduced information-processing rate: difficulty in doing two things at once

We have seen that impaired sensory efficiency and slowed information-processing rates interact to reduce the efficiency with which older people can obtain information from single sources, such as loudspeakers or monitor screens. In most everyday situations, however, information from two or more sense organs must be received and integrated in order to make sense of what is going on in the world. An obvious example is watching television, where correct inferences about what is happening require integration of what is seen of the action with what is heard of the soundtrack and dialogue. As the range of multi-media communication devices is extended, this will certainly become an increasingly important consideration for designers.

Obviously, the necessity to monitor and integrate data from more than one sensory modality must increase the total amount of information that has to be processed in unit time. We would expect this to place a relatively greater strain on people whose information-processing rate has been reduced by old age, and who may be further handicapped by memory impairments. Unfortunately, the balance of the evidence suggests that humans of any age have difficulty in processing events that occur at precisely the same instant on separate sensory channels. It would be misleading to suggest that we yet adequately understand precisely how people attend to, select between and integrate simultaneous input from separate sensory channels. In particular, there is comparatively little information on how these capabilities are affected by aging. Within all of the four different frameworks in which these processes have been discussed we can, however, make useful guesses as to how old age may reduce efficiency.

The earliest model for selection between information channels was proposed by Broadbent (1958). He compared recall of separate short messages simultaneously and separately presented to each ear, or to the eyes and ears (Broadbent and Gregory 1961). He interpreted his results as evidence that humans cannot jointly process both members of a pair of messages that are presented to two different sensory channels simultaneously or in rapid succession. They must, rather, process information presented to one channel while holding information presented to the other in a limited capacity short-term memory system with a very brief storage half-life. Thus, if too much time is spent processing information from the momentarily selected channel, the stored information from the other rapidly degrades, or may be entirely lost (Broadbent 1958, 1971, Broadbent and Gregory 1961). This theoretical framework was used in early studies carried out to examine the effects of old age on selective attention. Older age did not appear to impair the accuracy of messages that were selected and reported first but did increase the rate of loss of information from messages that were reported second. This suggested that old age does not affect the ability to select between messages but it does accelerate the rate of loss of information from temporary storage and so reduces the total amount of information that can be processed from two independent sources (Broadbent 1958).

A second model for selective attention has an equally long pedigree (Treisman 1969). This suggests that while it is indeed possible to process data from two independent channels or information sources simultaneously, there is an overall limitation of resources so that input from one channel is favoured and that from the other is 'degraded' so that information from this source is lost. Within this framework we would again expect that attenuation of information-processing resources with increasing age would lead to progressively greater degradation of information from the channel to which lower priority is given.

A more recent model associated with the work of John Duncan is that it is possible simultaneously to process input of two or more kinds, or from two or more sensory modalities. The efficiency with which this can be done, however, and, in particular, the efficiency with which these different kinds of data or sensory evidence can be integrated into meaningful percepts depends on the availability and efficiency of deployment of central attentional resources.

All these models would predict that elderly people will be handicapped in their ability simultaneously to select and integrate sensory information of more than one kind, and from more than one source. This is because they process information more slowly, have reduced attentional resources and also have a reduced short-term memory capacity, and more rapidly forget information from an unattended channel that must wait its turn to be processed.

Designers of communications and information-handling systems which may be used by the elderly should be aware of these limitations. They should also consider the probable impact of another, as yet undocumented, source of difficulty for the old. The efficiency and speed with which people can select between information sources depends on the ease with which they can discriminate between them. Thus people can easily, accurately and rapidly select between speech messages which are in markedly different voices. As the mutual discriminability of messages declines, however, selection becomes increasingly effortful, slow and inaccurate. We must bear in mind that the reduced efficiency of older eyes and ears will tend to reduce discriminability between separate samples of speech and text and so tend to slow, and to complicate the task of attentional selection between them. These difficulties will be greatly increased, and the gaps in efficiency of younger and older operators will correspondingly widen if the quality of signals over communications equipment is also degraded.

Empirical evidence that the elderly have difficulty integrating information from multiple sources

Several studies have found that older people have difficulty integrating and remembering information from more than one channel (Broadbent and

Gregory 1965, Caird and Inglis 1961, Inglis and Caird 1963). Recent work suggests that this may sometimes be true even when the information on one channel simply replicates that given on the other. For example, Wingfield and Stine (1986) and Stine and Wingfield (1987) found that young adults recalled more information from a passage of text when they simultaneously read it and heard it read aloud. However, their older subjects did not benefit from this replication of information. Another example that older people have difficulty integrating visual and auditory information was found by Rabbitt (1976). He required volunteers to watch TV recordings of four successive statements, which were either all made by the same person, made by two different people speaking in turn or each made by one of four different people. Volunteers of all ages could recall the content of the four statements equally well but, in the two- and four-speaker conditions, older volunteers had difficulty in remembering correctly which speaker had made which statement.

A further demonstration of this difficulty comes from a study recently carried out in our laboratory in support of the RACE 'Tudor 1088' project to evaluate some possible benefits of videophones. An obvious benefit is that because videophone users can see strangers to whom they talk they will be able to recognize them when they meet them. We thought that another benefit might arise from the fact that people's recognition memory for faces is usually better than their memory for other kinds of information about them, such as their names. Thus memory of the face of someone we have spoken to may act as an exceptionally effective retrieval cue to call to mind other information about them.

The studies we have described above also suggested, however, that older people might be less able to benefit from this additional information because they process information more slowly and find it increasingly difficult to integrate and hold in memory different kinds of information from more than one source. We tested this by showing volunteers aged from 55 to 83 years slides of men's faces and verbally 'introducing' each by a name and three items of biographical information, such as their occupation, place of residence and a hobby or pet. After each set of four different pictures had been presented and described, our elderly volunteers were cued to recall as much biographical information as possible about the imaginary personalities described. Their recall was triggered by three different types of cue, i.e. the people's names alone, their faces alone, or both their faces and names. This cued recall occurred under two conditions. In one subjects were informed prior to presentation as to which of the three cue types would be used for recall of that set; in the other, cue types varied unpredictably. We had expected to find a hierarchy of cue efficiency, with the most powerful cue being the 'face plus name' cues, followed by 'face only' cues, with 'name only' cues being the least effective. To our surprise, regardless of whether or not subjects knew in advance which type of cue to expect, recall was significantly better for 'face only' cues than for 'face and name' cues, although, as expected, 'name only' cues were least effective of all. It would seem that the additional demand of remembering both faces and names as cues reduced the total amount of information which could reliably be reported. That is, we found the same cue effects across the predictable and unpredictable cueing conditions with only the magnitude of recall to all cue types being reduced in the last case. This suggested that, in this last condition, subjects attempted to learn both cues 'just in case', thus reducing their capacity successfully to store the items to be remembered.

It would seem that these results do not bode well for any attempt to enhance elderly visually impaired people's enjoyment and understanding of TV programmes by adding yet another source of information to densely packed and rapidly changing visual and auditory information. For those with a degree of visual impairment so severe that they lose virtually all visual information on the TV screen we might expect that replacing lost visual information, even in a limited and time-consuming way, may bring some benefits. However, only a very small minority of the visually impaired population are totally blind; most have some useful vision and, because most of these are also elderly, they run the risk of being 'overloaded' by information in different modalities. We tested this by getting non-visually impaired elderly people, aged between 70 and 86 years, to watch an episode of the popular police series 'The Bill', with and without audio description. This programme was chosen because its fast-moving action, rapid scene changes and presentation of complex scenes in which more than one thing is going on at any given time might be expected to make relatively severe demands on the information-processing rates of older viewers. Two groups of subjects were matched for age and IQ attainment: one group watched the programme as it would normally be transmitted and the other a version accompanied by audio description. All then completed a battery of questions probing various aspects of their comprehension of and memory for the programme. The main performance measure was a composite score of their ability to identify and describe characters and locations which had appeared in the episode. We were pleasantly surprised to find that those who had experienced the audio-described version performed significantly better. There were effects of both age and performance on IQ tests but these were independent rather than interactive. Figure 14.5 illustrates that groups who saw the audio-described version remembered more of what had happened. It also shows that while older volunteers and those with poorer scores on IQ tests remembered less of both the audio-described and undescribed versions, they derived equal benefit from audio description, as did younger volunteers and those with higher IQ tests scores.

This gain from audio description seems surprising in view of the literature that suggests that older people do not benefit, and may even suffer from presentation of additional information (e.g. Stine and Wingfield 1987, Wingfield and Stine 1986). However, a likely explanation is that because episodes of programmes such as 'The Bill' tend to have very straightforward and basic main plots, their producers try to disguise their potential banality by including irrelevant, or even misleading, sub-plots and redundant detail. It seems likely that while audio description may have filled some gaps in understanding, its main effect was emphasis of relevant, as opposed to irrelevant or misleading,

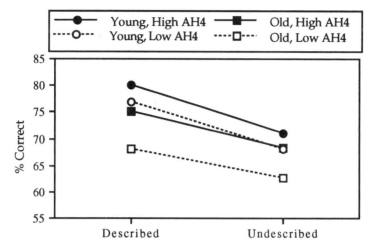

Figure 14.8. Comprehension performance in 'The Bill' experiment, by age and AH4 attainment.

information. Because audio describers cannot describe all, or even most, of the visual information available, they focus on the most important aspects. This unobtrusively helps the elderly by focusing their attention on salient events rather than on distracting irrelevancies.

Conclusions

These experiments provide at least some preliminary guidelines for those concerned with design of communication systems for elderly and disabled users. They make the general points that even mild sensory deficits may cause problems for elderly users because they interact with each other and because they exaggerate the additional problems caused by a general reduction in information-processing capacity. These interactions are seen to be subtle and sometimes counter-intuitive. For this reason, wherever possible, human-factors engineers should not assume that the systems they develop will necessarily be acceptable to the majority of users for whom they are intended. Empirical trials on elderly and disabled individuals should be carried out as a matter of course. These trials can be very usefully supplemented by questionnaire surveys but, as we have seen, information from these surveys should be evaluated with caution.

References

Bergman, M. 1971. Hearing and aging, *Audiology*, **10**: 164–171.
Broadbent, D. 1958. *Perception and Communication* (Pergamon, Oxford).

Broadbent, D. E. 1971. *Decision and Stress* (Academic Press, London).

Broadbent, D. E. and Gregory, M. 1961. On the recall of stimuli presented alternately to two sense organs, *Quarterly Journal of Experimental Psychology*, **13**: 133–140.

Broadbent, D. E. and Gregory, M. 1965. Some confirmatory results on age differences in memory for simultaneous stimulation, *British Journal of Psychology*, **56**: 77–80.

Broadbent, D. E., Cooper, P. F., Fitzgerald, P. and Parkes, K. R. 1982. The Cognitive Failures Questionnaire (CFQ) and its correlates, *British Journal of Clinical Psychology*, **21**: 1–16.

Carmichael, A. and Rabbitt, P. M. A. 1993. Human factors design for the elderly in the 'Audetel' project. In IEE colloquium 'Information access for people with disability', Digest reference number 93/103.

Caird, W. K. and Inglis, J. 1961. The short-term storage of auditory and visual two-channel digits by elderly patients with a memory disorder, *Journal of Mental Science*, **107**: 1062–1069.

Corso, J. F. 1981. *Aging, Sensory Systems and Perception* (Praeger, New York).

Dickinson, C. M. and Rabbitt, P. M. A. 1991. Simulated visual impairment: effects on text comprehension and reading speed, *Clinical Vision Science*, **6**(4): 301–308.

Gould, S. J. 1991. *The Mismeasure of Man* (Norton, New York).

Herrman, D. J. 1979. Know thy memory: the use of questionnaires to assess and study memory, *Psychological Bulletin*, **92**: 434–452.

Herrman, D. J. and Neisser, U. 1979. An inventory of everyday memory experiences. In M. M. Gruneberg, P. Morris and R. N. Sykes (eds) *Practical Aspects of Memory* (Academic Press, London).

Holland, C. A. and Rabbitt, P. M. A. 1993. People's awareness of their age related sensory and cognitive deficits and the implications for road safety, *Applied Cognitive Psychology*, **6**: 234–245.

Hulsch, D. 1993. Paper presented at the International Conference of Gerontology, Budapest, July 1993.

Inglis, J. and Caird, W. K. 1963. Age differences in successive responses to simultaneous stimulation, *Canadian Journal of Psychology*, **17**: 98–105.

Rabbitt, P. M. A. 1968. Three kinds of error-signaling responses in a serial choice task, *Quarterly Journal of Experimental Psychology*, **20**: 179–188

Rabbitt, P. M. A. 1976. Talking to the aged. *New Society*, July 1996. p. 123–125.

Rabbitt, P. M. A. 1988. The faster the better? Some comments on the use of information processing rate as an index of change and individual differences in performance. In I. Hindmarch, B. Aufdembrinke and H. Ott (eds) *Psychopharmacology and Reaction Time* (John Wiley, London).

Rabbitt, P. M. A. 1991. Mild hearing loss can cause apparent memory failures which increase with age and reduce with IQ, *Otolaryngologica* (Stockholm), Suppl. **476**: 167–176.

Rabbitt, P. M. A. 1993. Does it all go together when it goes? The nineteenth Bartlett Memorial Lecture, *Quarterly Journal of Experimental Psychology*, **46A**: 385–434.

Rabbitt, P. M. A. 1993. Crystal Quest: an examination of the concepts of 'fluid' and 'crystallised' intelligence as explanations for cognitive changes in old age. In A. D. Baddeley, D. Jones and L. Weiskrantz (eds) *Festschrift for Donald Broadbent* (Oxford University Press, Oxford).

Rabbitt, P. M. A. and Abson, V. 1990. Lost and found: some logical and methodological limitations of self-report questionnaires as tools to study cognitive ageing, *British Journal of Psychology*, **81**: 1–16.

Rabbitt, P. M. A. and Abson, V. 1991. Do older people know how good they are? *British Journal of Psychology*, **82**: 137–151.

Rabbitt, P. M. A., Cumming, G. and Vyas, S. 1978. Some errors in perceptual analysis in visual search can be detected and corrected, *Quarterly Journal of Experimental Psychology*, **33A**: 223–239

Rabbitt, P. M. A. and Maylor, E. A. 1991. Investigating models of human performance, *British Journal of Psychology*, **82**: 259–290.

Reason, J. 1979. Actions not as planned: the price of automatization. In G. Underwood and R. Stevens (eds) *Aspects of Consciousness* (Academic Press, London).

Reason, J. 1993. Have Self Repair Questionnaires delivered the goods for cognitive psychology? pp. 302–349 in order of editors as stated. In A. D. Baddeley, L. S. Weiskrantz and D. Jones (eds) *Festschrift for Donald Broadbent* (Oxford University Press, Oxford).

Reason, J. and Mycielska, K. 1982. *Absent Minded? The Psychology of Mental Lapses and Everyday Errors.* (Prentice-Hall, Englewood Cliffs, NJ).

Salthouse, T. A. 1985. *A Theory of Cognitive Aging* (North-Holland, Amsterdam).

Salthouse, T. A. 1991. *Theoretical Perspectives on Cognitive Aging* (Lawrence Erlbaum, Hillsdale, NJ).

Stine, E. A. L. and Wingfield, A. 1987. Process and strategy in memory for speech among younger and older adults, *Psychology and Aging*, **2**: 272–279.

Treisman, A. M. 1969. Strategies and models of selective attention, *Psychological Review*, **76**: 282–299.

Winder, B. 1993. Intelligence and expertise in the elderly. Unpublished MSc thesis, University of Manchester.

Wingfield, A. and Stine, E. A. L. 1986. Organizational strategies in immediate recall of rapid speech by young and elderly adults, *Experimental Aging Research*, **12**: 79–83.

Training the middle-aged in new computer technology: a Pilot Study using signal detection theory in a real-life word-processing learning situation

B. Baracat and J. C. Marquié

Summary. New computerized techniques can be mastered more easily by older workers if the teaching methods used are appropriate for this population and if the necessary measures are taken in the area of work organization within the company. The first part of the paper discusses the results of a recent survey which pointed out some of the difficulties older employees encounter when faced with computerized work situations. The second part analyses how some of these difficulties may be manifested in a real-life training situation where new computer techniques are being learned. Based on a method derived from signal detection theory (SDT), this study revealed that less knowledge was acquired by older trainees (aged 36 to 52) learning a word-processing program (Word, version 5) than by younger trainees (aged 20 to 35). Old trainees had also stricter decision criteria: they appear more conservative in their responses and more reluctant than younger individuals to commit to a response when they are not completely sure. Future research should be conducted to validate the hypothesis set forth concerning the significance of this attitude, its origin and its consequences on learning.

Keywords: age, older workers, computerization, training, signal detection theory.

Introduction

Owing to rapid changes in the work environment, it has become obvious that an individual can no longer spend his or her entire working life in the same occupation or profession, or at least cannot continue to use the same work methods and techniques. An increasing number of employees will thus be required at some time during their career and perhaps even several times, to receive occupational or professional training. Because demographic changes have caused economic activity to rely more and more heavily on an increasingly older working population, older workers cannot avoid this retraining.

The answer most often given to cope with our society's economic problems is new technology and higher qualifications. But in order for these means to be effective not only for economic development but for social development as well, they must be accessible to all users so that certain groups are not excluded, namely the older working population. Initial training and education no longer suffice in this respect. For centuries, the transmission of knowledge and know-how took place in a stable environment: the occupation first learned was practised throughout life. The daily exercise of the profession was enough in many fields for workers and professionals to develop specialized skills for which they would be appreciated and valued until retirement, and they could transmit their knowledge to the younger generation. But we now live in an ever-changing environment. Sometimes the changes are gradual, but at other times they occur in technological 'leaps and bounds', requiring individuals to update their skills completely and modify their habitual work procedures.

Even more than before, this new situation involves making room for *truly ongoing education* which not only imparts immediately useful knowledge and know-how but also capitalizes on prior skills, prepares for and accompanies change and maintains and develops the ability and motivation to learn throughout professional life. Training cannot, however, be planned without also considering the setting in which the work activity is carried out. The work environment is a place where various types of transformations must be achieved as harmoniously as possible. These transformations include changes in the individuals themselves, in their basic capacities and knowledge and in the technical systems used. The changes must be foreseen and the workplace must be organized accordingly (what must be done and how to do it), so that the work itself can contribute to development and adaptation. As Leplat said, 'It is [. . .] important to avoid over specialization of operators and to place them in a variety of work conditions which will allow them to apply their knowledge to a broader and more diversified field and build extended, more highly adaptive abilities' (Leplat 1988: 154). But other aspects of the organization of work (in the broadest sense) can prepare employees for a more successful experience of the inevitable changes that will occur during their working life (see Chapter 18 by Boerlijst).

This chapter attempts to point out some of the difficulties likely to be encountered by older employees facing the computerization of the workplace and analyse how these difficulties may be manifested in a real-life training situation where new computer techniques are being learned.

Older workers in a computerized setting

Evidence of the problem of the participation of older workers in new technological advances was provided by a study on 20,000 people conducted in
France by the National Institute of Statistics and Economic Studies (Bué and
Gollac 1988). This study clearly revealed that workers aged 45 and over were
grossly underrepresented among computer users. The most extensive use of
the computer was observed in the younger age groups, in approximately equal
proportions for males and females. Among the various computer tools, the
only one which was the same for all age groups was the consultation of
printed lists, which is not really a computer task since it does not involve
direct interaction with computers. A similar state of affairs was reported by
Huuhtanen (1988) for insurance companies and banks.

How can this phenomenon be explained? Why does new computer technology lead to the marginalization of some categories of workers, particularly
older ones (Marquié and Gollac 1989)? What specific difficulties do these
employees have when confronted with technological changes in their work?
Finding an answer to these questions is crucial at a time when computer processing has invaded all sectors of professional and private life, especially if we
agree that generation effects do not satisfactorily account for the entire phenomenon. A recent publication by Cezard et al. (1992) revealed that the curve
representing the percentage of workers using computer tools had gone up
slightly since the last survey, for all age groups, but the slope after the age of
45 was approximately the same as that previously obtained by Bué and Gollac
(1988). Thus, a generation effect certainly exists, but it does not explain everything. Rather than hastily dismissing this phenomenon as temporary and
resulting from the poor preparation of the transitional generation for dealing
with the switch from an outdated system to a radically new one, it would be
preferable to focus on what might be the more durable aspects of the phenomenon. Such more lasting effects are a consequence of the rapid and continuous nature of the changes, which may be such that the differences between
the successive phases of computerization are as great as the recognized differences between traditional and computerized workstations.

In a recent study (Marquié et al. in press) we attempted to determine some
of the reasons behind the less frequent use of computers by individuals over
45. A questionnaire was filled out by 620 office workers and executives
between the ages of 18 and 70. The questions dealt with their representations
of the consequences of computerization on work conditions, employment and
various other aspects of their career, their training on the computer, their
involvement and so on.

A first important finding from this study was that age itself is not the most
important factor associated with negative representations of computer work.
The least favourable opinions were expressed mainly by those who had not
used a computer at all. Attitudes about various aspects of computer work were
more favourable for light users and considerably more favourable for heavy

users. These results show that until the 'step' towards using the computer is taken, individuals generally have less positive attitudes about computers. In non-users, however, age (age 45 and above) and especially seniority (20 or more years in the same department) tended to be associated with greater reluctance concerning these new technologies. Once the step has been taken, the advantages of computerization become more apparent to workers and numerous fears held by non-users decrease greatly. This fact was also noted by Gillet (1990).

A second point worth mentioning concerns the socio-professional consequences of computerization in terms of its expected benefits for career purposes or, inversely, of its threat to employment. Unlike young people and more generally people who are newcomers in a department, senior workers do not have such high expectations for their career. This appears to be responsible in part for the low degree of motivation of some of them to make the necessary efforts to learn the new work techniques and methods. On the other hand, young people and more generally newcomers (even if they do not have a high level of training or qualifications) exhibit much more initiative with respect to computer work, because they see clearly the benefits of computer skills for negotiating a promotion or future employment. It is easier for them to consider these skills as an advantage that is likely to increase their qualifications and enhance their status with respect to their peers.

It appeared as well that a relatively significant proportion of working individuals consider computerization to be accompanied by threats to employment. This holds for all subjects, but is especially true for senior workers. For some of the latter, this fact may cause extreme anxiety, as older workers are the ones who suffer the most from restructuring or elimination of jobs following technological changes, either in the form of transfers to other work positions which do not necessarily correspond to their qualifications, or through forced early retirement or even dismissal. This was admitted by some of the department and company managers we questioned. A minority said that it was true, although the majority declared that one of the major objectives of computerization was to increase productivity without increasing the size of the staff, perhaps even decreasing it.

This survey revealed that all users, even heavy ones, exhibited some degree of realism, accompanied by reservations, concerning the improvement that computerization might bring to work conditions. But the greatest number of criticisms of the computer were found among the older staff members and those with more seniority, who stated that computer tools are too abstract: these individuals do not have a correct representation of how the machine operates or of how information is manipulated and stored. Those with the most seniority also said that the procedures were rigid and that certain techniques were inappropriate and lacked flexibility for moving around during the execution of computer tasks (backtracking, for example). It is natural that individuals who have been working for years with manual procedures that are well suited to their jobs feel more strongly disrupted than others by the transition to computerized techniques. This feeling may be even greater when standard program packages are used rather than custom-designed systems which are specifically adapted to individual needs and tasks.

The last point to be considered is consultation. It appears that consultation is low, that it applies mainly to the best trained and qualified people, and that it is more developed in small companies where communication is likely to be easier. But, apart from the few questions devoted specifically to this topic, the overall results of the survey pointed out the importance of careful planning of the successive phases of preparation, explanation, training and integration of these new techniques by the staff, to the success and general acceptance of computerization.

The major finding of this study was that resistance to change was due, above all, to the lack of knowledge of the computerization effects. Indeed, as mentioned above, most of the reservations were expressed by non-users, especially older ones with more seniority, but this reluctant attitude decreased or disappeared as soon as workers started using the computer. This is a clear demonstration of the essential role of information and the necessity of involving all staff from the beginning, in order to demystify the computer and give employees a means of properly assessing the personal consequences of the computerization. This approach should decrease anxiety, a natural reaction to changes likely to affect one's daily life. Anxiety of this sort was quite apparent in our interviews with both department heads and employees. Several of the department managers said that by the time they became aware of this need to inform and involve their employees, it was already too late. When asked, 'What would you do differently if you had to start computerization of your department all over again?', many replied that they would attempt to get more employee support by allotting additional time to preparing the staff and explaining the nature and consequences of the proposed changes. Such measures would promote a more confident and calmer state of mind among employees as they begin learning how to use the computer, especially older employees.

A study in a word-processing learning situation

It would be inaccurate to assume that age constitutes an obstacle to the learning of new technologies: many persons have learned to use the computer after age 40, demonstrating that it is indeed possible. But it would also be inaccurate to deny the fact that older adults are likely to encounter specific difficulties during training (Czaja et al. 1989). These difficulties must be identified not only in the phases preceding the decision to undergo training but also during the learning phase itself. In doing so, we could minimize the risk of failure and develop ergonomic training methods which consider the specific characteristics of older trainees.

Accordingly, the present study was designed to determine how the above representations and attitudes of older workers are manifested in a real-life learning situation. Data obtained in research on the psychology of aging allow hypotheses to be made with respect to age-related problems in the acquisition and memorization of new information. But very few studies have looked at the repercussions on learning of non-cognitive factors like the trainee's attitude about his/her own skills during the training period, particularly those of older learners.

A method derived from signal detection theory (SDT, initially developed by Tanner and Swets 1954; see also Baracat 1992) was used in this study, which took place during a word-processing course (Word 5 for DOS). Subjects were asked to take true–false tests about some of the main functions of the word-processor. This enabled us to make separate assessments of the formal knowledge the trainees had of the program functions (A' index, a non-parametric analogue of d' which measures sensitivity) and of the ease with which they used their newly, still uncertain, acquired knowledge (B", a non-parametric analogue of β which measures decision criteria).

(a) The A' index could show us whether one source of difficulty in adult learning comes from age-related effects on new knowledge acquisition and memory processes, as mentioned above. Age effects are observed in laboratory learning tasks quite early in adulthood, although they are moderate before the age of 60 (Arenberg and Robertson-Tchabo 1977). One characteristic of older learners is that they are no longer used to learning and dealing with formal information, a factor which makes learning more difficult, particularly in a context which still frequently resembles a school-learning environment. These trainees are not only older, they also have not been in an academic environment for quite some time. Finally, integrating new information and skills into pre-existing ones seems to be increasingly more demanding with age: the links between new knowledge and previously acquired skills are by nature more numerous and difficult. This could explain why 'older trainees have more difficulty than younger ones in moving within frames of reference other than their own', as reported by computer trainers. While their skills, crystallized from long experience, usually ensure that these individuals are well adapted to a familiar, unchanging environment, a deeper and more costly reorganization of knowledge is required when new methods and techniques are introduced. For all of these reasons, the older trainees in the present study were expected to have a lower sensitivity index (lower A' values) than the younger subjects.

(b) The other dependent variable analysed in this study was the older trainees' decision criteria. Learning is itself a source of anxiety, as it disrupts cognitive and non-cognitive equilibria acquired in the past, sometimes at a high cost. But the anxiety may be even greater in the computerization environment described above. Indeed, our prior survey revealed that when faced with computerization or at the onset of training in new computer tools, older workers often fear having to start all over and are reluctant to accept that the skills and know-how they have been using all these years have become obsolete. They feel they are at a disadvantage relative to younger people whom they consider more skilful at using the new tools; they are also afraid they will be unable to learn how to use the computer and will make serious mistakes like deleting a program or even damaging the hardware by incorrect operation (Stewart 1992). In a training situation, these fears and feelings of inadequacy may show through as a lack of self-confidence and difficulty in appropriating and applying new knowledge. Thus the second and most important hypothesis in the present study was that such a context is likely to increase the level of caution of the older trainees, older people generally being expected to be more cautious than younger ones and to make more conservative decisions (Craik 1980). In other words, stricter decision criteria were predicted for older trainees (higher cost attributed to false alarm responses). The questionnaire was worded in such a way that attitudes reflecting reluctance to use a given word-processing function if it might lead to an error were shown by High B" values (cautious attitude).

Method

Subjects

Fifty professional women typists who were novices at using a computer were taking a computer training course for professional reasons. Given the limitations imposed by this real-life training situation (namely the large number of young trainees), it was not possible to obtain balanced age groups. Subjects were divided into three age groups: 14 trainees between 20 and 25 (mean age 23), 21 trainees between 26 and 35 (mean age 31) and 15 trainees between 36 and 52 (mean age 43).

The training situation

The word-processing course (Word 5, IBM-compatible PC) was paid for and lasted for eight full days (6 hours per day). The course material included a presentation of the hardware and operating system; the basic functions required to move the cursor and format a text; how to record, open, close and print a file; the ruler functions (margins, tabs, indentation and so on); creating tables; managing several windows; glossaries; form letters; outline mode and MS-DOS functions (Microsoft—Disk Operating System). The training method consisted of alternating between group training sessions and longer, individual exercise sessions. Individual assistance was available when needed.

Procedure

Three program functions were analysed in this study: (1) text formatting, introduced essentially during the first two days of training but applied throughout the course, (2) glossaries, taught on the third day and used for the rest of the course, and (3) form letters, studied on the last three days. At the end of the second, fifth and eighth days, tests containing a series of statements about the word-processing functions learned so far were passed out (e.g. 'To justify a portion of text, you must use ⟨FORMAT⟩ ⟨PARAGRAPH⟩ or ⟨ALT + J⟩'). The trainees were asked to indicate 'true' or 'false' for each statement on the test. There was an equal number of true and false statements. To test for a practice effect, certain functions were tested several times during the course (second, fifth and eighth days for text formatting and fifth and eighth days for glossaries).

The trainees were asked to consider the statements as potential actions, so that whenever they were in doubt about the answer, they should act as if they were deciding whether or not they would actually perform the action. The tests were taken individually, without documentation.

Results

The data were processed in ANOVAS, using as factors age (3) and, depending on the case, word-processing function (3) or practice (2 or 3). The results revealed an effect of age on the value of A' (figure 15.1).

This means that the error rate for the older trainees was higher for all three program functions studied ($F[2,47] = 6.1$, $p \leq 0.005$) as compared with the younger trainees. Likewise, A' varied significantly across functions ($F[2,94] = 18.8$, $p \leq 0.001$), although identically for all three age groups. As to the decision criteria (B"), the only significant effect was age ($F[2,47] = 20.7$, $p \leq 0.001$).

As figure 15.1 shows, the older group differed from the other two by its stricter decision criteria (higher B" values): the older trainees produced proportionately more omissions than false alarms compared with the younger subjects. The A' values obtained for the glossaries on the fifth and eighth days (figure 15.2) showed no significant effects of age or practice, although the age effect nearly reached the significance level on the first test ($F[2,47] = 2.9$, $p = 0.06$).

The achievement levels (knowledge acquired), as a whole quite low, were very similar in all three age groups and did not progress significantly between the fifth and eighth days.

As to the decision criteria, age was associated with higher B" values ($F[2,47] = 8.5$, $p = 0.01$). The repetition of the test three days later did not have a significant effect on the decision criteria of either group, despite the tendency indicated in figure 15.2 for the older group.

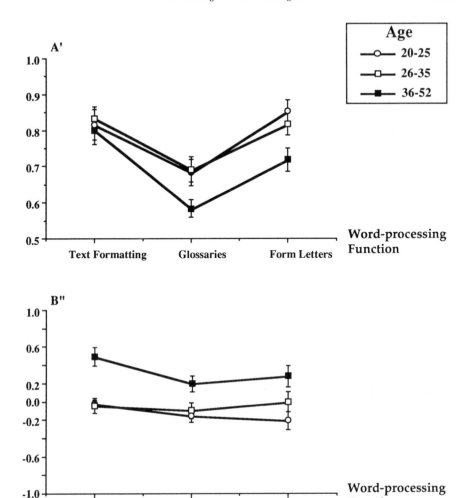

Figure 15.1. Mean A' and B'' values (and standard error) for three word-processing functions, by age.

A comparison of the three age groups for the questions on text formatting at three different times during the training (2nd, 5th and 8th days) showed (figure 15.3) that age ($F[2,47] = 4.7$, $p \leq 0.02$) and practice ($F[2,94] = 8.1$, $p \leq 0.01$) had statistically significant effects on A', with older trainees showing lower overall achievement levels than younger ones.

Achievement improved with practice, but did so equally in all three age groups, although the age × practice interaction was close to the significance level ($F[4,94] = 2.3$, $p = 0.06$). The results obtained for the B'' values (figure 15.3) showed that the older trainees were significantly more cautious in their

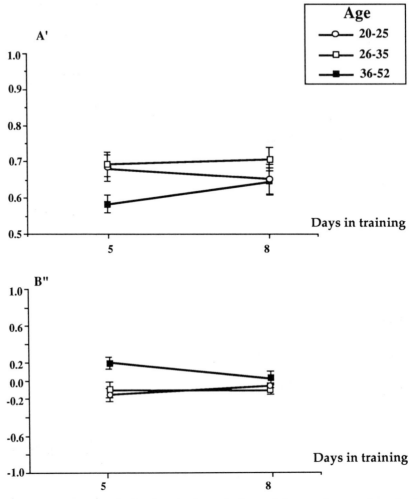

Figure 15.2. Mean A' and B" values (and standard error) for questions on glossaries, by age and day of training (5th and 8th days).

responses (high B" values) than the younger ones (F[2,47] = 3.6, $p \leq 0.05$) and that their decision criteria varied with practice (F[2,94] = 4.4 $p \leq 0.02$), although in similar proportions, as indicated by the lack of a significant interaction. However, such a significance appeared when the two younger groups were pooled and compared with the older group (F[1,76] = 3.8, $p \leq 0.05$), suggesting, as the figure indicates, that the older subjects differed more from the younger subjects at the beginning of the training that at the end.

Analysis of the individual values can provide us with information about the variability of index B", of particular interest to us here, in addition to furthering our understanding of the variation trends during the training. Regres-

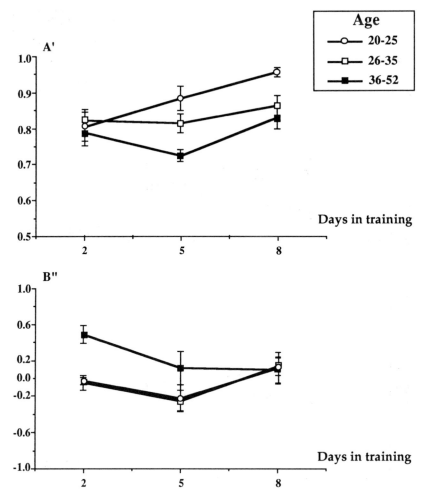

Figure 15.3. Mean A' and B'' values (and standard error) for questions on text formatting, by age and day of training (2nd, 5th and 8th days).

sion analyses between age and B'' for the text formatting operations were computed for each of the three testing times (figure 15.4). A significant correlation between age and B'' was obtained for the second and fifth days $(r^2 = 0.19, \ t[48] = 3.38, \ p \leq 0.001;$ and $\ r^2 = 0.11, \ t[48] = 2.38, \ p \leq 0.05,$ respectively), but not on the eighth day $(r^2 = 0.003)$.

Visual inspection of the figure shows that older subjects used stronger decision criteria than younger ones at the beginning of the course and that this tendency persisted until the fifth day but with a wider spread of values. By end of the course, a majority of subjects (younger and older) used more neutral criteria (close to zero) and age was no longer a predictor of the subjects' attitude.

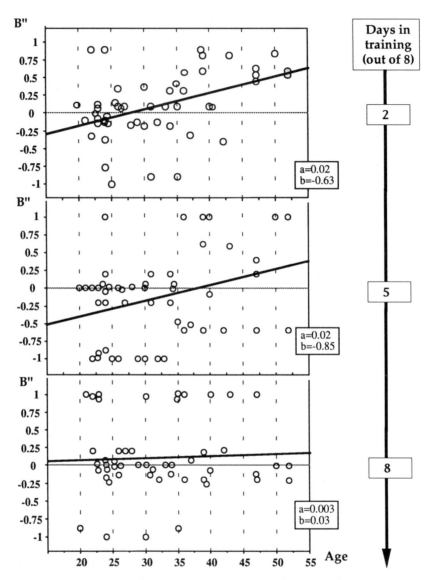

Figure 15.4. Individual B" values by age and regressions between age and B" for questions on text formatting, at three different times during the training (2nd, 5th and 8th days). Slopes (a) and intercepts (b) are displayed in the figure.

Discussion

SDT can be applied to learning situations in order to gain insight into two complementary aspects of the learning activity: a quantitative aspect, which informs us about the amount of formal knowledge acquired at different points in the training process (sensitivity); and a more qualitative aspect, which informs us about the attitudes of the trainees confronted with their knowledge (decision criteria).

In this study the analysis of the sensitivity index revealed as a whole that the older trainees did not acquire as much formal knowledge about the word-processor as the younger ones did. The older subjects seem to have been less able to memorize the procedures they learned and practised during the exercises. This age difference is consistent with the data in the literature on adult learning at this age.

Indeed, middle-aged trainees appear to be slower on average than younger ones and need more time to achieve the same level of performance (see Paumès and Marquié in press). Czaja et al. (1989) obtained the same results in a situation similar to our own. These authors showed that when learning how to use a text-editing system, the younger group (ages 25 to 39) made fewer errors and were quicker than the older groups (ages 40 to 54 and 55 to 70).

It is nevertheless very difficult on the basis of the available results to determine whether these differences between age groups are a result of a decline in the efficiency of memorization processes with age, or because of other factors such as the older generation's potential lack of exposure to the 'computer culture', or some form of learning disuse. The main area in which practice effects significantly appeared during the training course was text formatting (this operation was taught and practised right from the beginning of the training). The results do not, however, allow us to conclude that the young progress faster than their elders.

The most original aspect of this study pertains to the decision criteria. As predicted, the older subjects did not make the same type of errors as the younger ones: they produced proportionately more omissions than younger subjects. The stricter decision criteria of the older subjects indicate greater caution, which in the present case means that when in doubt, they were more inclined to say that a true statement was false than to say that a false statement was true. Older trainees therefore appear to be more reluctant than younger ones to commit to an action when they are not completely sure that it is correct.

This difference in attitude seems to reflect fear of making a mistake on the part of older trainees. Indeed, with the instructions given here, only false alarms in reality could lead to the use of an erroneous procedure in practice. It is possible that this more conservative and less experimental attitude contributed in the present study to slowing down the learning process in the older trainees, since such an attitude may have put them in a less active environment for empirically manipulating and discovering how the word-processing system works. But perhaps the more cautious attitude of the older subjects also reflects a strategy used to guard against certain difficulties: as shown in Belbin and Shimmin's study (1964), older individuals seem to have more trouble eradicating errors made during the first stages of the learning process.

This more cautious attitude appears to be a frequent characteristic of older adults (Craik 1980), although not consistently so (Baron and Le Breck 1980, Baracat and Marquié 1992). It was already observed by Belbin and Shimmin (1964, data reanalysed by Craik 1980 using SDT) in quality control learning that the tendency of the older trainees to be too strict in their decision criteria led them to reject many items as incorrect when they were, in fact, correct. The use of teaching methods that were more suitable for these older adults allowed them to bring their decision criteria back down to the level required by the instructions.

The results of the present study also suggest that the more cautious attitude clearly exhibited by older individuals at the beginning of the training course decreased with time: for most older subjects, the decision criteria tended to become neutral as the course progressed and in the end did not differ from those of the younger subjects. It would be worthwhile in future research to find out if this tendency exists for other types of training and whether it is associated with an increase in specialized knowledge, or simply with the subjects' familiarization with the situation. This could be of some interest for future teaching applications.

Conclusion

Whatever the case may be, older trainees appear to approach learning with a different, apparently more worrisome, less facilitating attitude than younger individuals. Any teaching method that reduces this anxiety could be particularly beneficial to middle-aged and elderly trainees. Future research should attempt to determine the significance of the attitude of older trainees, as well as its origin and consequences on learning. Data collected on the attitudes and representations of older individuals faced with the computerization of their work environment (Marquié et al. 1994) suggest that various measures concerning the organization of work within the company (e.g. consulting employees during the implementation of technological changes, keeping them more informed, improving the ergonomics of computer tools, more regular training and so on) could be beneficial and better prepare the staff for learning new skills.

Acknowledgements

This work was supported by grants from Le Ministère de la Recherche et de la Technologie and Le Ministère du Travail, de l'Emploi et de la Formation Professionnelle. We wish to thank the Institut de la Promotion Supérieure du Travail in Toulouse for having made this study possible and especially D. Barbarou for her contributions.

References

Arenberg, D. and Robertson-Tchabo, E. A. 1977. Learning and aging. In J. E. Birren and K. W. Schaie (eds) *Handbook of the Psychology of Aging* (1st edn) (Van Nostrand Reinhold, New York) 421–449.

Baracat, B. 1992. Changements liés à l'âge dans les processus de prise de décision. Application de la Théorie de la Détection du Signal chez l'homme adulte. Thèse d'Université. Université P. Sabatier, Toulouse.

Baracat, B. and Marquié, J. C. 1992. Age related changes in sensitivity, response bias and reaction time in a visual discrimination task, *Experimental Aging Research*, **18**(2): 59–66.

Baron, A. and Le Breck, D. B. 1987. Are older adults generally more conservative? Some negative evidence from signal detection analysis of recognition memory and sensory performance, *Experimental Aging Research*, **13**: 163–165.

Belbin, K. and Shimmin, S. 1964. Training the middle aged for inspection work, *Occupational Psychology*, **38**(1): 49–57.

Bué, J. and Gollac, M. 1988. Technique et organisation du travail. Premiers Résultats, no. 112 (Insee, Paris).

Cezard, M., Dussert, F. and Gollac, M. 1992. La percée des nouvelles technologies, résultats de l'enquête 'conditions de travail' de 1991. Premières Informations, 266 (Ministère du Travail, de l'Emploi et de la Formation Professionnelle, SES, Paris).

Craik, F. I. M. 1980. Applications of signal detection theory to studies of aging. In A. T. Welford and J. E. Birren (eds) *Decision Making and Age* (Arno Press, New York), 147–157.

Czaja, S. J., Hammond, K., Blascovitch, J. J. and Swede, H. 1989. Age related differences in learning to use a text-editing system, *Behaviour and Information Technology*, **8**(4): 309–319.

Gillet, B. 1990. De l'usager-type aux types d'usagers, ou l'influence des différences individuelles dans l'adaptation homme-ordinateur, In Ouvrage Collectif (ed.) *Informatique et Différences Individuelles* (PUF, Lyon), 111–123.

Huuhtanen, P. 1988. The aging worker in a changing environment, *Scandinavian Journal of Environmental Health*, **14**(1): 21–23.

Leplat, J. 1988. Les habiletés cognitives dans le travail. In P. Perruchet (ed.) *Les automatismes cognitifs* (Pierre Mardaga, Liège), 139–172.

Marquié, J. C. and Gollac, M. 1989. Caractéristiques des populations au travail et nouvelles technologies. In *Evolutions technologiques et ergonomie. Actes du XXV-ème Congrés de la Société d'Ergonomie de Langue Française*, Lyon, 107–118.

Marquié, J. C., Thon, B. and Baracat, B. 1994. Age influence on attitudes of office workers faced with new computerized technologies. A questionnaire analysis, *Applied Ergonomics* **25**(3): 130–142.

Paumès, D. and Marquié, J. C. 1995. Travailleurs vieillissants, apprentissage et formation professionnelle. In Ouvrage Collectif (ed.) *Le travail au fil de l'âge* (PUF, Paris) (in press).

Stewart, T. 1992. Physical interfaces or 'obviously it's for the elderly, it's grey, boring and dull'. In H. Bouma and J. A. M. Graafmans (eds) *Gerontechnology: Studies in Health, Technologies and Informatics* (IOS Press, Amsterdam), 197–207.

Tanner, W. P. and Swets, J. A. 1954. A decision-making theory of visual detection, *Psychological Review*, **61**(6): 401–409.

16

Working conditions: problems ahead for workers over the age of 40

A.-F. Molinié and S. Volkoff

Summary. Three factors concerning the increasing problems with the assignment of employees over 40 years are discussed. These factors are shift- or nightwork, production pace or time pressure and stress from new technologies. In addition, there is a relative shortage of younger people, who may replace the age group of 40–50 years. Anyhow, the change to 'normal' schedules, because of reasons such as health or changes in the personal sphere, is hampered by such causes. In our modern society, for the near future, shiftwork will tend to increase, there is much pace-dictated, assembly-line work and new technologies are introduced continuously. Taken together, these factors may indicate the increasing difficulty of placing workers over 40 years, since they seem to be, and feel, out of place in this changing, technological world.

Keywords: job assignment of employees over 40, shiftwork, time constraints, new technology.

Introduction

The working population is aging and will continue to age in the coming decades. The reasons are well known: the numerous generations born after the war are already 40 years old or older; they tend to end their education at a later age; the percentage of working women between 40 and 60 years old is increasing; finally, it is unlikely that they will retire at an earlier age or will take an early retirement.

Such developments should be taken into account in designing the working environment. The relationship between workplace characteristics and the age breakdown of the worker population is changing. It is necessary to determine

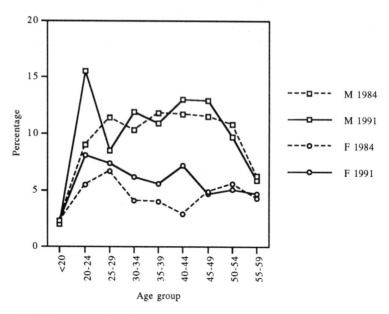

Figure 16.1. Percentage of employees affected by shiftwork in 1984 and 1991.
Source: Ministry of Labour (France), National Surveys on Working Conditions, 1984, 1991

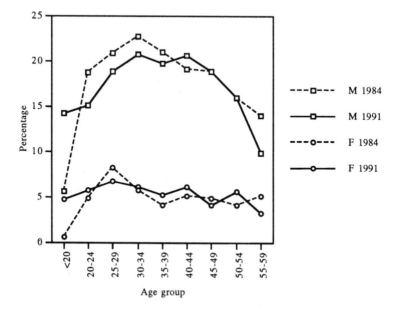

Figure 16.2. Percentage of employees affected by nightwork at least once a year in 1984 and 1991.
Source: Ministry of Labour (France), National Surveys on Working Conditions, 1984, 1991

the nature of these changes. There are three areas in which selection processes specifically based on age have been observed: shiftwork or nightwork, constraints imposed by work pace, and the use of new technologies.

Returning to a dayshift after the age of 45

The proportion of employees who work on a shift decreases after the age of 45. Such a decrease even begins after the age of 40 among nightshift workers (see box 16.1 for a description of the survey).

This becomes even clearer when applied to a single sector and to one category of workers. Thus, in industries of semi-manufactured goods where there is a high rate of shiftwork, the proportion of workers on a shift decreases continuously after age 45, from 53 per cent between 40 and 44, to 24 per cent between ages 55 and 59. In addition, for the same age groups, the percentage of people who work in three or more shifts—which implies a nightshift—decreases from 30 per cent to 13 per cent.

These results show a marked trend to returning to 'normal' schedules after age 40—a well-known fact among personnel directors and occupational physicians, who are frequently called on to make necessary reassignments. The

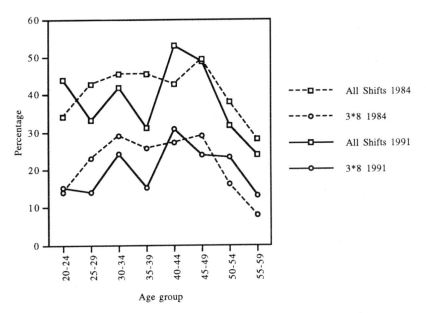

Figure 16.3. *Percentage of blue-collar males affected by shiftwork in semi-manufactured goods industries in 1984 and 1991.*
Source: Ministry of Labour (France), National Surveys on Working Conditions, 1984, 1991

request for changes can be a result of the health of shiftworkers, which can deteriorate because of the long periods of early and late schedules (Bourget-Devouossoux Volkoff, 1992); or it can be a consequence of sleeping disorders that have come on with age; or because of a change in family and social situations which no longer require the extra pay generated by the nightshift. All these factors, which can have a cumulative effect, are persistent, as is evident from the similar results gathered in 1984 and 1991.

Assuming that shiftwork continues at the current rate or increases in the years to come, however, one can wonder whether the aging of the working population is not likely to make reassignments more difficult than today. This concern could modify the criteria that determine the volume of shiftwork and its organization.

Dictated speed for young female workers

Time constraints give rise to hardships that worsen with age. When time constraints are overly demanding, workers have less and less freedom to manage their work. This is especially so for aging employees, namely because it restricts their capacity to anticipate their next move (Laville, 1989).

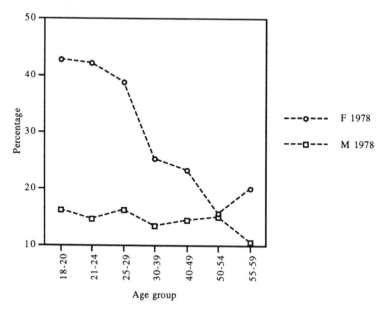

Figure 16.4. Percentage of factory workers subjected to tight pace constraints in 1978 (old nomenclature*).
* *The category of factory workers is defined here according to the old French nomenclature of socio-professional categories (SCP), whereas figure 16.5 refers to the new nomenclature.*
Source: Ministry of Labour (France), National Surveys on Working Conditions, 1978.

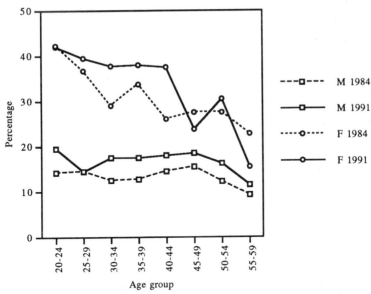

Figure 16.5. Percentage of factory workers subjected to tight pace constraints in 1984 and 1991 (new nomenclature).
Source: Ministry of Labour (France), National Surveys on Working Conditions, 1984, 1991

Ergonomic and statistical studies have shown that, when time pressure is heavy, there are 'age limits' beyond which it is difficult to be at certain work-stations or to be subjected to certain work constraints (Teiger, 1989, Volkoff 1989).

The results of the 1991 survey on working conditions confirm the analyses made to that effect in previous surveys. In 1991, as in 1978 or 1984, factory workers experience the tightest pace constraints (see box 16.2) and particularly women, especially those under 30. Four out of 10 of these young women work on assembly lines or at a pace imposed on them by the automatic motion of a product or a part, or by the automatic movement of a machine.

Along with these constraints comes the sense of not having enough freedom to manoeuvre within the given time. The youngest women are also those who most frequently cite 'being not able to change the time limits': more than 60 per cent of female workers between age 20 and 34. They also form the largest number of those who cannot 'interrupt their work': approximately 40 per cent of female workers between 20 and 34 years old.

These selection phenomena related to pace constraints can also apply to less apparently tight constraints, such as the compliance with production stan-dards, deadlines, or the response to outside requests. Here again, the 'tighter' the constraints, the more they seem to entail an age-based selection. For example, the proportions of employees who 'must meet production standards or deadlines at most within an hour', or those who must 'answer immediately' a request from clients or from the public, decreases sooner and more sharply with age than in cases where the time limits are longer ('production standards or deadlines to be met in at most one day' and a request 'not requiring an immediate response').

Aging on the assembly line

Tight time constraints have increased substantially between 1984 and 1991 while the working population in the age category of about 40 years old was growing and the category of youngest employees was diminishing. The proportion of female factory workers between 20 and 25 years old subjected to tight pace constraints has remained constant since 1978. In older age groups, however, this proportion has increased considerably. In 1991, having to work at a speed imposed by the automatic pace of a machine is almost as frequent among female workers between 40 and 44 years old as it is among female workers under 30. Workers of all ages depend much more often on the automatic movement of a product or a part nowadays than in 1984, even though the proportion of female workers affected by this constraint diminishes more or less constantly starting at age 35.

The spreading of these tight constraints has also been noticeable for male workers. Neither in 1991 nor in 1984, however, does the age-based selection occur in men as early or as markedly as it does in women.

Would tight time constraints have become less selective, either because of adjustments made to ease hardship, or because the absence of young workers makes a selection based on age harder to effect? Statistics alone cannot answer this question.

Two extreme examples illustrate the variety of situations encountered. In the 'meat and milk industries' there has been a profound change in the way work is organized, especially in the meat industry. Tight time constraints have become common, in particular in assembly-line work, which affected 43 per cent of factory workers in 1991 against 20 per cent in 1984. These time constraints are added on to others such as the carrying of heavy loads. Age selection is very pronounced in this sector. The age pyramid is very young in 1991, as in 1984, and the proportion of assembly-line workers decreases drastically with age.

Conversely, in the industry of 'land transport material' there has been a marked aging of workers and a drop in staff between 1984 and 1991. The percentage of assembly-line workers has gone from 25 per cent to 32 per cent and the percentage of workers subjected to constraints related to the automatic movement of a product or a part has risen from 17 per cent to 31 per cent. This increase has affected mainly workers between 40 and 49 years old. It is probably not so much an increase in the number of workstations as the result of aging of this sector—there aren't enough youths to fill the workstations—and of the elimination of low-constraint jobs. Such jobs include light preparation off the assembly line, maintenance of premises, warehouse

work, and surveillance—all of which employed older workers or those who
had difficulties 'bearing' assembly-line work. But, for aging employees to be
able to fill the workstations, some of the hardships associated with chain work
and particularly selective work requirements can both allow for the assignment of more workers on the stations and render the workstations less
harmful.

Despite these changes, the existence of tight pace constraints on the older
age group, in this sector as in others, could generate increased difficulties in
the long run, possible health problems or new exclusion processes. In designing automobile manufacturing plants it remains unquestionably imperative to
take into consideration the aging of the workers (Sailly, Volkoff; 1990).

Older workers cast aside from computers

In 1987, according to the survey on technologies and the organization of
work, the proportion of computer users peaked between ages 25 and 44 and
clearly dropped after age 45 (Gollac, 1989). Since then, computers have spread
tremendously and have affected all ages; so much so that in 1991 the percentage of workers aged 50 to 54 who used a computer at work was similar to that
observed in 1987 for ages 25 to 44. But, in 1991, as in 1987, computer usage
was less and less frequent for those over 45.

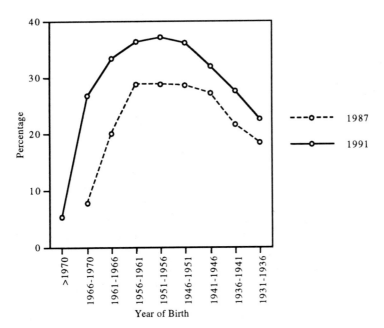

*Figure 16.6. Percentage of workers using a computer, per generation, in 1987 and 1991.
Source: Ministry of Labour (France), National Surveys on Working, Technics and
Organization 1987, on Working Conditions 1991.*

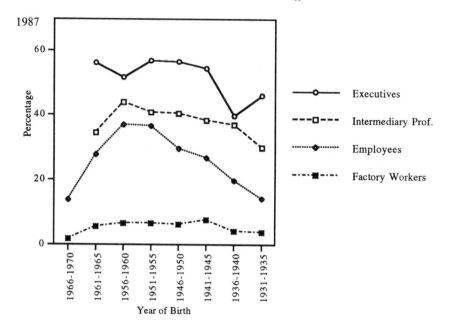

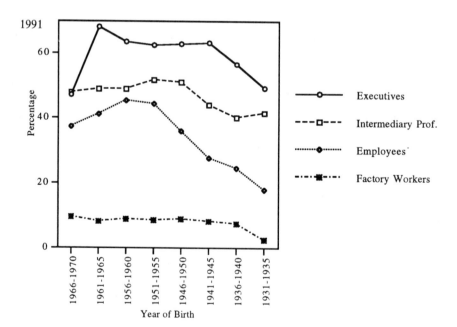

Figure 16.7. Proportion of workers using a computer by socio-professional category, per generation, in 1987 and 1991.
Source: Ministry of Labour (France), National Surveys on Working, Technics and Organization 1987, on Working Conditions 1991.

This relative casting aside of older workers when it comes to computer usage is evident in all social categories. But the decrease with age in the proportion of computer users occurs all the earlier and is all the more rapid when computer usage is more intensive: as of age 50 with executives, age 45 for intermediary professions and age 40 for employees.

The same applies to the use of word-processors, which decreases with age, especially after age 50. For the employees who use this equipment for much longer, this decrease is markedly faster and appears as early as age 30.

These selection phenomena based on age are harder to highlight for new industrial technologies. They affect a much narrower population: less than 2 per cent of employees use a robot or a digitally controlled machine tool. Their use appears to be less frequent among older factory workers, starting at age 45 for robots and at age 50 for digitally controlled machine tools.

A generation effect

In the oldest age groups, the spread of computers between 1987 and 1991 seems broadly to reflect a generation effect. It is, of course, impossible to ascertain this with the rigour of a longitudinal study. But one can move near this concept of generation by comparing the proportion of computer users in 1987 and 1991 for the same birth year group. It is as if employees who used a computer at work in 1987 continued to use it in 1991 however old they were, but that being a 'new user' would dwindle among the oldest generations. And, the more intensive the computer work, as in the case of employees, the less likely the employee is to 'begin' using the computer at work at an older age.

Computer usage seems to depend greatly on the age of workers at the time computers were introduced, as well as the conditions in which technology was introduced. Software design, means of consultation, training policies and methods should also play, for many years to come, an essential role in easing the access to data processing for the older employees.

Conclusion

In conclusion, whether it be shiftwork, pace or new technology constraints, a common concern arises: the assignment of employees over age 40 in the coming years. If certain, particularly harsh constraints on workers, such as assembly-line or a fast pace of work, persist or increase, and if, at the same time, their inclusion in the modernization of production means remains problematic, assignment difficulties could become greater; and with them would follow a loss of know-how and skills as a result of older employees' exclusion. This issue could turn out to be thornier for the 40–50-year-old age group than for the older one, because of its larger number and because the management of the professional career of this group will not be solved by a forthcoming retirement.

Box 16.1 Sources.

(1) *Surveys on 'Working conditions'*

The 1991 results stem from the third survey on working conditions. The two previous ones were carried out in October 1978 and March 1984. This survey was organized and operated by the Department of Studies and Statistics of the Ministry of Labour, Employment and Vocational Training, as a supplement to the Insee Employment survey. Its scope is therefore that of the Employment survey. The questionnaire was submitted to all the employed members in one out of three households covered in the Employment survey sample, that is about 21,000 people. The results produced here only concern factory workers. The questionnaire of the supplementary survey was submitted to each employed member in the household, who was required to answer it individually. The answers collected thus refer to the working conditions as perceived by the individual employee.

The 1991 survey tackles the following broad themes:

— risks, nuisances, work hardship; new questions covered aspects such as the work-related mental burden and 'stress' factors;
— organization of work, autonomy;
— machines and technologies used;
— work hours and organization of work time, shiftwork and work on weekends and at night.

(2) *Survey on 'Technology and organization of work'*

The questions on the organization of work and the new technologies were not included in the 1984 survey, but in the 1987 survey on 'Technology and organization of work'. This survey was also a supplement to the Employment survey.

Box 16.2. Measurement of tight pace constraints in the Working Conditions survey

An employee is said to be subjected to a 'tight pace constraint' if he or she mentions the presence of at least one of the following constraints: working on an assembly line, having 'a work pace imposed by the automatic movement of a product or a part', or having to 'keep up with the automatic pace of a machine'. The definition of assembly-line work is that used by the Ministry of Labour, Employment and Vocational Training in a number of surveys carried out on firms: 'any person fulfilling, at a given pace, a repetitive task on a product that is either moving in front of, or passed on to him by his neighbour, without intermediary stocks between them, is considered to be doing assembly-line work'.

The questionnaire brings to light the possible existence of certain pace constraints. The survey thus enables the measurement of the frequency of such constraints in the population, although it does not measure their intensity.

One could assume, however, that, by their very nature, the three above-mentioned constraints are particularly tight and demanding. In practice, compliance with standards and deadlines or the need to respond to a request (to cite examples of other constraints mentioned in the survey) can sometimes prove to be as demanding as a tight pace constraint.

References

Bourget-Devouassoux, J. and Volkoff, S. 1992. Bilan de santé des carrières d'ouvriers (Health check-up in factory worker careers), *Économie et Statistique*, **242**: 83–93.

Gollac, M. 1989. L'ordinateur dans l'entreprise reste un outil de luxe (Computers in firms are still luxury tools), *Économie et Statistique*, **224**: 14–25.

Laville, A. 1989. Vieillissement et travail (Aging and work), *Le Travail Humain*, **52**(1): 3–20.

Sailly, M. and Volkoff S. 1990. Vieillissement de la main-d'oeuvre et adéquation prévisonelle des postes, le cas des ouvriers de montage dans l'automobile (Aging and anticipating the adjustment of workstations; the case of assembly-line workers in the automobile industry), *Formation et Emploi*, **29**: 66–81.

Teiger, C. 1989. Le vieillessement différential dans et par le travail: un vieux problème dans un contexte récent (Differential aging in and due to work: an old issue in a recent context), *Le Travail Humain*, **52**(1): 21–56.

Volkoff, S. 1989. Le travail après 50 ans, quelques chiffres et plusieurs inquiétudes (Work after 50, a few figures and a number of concerns), *Le Travail Humain*, **52**(2): 97–116.

Conclusion

In part II, the relevance of age-related differences in task performance for everyday work was discussed, as well as age-related adaptations in work conditions for both employees and employers.

In chapter 11, Zeef, Snel and Cremer reported a study in which a group of old subjects underestimated the velocity of a moving object that has a partly invisible trajectory to a greater extent than a group of young subjects. It was noticed that practice on the task had no effect on the performance of either age group. The authors explained their results by adopting a model of cognitive-processing rate, indicating that old people tend to process a smaller amount of information per time unit than young people. Given that participation in road traffic requires frequent updates of spatial schemata of movements of objects in relation to the movement of the participant himself, this study may have some diagnostic value.

Kok and colleagues indicated in chapter 12 that in reaction-time tasks old subjects performed more slowly and less accurately as a function of task complexity than did young subjects. This was taken as a manifestation of reduction in the amount of available resources with advancing age. The combined effects of task complexity and conditions such as sleep deprivation and auditory noise showed, however, that the impact of the last conditions was stressor-specific. That is, auditory noise accentuated the interaction effects between age and task complexity, as suggested by the Brinley plots, whereas sleep deprivation alleviated these effects. Furthermore, the hypothesized beneficial effect of caffeine intake on attenuation of age-related differences in processing resources was not found. Given that the regression lines were not statistically tested, these results should be interpreted cautiously. In extrapolating these results to real-life working conditions, it can be concluded that old workers are apparently less vulnerable from lack of sleep than in a noisy environment. This may, however, be relative, as it is likely that other underlying variables are at stake.

Rudinger et al. showed in chapter 13 that a mere formal change of instructions for use may have tremendous improvement on the use of well-known everyday gadgetry such as ticket-vending machines and video-recorders. Simplification of both the display and the instructions resulted in 15 per cent fewer usage errors and a 40 per cent decrease of operation time in elderly people. This study hypothesizes plausibly that the main cause of problems of use of modern technology is the poor fit of the design of the devices and the preconceptions about the operational aspects and the function of the machine. The inadequate fit of the 'design model' and the 'user model' is easily improved by following some suggested guidelines on how to improve the design of user interfaces. By following such behaviour-oriented procedures it must be possible to develop a kind of 'technical grammar' or universal directions of use. When designers follow these criteria, it may result in technical devices that are more user-friendly than formerly.

Rabbitt and Carmichael presented in chapter 14 a valuable methodology which indicates that people with different types of disabilities have different attitudes to new technology. That is, the extent to which people look forward to new communication systems seems to correspond with their ability to make use of them and to their willingness to invest resources to acquire and master them. The authors emphasized, however, that people's knowledge of their own abilities depends on the degree of experience with a particular situation and particular tasks. It also depends on the extent to which these tasks or situations can give people accurate feedback about their own level of performance, as well as the ability of people to monitor and use feedback. Furthermore, they stressed that the use of new communication systems should be tested by people suffering from multiple minor disabilities. Disabilities such as loss of vision, hearing or taste produce greater restrictions in performance together rather than apart. These restrictions are even more true in old age, because information-processing resources reduce with age.

The topic of experience is also addressed in chapter 15 by Baracat and Marquié. They indicate that career expectations rather than age is the most important factor associated with negative representations of computer work. They designed a word-processing learning study, in which old trainees acquired less knowledge than young trainees. The old trainees had also stricter decision criteria because they responded more conservatively and were more reluctant when they were not completely sure. The authors suggest that teaching methods should be aimed at reducing the level of anxiety, especially in old trainees. This will be more effective when trainees are kept well informed.

That, in the near future, a common concern of employers will be the assignment of employers about 40 years is emphasized by Molinié and Volkoff in chapter 16. Realizing that it is unlikely that this working population will take early retirement, they advise that the work environment should be adapted to the possibilities of an aging working population. Areas of concern are shift or nightwork, pace constraints and training on new technologies.

Thus, in this part II, an age-related reduction in processing rate is emphasized, as well as an age-equivalent level of adaptability to new technologies. It is indicated that, for the sake of work efficiency, both young and old workers should be involved in training programmes and informed about new technologies.

Part III
Training and Educational Programmes

Introduction

Our course of life is subject to a culturally determined process of chronological standardization. This process is steered to a considerable extent by the organization of labour in our society. This results in a 'normalized biography' ('Normalbiographie') consisting of three phases: preparing for, participating in and resigning from labour (Kohli 1985, in Baars 1988). Older employees are considered to be in a kind of transition phase between participation and resigning. In this transition they suffer from a loss of (parts of) their roles, while there are no obvious immediate alternatives to replace this loss (Schuyt and Van de Klinkenberg 1988). As it is, their integrity of functioning is infringed on and they experience a loss of power and identity. Apart from that, many older employees undergo a decline of their physical and mental capacities. Although there is a lot of evidence that for most older employees this decline does not greatly affect their productivity, as a group, older employees stereotypically are perceived as deficient in this respect.

The perceived deficit functions as a justification for the application of an essentially medical frame of reference, from which all kinds of corrective measures can be taken (Baars 1988). This frame of reference has two important drawbacks here. First, it is based on the denial of the great variation in health and capacities among the elderly. As such, it reinforces the stigma of the elderly as a more or less uniform group of weak and dependent people. After all, medicine is a science and scientists know what they are talking about. Second, by relying on a curative approach, as is the normal way of doing things in medicine, the possibility of more preventive measures is obscured.

In this part III, which deals with preventive and curative measures, a somewhat different perspective is offered. It is shown that, though there is some loss of physical and mental capacities, the main source of problems must be sought in the loss of integrity of individual functioning, and the loss of power and identity resulting from it. Interventions have to be directed at preventing, neutralizing and mending these infringements. The main way of doing this, advocated here, is by (re)empowering employees: improving their capabilities by training them and giving them more autonomy to use these capabilities and to make choices about their own destinations. It is argued that by empowering employees of all age groups, it would be possible to prevent most of the problems that older employees now have to face. This is what we call an 'age-conscious' personnel policy. An additional advantage of such a general age-conscious personnel policy is that it does not set apart the elderly as a separate, stigmatized group. At the same time, however, we have to confront the problems that older employees have right now. This requires some additional policies in order to solve those problems and to counteract the effect of existing stereotypes.

In chapter 17, Schabracq gives an overview of the backgrounds of several problems of older employees. The following topics are dealt with: 'concentration of experience', the 'glass-ceiling' effect, lack of formal schooling and additional training, changes of networks and personal work objectives, managerial uneasiness, the generation gap, the organization of production processes, and the increase in change. Schabracq shows how these problems can interfere with one's integrity of functioning, resulting in loss of power and identity.

Boerlijst, in chapter 18, sketches the relative neglect of the development and growth of older employees (40 +) in organizations. The overview is based on the results of his research project into the situation of middle- and higher-level employees in 10 exemplary Dutch organizations (Boerlijst et al. 1993). Boerlijst presents data about the development of the quality of functioning of over-40s, their mobility, their being embedded in social networks, their participation in training and developmental programmes and the stimulation they get from management. Discussing these data, he distinguishes between the utility value and the learning value of functions, and demonstrates that the most serious problems centre around the latter.

In chapter 19, Cremer sketches the functional and psychological effects of aging on work performance from a psychonomic point of view. Special attention is given to the effects on the ability to learn. These effects have important consequences for the design of training programmes for the elderly. Some designing principles for such programmes, as described by Thijssen (1988), are introduced. Cremer concludes his article by stressing the overriding variability of older employees and by emphasizing the importance of tailor-made training programmes.

Holms presents an interesting case-study of older employees engaged in a reorganization programme of a Danish company, in chapter 20. A far-reaching technological innovation in the production process was combined with the introduction of production groups with extended autonomy. Whereas this kind of technological innovation often marginalizes older employees, this was not the case here. The reorganization resulted in active participation of older employees as equal members of the production groups. As it turned out, they were able to play all roles that members of such a group are supposed to play. In a way this article can be read as a kind of existential proof that older employees can be very mobile and adaptive indeed.

Chapter 21 by Schabracq describes an age-conscious approach to personnel management that focuses on the self-management of individual (older) employees. This approach on the one hand tries to boost the potential of individual employees to cope and adopt by stressing the importance of certain values and motives, and by teaching them better planning and managing skills. On the other hand, the approach emphasizes changes of the organization as a whole as well. The main objective here is to create enough opportunities for individual adaptive change. This requires—apart from neutralizing certain stereotypes and improving the quality of jobs (especially in the direction of a greater job decision latitude)—organizations to provide their (older) employees with the chance to change jobs in ways that are rewarding, both to these employees and to the organization. To facilitate the process, the organization has to become more transparent to all parties involved.

Warr indicates in chapter 22 that there is no significant difference between job performance of old and young workers. In seeking to identify aspects of jobs that are likely to remain stable or to change over the years, Warr presents a four-category framework of job activity.

References

Baars, J. 1988. De sociale constitutie van de ouderdom (The social constitution of old age). In C. P. M. Knipscheer, J. Baars and M. Severijns (eds) *Uitzicht op ouder worden: een verkenning van nieuwe rollen (A view of growing old: an exploration of new roles)* (Van Gorcum, Assen), 21–36.

Boerlijst, J. G., Heijden, B. I. J. M. van and Assen, A. van 1993. *Veertig-plussers in de onderneming (Over-forties in Organizations)* (Van Gorcum, Assen).

Kohli, M. 1985. Die Institutionalisiering des Lebenslaufs (The institutionalizing of the course of life), *Kölner Zeitschrift für Soziologie und Sozialpsychologie*, **37**: 1–29.

Schuyt, T. and Klinkenberg, T. van de 1988. Helpen en de zuigkracht van de macht (Helping and the pull of power). In C. P. M. Knipscheer, J. Baars and M. Severijns (eds) *Uitzicht op ouder worden: een verkenning van nieuwe rollen (A View of Growing Old: An Exploration of New Roles)* (Van Gorcum, Assen), 53–64.

Thijssen, J. G. L. 1988. Bedrÿfsopleidingen als werkterrein (In-company training as a field of work). VUGA, Den Haag)

17

Motivational and cultural factors underlying dysfunctioning of older employees

M. J. Schabracq

Summary. The way elderly people are seen and treated in the work environment is determined partly by motivational and cultural factors. In the present chapter, attention is paid to 'concentration of experience', the overspecialization of the elderly worker. Plausible explanations are his or her reluctance to change and the continuous, rigid way the work is done. The restricted specialization of the elderly worker on only some aspects of the job is further strengthened by a slow though ongoing decay of physical and mental abilities. This change may enforce the rigidity in behaviour and induce a loss of flexibility to adapt to changing work conditions and the introduction of new technologies. The consequential loss of interest and the associated demotivation form a serious threat to the health of the elderly worker and to the work organization as well. Although such a situation may develop in both men and women, the latter more often than men may suffer from the so-called 'glass-ceiling effect'. The glass-ceiling phenomenon is characterized by a block, imposed by the organization itself, to the attainment of the appropriate occupational level in the organizational hierarchy, in spite of the proven capability and aspirations of the employee involved. Several other factors that contribute to the health-threatening dysfunctioning of elderly workers are discussed, such as lack of formal schooling and opportunities for additional training, the decay of social networks within the work environment and, as is true for some work environments, a lack of autonomy of the elderly worker owing to the way the production process is organized.

Keywords: concentration of experience, glass-ceiling phenomenon, lack of training, aims of organization, personal work objectives, work stress, health, cultural factors.

Introduction

Cicero mentions in his *De senectū* (*About old age*) several qualities of old age which make this life stage, supposedly, an unhappy one. Old age interferes with the management of one's affairs; it weakens the body; it deprives one of almost all bodily pleasures; and it brings one close to death. Having described those qualities he shows that these ideas are mistaken. Older people should rely more on their skill and consideration, should give themselves other tasks, should enjoy and use the freedom that results from the waning of carnal desire, and they should not worry about dying: dying can occur at every age, and imminent death can be a good reason to use the remaining years in a better-considered and more meaningful way. Taken together, the elderly should be wiser and more virtuous. Though evidently sound advice, this line of reasoning rather easily passes over the problems that confront aging employees in their everyday reality.

In everyday life, people differ in the way they grow older. Some of the elderly do develop a form of wisdom that enables them to cope graciously with changing circumstances, while at the same time keeping up the capacity to pursue the goals that matter to them. They show the proverbial kind of wisdom that is supposed to come with the years. Other elderly people, however, feel that it becomes increasingly difficult to adapt to the changes to which they are exposed.

Unfortunately, within organizations the second category of elderly people, the ones who adapt less easily, has come to determine the overall picture. Tendencies exist to treat older employees as somewhat unwanted persons. As such, the elderly are not seriously opposed when they want to leave. Personnel managers often consider their older employees as sources of problems. They are thought to be less productive, less flexible and more liable to fatigue, stress and illness. Though partly an obvious case of stereotypical exaggeration, it is also known that some older employees do experience serious difficulties in these areas.

This chapter is concerned with underlying motivational and cultural factors, leaving aside all somatic factors, notwithstanding their evident importance in this respect. For these somatic factors I refer to part I.

Describing the difficulties of and with older employees may be a risky undertaking. It seems only one step away from passing the responsibility for these problems on to the older employees involved. Obviously, this is not what is intended here. As a matter of fact, the two factors which are presented as the most influential ones, the emphasis on change in our culture and the way in which we organize our production processes are purely cultural in nature, for which no individual employee can be held responsible.

As mentioned before, the factors to be described are partly motivational and partly cultural in nature. These two kinds of factors are difficult to separate, and I make no attempt to do so.

'Concentration of experience'

Employees show a tendency to become specialists at their jobs. They invest in the development of necessary skills and become good performers. This is, of course, a highly conceivable and normal process: it is nice to be good at something. It provides employees with a stable repertoire of successful behaviour, adequate thoughts and comfortable feelings, which are utterly their own. In short, they become one with their jobs: their jobs become part of the integrity of their functioning. When their jobs are questioned, they hold on to them. And, if necessary, they are inclined to defend them. As such, this is a familiar and seemingly harmless series of events.

As employees grow older in their jobs, however, the process of specialization loses its seemingly harmless quality. Their knowledge and skills only develop within the narrow limits of their specific jobs; they become overspecialized, a process called 'concentration of experience' (Thijssen 1988). At the same time, capacities necessary for the development of skills outside the specific job domain atrophy and waste away. Furthermore, thoughts of possible loss activate an emotional process (Frijda 1986).

Generally speaking, the effects of this 'concentration of experience' process are enhanced with aging. The process fits in with the relatively strong needs for conservation existent in older employees: as the younger are more focused on change, progress and advancement, the elderly are, on average, more directed towards the maintenance of a stable present.

In addition, when employees grow old in their jobs it is as if they become less visible to their superiors. Except for the familiarity of their presence, this also has to do with the work ethics of the elderly: they work hard, are not very critical of authority, do not complain much, and are not absent frequently. As a result their being there becomes a self-evident datum, similar to the presence of a familiar piece of furniture.

'Concentration of experience' has some important drawbacks.

(1) Careers of persons who are subject to 'concentration of experience' show a definite ceiling effect: they cannot rise higher than their jobs or, in the best of cases, not higher than the top of their department.
(2) Employees become less mobile and flexible when it comes to transferring them to other jobs or departments, even within the same organization. This effect becomes especially consequential when the organization changes in a way that affects their positions or when these jobs disappear altogether.

As a consequence of these two processes, older employees often end up being stuck in their jobs. This state of affairs holds a serious threat for the well-being and health of those directly involved: working in the same, unchanging job for a long time can easily lead to qualitative task underload.

Such a chain of events has several links:

—As the job does not any longer appeal to the capacities of the person involved, it becomes less challenging. The employee becomes bored and feels somewhat alienated. It is as if the meanings inherent in the job dissolve. Still, he or she is supposed to pay attention to the task. This demands a form of forced attention, which can become very straining and tiring. Concentration disturbances may arise: suddenly the task seems less interesting than everything else. Emotional processes are activated, aggravating the situation further. In the long run this may turn into a pernicious stress process, which interferes with the employee's well-being and health. Processes such as these are responsible for a large amount of sick leave, particularly among white-collar workers.

—When the employees involved have made some advances within their department, the effects of the 'concentration of experience' process are even more enhanced, particularly when these advances have led to a substantially higher income. As it simply becomes impossible to earn the same amount of money somewhere else, this leads to further reduction of mobility. This so-called 'golden cage syndrome' may occur in combination with 'structural plateauing': the impossibility of further advancement owing to a lack of room in the organization. The problem can be aggravated by non-transferable pension rights; a valid reason to stay, even with a preferable job available elsewhere.

—A related problem is created by pension schemes in which the amount of the ultimate payment depends on the salary earned during the last working years. Schemes like these effectively block the acceptance of a less well paid, but less demanding job for the last working years.

—'Automatic' promotion policies may form a source of trouble in the later stages of careers. In some organizations employees are promoted time after time, eventually beyond their capability (Peter's principle). Elsewhere, incompetent employees are promoted to a formally higher, but in fact empty function (being 'kicked upstairs').

The essence is that people grow old in functions that do not appeal to them (any longer), a state of affairs that represents serious health risks.

The 'glass-ceiling' phenomenon

Though nowadays more women are employed and though some have made significant advancement towards higher organization levels, female employees are confronted by much resistance from other organization members when they try to advance above a certain level in the organizational hierarchy. In addition, many women employ tactics and courses of action that can be characterized as somewhat self-defeating. As a result, many female employees can clearly see the levels above them that they want to reach, without being able to do so. Such a predicament is called the 'glass-ceiling' phenomenon.

Essentially, all of this has to do with assumptions concerning the division of power and the division of roles between men and women. These assumptions have a limiting effect on what is possible, thinkable and speakable. (Schabracq

1991). For the greater part, these assumptions operate at a subconscious level, in men and, partly, in women as well (Meyer 1983, Davis 1988) and lead to various self-enhancing processes and distorted, stereotypical perceptions in everyday reality.

The 'glass-ceiling' effect often interferes even more with the health and well-being of female employees than the effects of mere 'concentration of experience'. The point is that those involved often have both the capability and the ambition to rise to a higher organizational level. At school, and even in the organization, they are told that everyone, male or female, has the same opportunities: just a matter of hard work and achievement. So, they work hard, often very hard, to find out that their achievements are not as important as they were told that they would be. In addition, these women see how their male colleagues and subordinates often are making much quicker progress in their careers than they themselves are making. Many of these female employees become frustrated, feel themselves betrayed, become cynical and show all kinds of burn-out symptoms (Schaufeli in press).

All these ways of getting stuck in inappropriate and unsatisfactory jobs have detrimental consequences, for the employees involved as well as for the organization. As the fit between the job demands and the capacities of the employee is poor, without much chance of a fundamental change, these jobs may activate serious stress processes and may induce damage to health. The organization becomes bewildered by an impairment of overall flexibility, costs of inefficient functioning, sick leave and the replacement of the employees involved. In addition, younger employees see the advancement of their careers blocked and may leave the organization to try their luck elsewhere.

Lack of formal education

Compared with younger employees, relatively many older employees show a low level of formal schooling. This applies particularly to the oldest category, born in the late 1920s and 1930s, as well as to the older immigrants from economically less developed countries.

A low level of formal schooling often goes together with a poor economic background and lack of possibilities in many other areas. People from such a background do not dispose of the cultural capital (social skills, attitudes, ambition, models; Bourdieu 1986) needed for the successful capturing and keeping of an interesting job.

Such a state can start a self-reinforcing process. People with a low level of schooling get uninteresting jobs without many possibilities of personal development. Consequently, they are less motivated, which makes the job even less interesting and so on. At the same time people with a high level of schooling can get highly stimulating jobs, which offer plenty of opportunities for further personal development. As a result, the gap between the possibilities for both groups widens more and more (Kohn and Schooler 1983, Schaie and Schooler 1989, Lee 1991). In addition, a poor economic background is often visible (style of clothing, speech style, dialect, nonverbal behaviour). Employees displaying such a background, from the very start, are treated by superiors as if they were inferior, verifying the self-fulfilling prophecy of poor functioning (Vrugt and Schabracq 1991). This process results in a considerable group of poorly schooled and poorly motivated workers. This is even more pronounced when they grow older, because a lot of work that they have been doing has often been very damaging to their health, resulting in a lot of wear and tear on their physical make-up.

Lack of additional training

As Boerlijst describes in chapter 18, many older employees get little or no additional training. At the same time, technology at the workplace is changing faster than ever, and so are the ways in which work is organized. Computers have been introduced and production processes are reorganized frequently. Moreover, it is known that elderly people have difficulties keeping up with these changes. As a result the skills of older employees are becoming obsolete at a fast rate. Although the obvious thing to do is to give these older employees additional training, the question arises as to why this does not happen on a wide scale.

Partly, the omission of additional training is a matter of comparable self-fulfilling prophecies (see above). This is not only a matter of near sighted management. Among older employees there does exist also a certain persistent reluctance against additional training. Several motives can be mentioned.

—Older employees often have learned to mistrust the motives of the management of the organization. Though in itself clearly not an irrational line of thought, it obscures the fact that additional training could be to the benefit of the employee.

—Additional training is often seen as a start of some change in work design and in the work to be done. Introducing change without consulting the elderly worker is often conceived as a threat to their integrity of functioning and may evoke emotional and tenacious resistance. As a consequence these people are hardly motivated to follow the needed additional training. They feel that there is nothing to be gained by it and consider it an extra, unpaid for, work effort, which the organization is not allowed to ask from them. Often such resistance is a group feeling: old hands supporting each other in resisting the change forced on them.

—Sometimes older employees, rightly or not, see themselves as lagging so far behind that it will be impossible to catch up. Lack of formal schooling and a low perceived self-efficacy in learning are of relevance here.

—Other motives have to do with the nature of the training itself and the way it is given: not adapted to older employees, in conflict with their value systems, seen as completely useless and with insufficient practice opportunities (see chapters 21 and 22).

A lasting lack of additional training will initially result in some form of qualitative overload: the work has become too difficult. Mostly, there will be someone around for the new difficult parts of the task, leaving the older employee with some easy, but not so interesting work. Such impoverished function, stripped of much of its meaning, will lead eventually to a situation of qualitative, if not also quantitative underload, which may provoke stress.

Aging of networks

When people work in an organization they develop a social network. Personal preferences and contacts may evolve in personal relations and friendships. One finds oneself protectors, favourite subordinates, colleagues and favourite contacts elsewhere, inside and outside the organization. Such a network is very useful for giving and receiving information, warnings, advice, emotional support and factual help. A good network is important for the development and maintenance of a power position in the organization. It helps when it comes to promotion, and it can also help to prevent dismissal. Moreover, evidence shows that a good network helps to prevent the activation of stress processes, that it makes stress less harmful and that it facilitates the recovery from stress-related complaints.

Social networks however, decay over the years. People leave, move elsewhere, they die: older employees are confronted with the loss of emotional support of the ones who are important for their position. Given that many elderly people do not make new friends as easily as when they were younger, it implies that the positions of older employees may become more isolated. As a consequence, they lose the yields of social networks. This loss is an emotional one, involving feelings like grief, anxiety, depression and bereavement, feelings that can interfere with their functioning. In addition, this process may negatively affect the communication within the organization.

Change of personal work objectives

When employees grow old their outlook on life becomes different. Personal work objectives change and may interfere with organizational interests.

Many older employees are not so focused on increasing their income as younger ones are. A further substantial increase is unlikely, and a stabilized salary-adapted lifestyle has been developed. Generally speaking, many older employees have grown tired of postponing their rewards to an infinitely regressing future. They have learned that a further progression in their careers is limited, realize that the time to go is finite and makes them focus less on objectives like increases in money and power. By implication, they are more directed to the present. For these reasons, they demand more from their present jobs in terms of meaning and usefulness, freedom to arrange their work, and the quality of their work life, including its social contacts.

As their work goals have become more focused on the present, there is a tendency to become more conservative. Consequently, they often are rather weary of the prospect of any organizational change and the required adaptation.

Unfortunately, many jobs do not have much to offer when it comes to the realization of such objectives like freedom and meaningfulness of tasks and products. When older employees realize that the actualization of objectives within the work context is too difficult, they usually turn more to their leisure-time activities for their rewards. In the view of the organization, these older employees show signs of demotivation.

All in all, one can say that, by changing the work objectives, older employees exhibit the kind of wisdom that is supposed to come with the years.

Managerial uneasiness

Many young managers experience some uneasiness in dealing with their older employees. Since older employees have often worked for a long time for the organization, they have built up a certain position. They have acquired privileges, have a thorough knowledge of the organization and are not as easily impressed or fooled by management schemes as younger employees are. Moreover, as they have probably received several pay rises, they earn relatively high wages. Their seniority, their proven contributions to the organization and their age entitle them to a certain respect from younger managers.

The due respect makes it difficult for younger managers to deal with the problems of and with older employees in the same ways as they are used to with younger personnel. Disciplining and correcting older employees can activate feelings of uneasiness and guilt. Managers may avoid doing this and, consequently, the older employees involved get little or no feedback about their performance and by that are exposed to a greater risk of role ambiguity and isolation.

Generation gap

Employees over the age of 45 are from different generations than employees under 35. As the first groups grew up in the era of the great depression, the second world war and the poverty of the first post-war years, the second group grew up in relative peace and wealth. In addition, during their formative periods, say from their tenth to their twenty-fifth birthday, these groups have been exposed to completely different realities, ruled by different value orientations. As a consequence, most employees of these groups have different outlooks on life, which are not always compatible. In order to support this view, some of those differences are described below.

—In general, older employees have been exposed to and have experienced more hardship and have had more models that taught them to cope by perseverence and persistence. So they seem more hardened than the younger ones who have much less experience with these matters. On average, older employees carry on longer when they do not feel too well. This is confirmed by statistics, showing a lower frequency of sick leave than younger employees. This is also a matter of honour and pride to them. Being independent and self-supporting means a lot to people who themselves or whose parents were witnesses or victims of the depression and the second world war. However, considering the equally well-known fact that when elderly workers are sick, the mean duration of sick leave periods is substantially longer, one can question the wisdom of this way of acting.

—The other side of this form of hardiness is that many older (male) employees often show more reluctance in showing emotions and, as they see it signs of weakness compared to employees from the younger generations.

—Individuality and identity are different concepts for the elderly and the younger generations. The older employees grew up in a world where they were born into an extended family, a social class, a church and a political party. Although the importance of those institutions has declined, for an important part by the doings of members of the older generations themselves, they still tend to live by the rules and values of these institutions. At the same time they are surrounded by younger people who have learned to make choices concerning their identity, such as for education and career, religion, sexual identity, social relations, life style. etc. This adds up to a more segmented or situated life, with rules that are valid in the chosen specific life domains. Having to choose one's own way of living presupposes that one lives in a world where these things are not fixed beforehand, within such a world such an approach is not so much a privilege, choosing one's own destiny has become a more or less necessary course to take in order to cope adequately with the ongoing changes of everyday. This proves to be a difficult point to accept for older employees whose outlook on life is shaped by a world in which fixed forms and meanings were much more prevalent, a world that did not ask for these kind of choices. Leading a life based on personal choices and preferences implies that one is more responsible for one's life, such as for example for one's own health and well-being and the actualization of one's possibilities in work as well as leisure time. Such a general attitude to life can be an effective way to adapt quickly to changing circumstances. However, such an attitude appears to be difficult to learn and to accept by older employees. Besides, they often have some vague feelings about such an approach; though obviously not a forbidden course of action, it is not felt to be completely right either.

—The division of roles between men and women is a cause of different opinions between employees of the older and younger generations. Though men and women are still far from being evenly represented in the higher levels of organizations (see the section on the 'glass ceiling'), it is nowadays not entirely uncommon to come across a female superior. On average however this is a rather recent phenomenon. Again, on average, older employees, male and female, find it more difficult to work under a female superior, especially when younger. Older subordinates often experience this as a threat. This threat goes back upon the implicit, deeply anchored assumptions, which were already mentioned, about the nature of reality. According to these assumptions, a female superior is a phenomenon which goes against the way in which reality is, or should be, constructed.

As for their attitude to authority, older employees have, on average, learned to accept authority without much questioning. Though most of the older employees have changed considerably in this respect, they still expect, to a certain degree, a corresponding treatment from their subordinates, and sometimes find it difficult to interact with younger employees who, in their opinion, always comment on what they have to do.

Age intrinsically deserves respect, just as authority does. Older employees often feel that they do not get enough respect from their younger colleagues, notwithstanding the fact that they already have proved their merits for a long time. The changed attitude to age is also related to the speed of society's change. In a time when change was less prominent and less fast than it is nowadays, old people were in a position to be wise and sensible. Nowadays the world changes so fast that older people lose their grip. Except for a few extraordinary individuals, they are at best seen as people trying their best to follow the developments.

Organization of production processes

In western cultures it is a normal way of thinking to make a strict distinction between goals and means. Subsequently, a rather absolute primacy is given to the goals: they justify the means.

One of the consequences of this distinction is that it is often tempting for those in charge to pay little attention to undesired side-effects of the used means. When nothing goes seriously wrong, they are not inclined to adjust these means; it would only distract them. The goal also is too important to act that way: one cannot make an omelette without breaking eggs. This way of thinking has been of crucial importance in the shaping of our production processes, industrial and otherwise, and becomes strengthened when the persons in charge are not present at the place where the production process and its side-effects take place. Being elsewhere, they are not in a position to perceive these side-effects. As a consequence, the process of realization of the goal becomes disconnected from restraining negative feedback from the environment, which may express itself in a certain indifference to the well-being of those directly involved in production.

This dissociation can become even more drastic when the persons in charge only are the owners of (part of) the means of production, without firm roots in the specific production process and without much affiliation for it. Moreover, such persons often see themselves as completely different from those who are executing the production process, thereby widening the (psychological) gap between themselves with the people on the work-floor. Although many things have been changed since the industrial revolution, certain principles introduced during the industrial revolution have remained influential.

For example the owners of the production means still determine the definite form and nature of the primary goal of the production process: the choice of the products to make. The next step also has not changed: the design of a production process, which consists of applying means to this end as effectively and efficiently as possible, though within the limits set by legislation.

The most important solutions that have been developed in our western world centre around a very strict division of labour, in which the manufacturing of one product is divided into a great number of separate, relatively simple and well-specified acts. Though this division of labour goes back a long time, its present forms have been mainly developed during the industrial revolution. Before that time production processes were also characterized by a definite division of tasks, but this one was of an entirely different quality. Before the industrial revolution the one in charge of production usually was the owner, most of the time a somewhat older person who was well trained and particularly skilful in the most difficult parts of the production process. The division of tasks was aimed at delegating the more easy and monotonous tasks to the younger and less well skilled workers in order to give the more well-trained ones the freedom to occupy themselves with the more difficult tasks. At the same time an apprenticeship system existed, aiming at educating the less well trained younger ones to gain higher skill levels. Obviously, such a system is much more favourable to the elderly than our present systems, even when sometimes the resulting production process resembled an assembly line, as it was the case in the 'Arsenale', the shipyards of Venice, during the sixteenth century.

The introduction of more extreme forms of division of the production process into a sequence of different acts, each performed by another person, served several purposes.

One of the first considerations to introduce such a division was one of standardization of the goods to be assembled. In this way it was possible to manufacture a completely standardized product, assembled from standardized modules, which in their turn were assembled from uniform parts. In this manner the product quality could be guaranteed and repairs could be easily done by replacing some parts.

Later on, when the mechanization of production automatically lead to more standardized products, the reason behind the more extreme forms of division of the production process changed, only to become more influential. The most important reason was that having to learn one simple act is much easier and takes much less time than having to learn a complicated skill. Its importance lies in three of its consequences.

—It sufficed to hire unschooled workers. Since nearly everyone can act as an unschooled worker, this implies that, given the 'right' circumstances, the availability of this kind of worker is almost guaranteed and their payment can be low.

—The fact that everyone has to do simple tasks implies that the organization does not need to invest in the training of new workers.

—By allowing people to do only one thing, they become easily exchangeable. In such a system no worker has any skills that make him or her indispensable. If a worker causes any trouble, (s)he can easily be fired and be replaced. So, fragmenting a production process into a series of simple acts, each to be performed by a different person, is foremost a power tactic.

This last aspect, control by the one in charge of the organization by diminishing the power of the individual worker is an important feature of many production processes. It is also a logical feature, since the designers of the production process knew well that it is not always pleasant to work in such a process. So, they knew that they could not depend on the wholehearted and voluntary cooperation of their workers.

A consequence of this institutionalized mutual low trust is a serious imbalance in the distribution of decision latitude. As the persons on the work-floor are not allowed to think for themselves, management is responsible for everything. Though managers are by no means the designated experts and often do not possess all the right information, they do have to make all the decisions. Their decision latitude is, in a way, too great. Except that this can easily activate a serious stress process in the managers involved, this can also lead to wrong decisions. Such faults are avoidable, because the required information is present on the work-floor, available to anyone who cares to listen to the people working there.

On the other hand, workers at the work floor are given little to none job decision latitude. They often have no control at all over the way their (simple) tasks are arranged. This holds in particular for employees in production processes which are characterized by short-cycled, often machine-paced tasks.

This can lead to a lot of time pressure for individual (older) employees who experience difficulties in keeping up with the pace. Also, the tasks have, as we have seen, little intrinsic meaning to the employees involved as they do not appeal to their specific skills and capabilities. Nor do these tasks in any way invite the employees involved to develop other skills. As such, many of these tasks result in a combination of serious qualitative under-load and quantitative overload.

Such a combination of work characteristics is particularly damaging to the worker's well-being and health (see, for example, Karasek and Theorell 1990). The impact on older employees is even more harmful because of their greater physical vulnerability, the decay of processing capabilities and their specific motivational characteristics (for example, their greater reluctance to do meaningless work, see the section on change of personal work objectives).

Since older employees are, on average, exposed to such harmful working circumstances for a longer timespan, they are more burdened with the wear and tear resulting from it. Continuous repetition of short, one-sided tasks, which always tax the same parts of their bodies, is likely to result in serious physical damage in the long run. At the same time, owing to disuse of their remaining physical and psychological skills, various forms of atrophy may occur.

The increase in change

The last few decades have been characterized by a rapid acceleration of societal, cultural, technological and economic change. In order to survive, organizations have to adapt to these changes. This implies organizations themselves have to change too. As we have seen before, older employees have more difficulties with this kind of change than younger ones.

Changes in an organizational culture imply changes in the prominence and priority of certain values and objectives. Objectives such as quantity of output, flexibility and focusing on certain markets may become more prominent, while values like technical perfection and respect for the professional freedom of employees may lose their dominant roles. However, these last objectives may well have been the reason why the somewhat older employees joined that particular organization. The change in organizational values leaves them with work which, in their eyes, has been stripped of most of its challenges and intrinsic meaning. Besides this loss and the resulting qualitative underload and stress, the demands to make drastic role transitions and to learn new skills are not appealing.

The emphasis on change is a peculiar feature of our culture. As most cultures are structured in order to provide their members with a stable environment, our culture has given its members environments which are changing at an accelerating rate.[1] A consequence is the phenomenon that the present generation sees the former generation as somewhat lagging behind and as old-fashioned. A second consequence is that the cultural forms, missions and role models, which parents and instructors have given us, are insufficiently geared to the present environment.

Another problem that many older employees experience has to do with the content of much organizational change. Organizations nowadays have to change to horizontal ones; to get more flexible, less product-oriented, more market-oriented, more transparent, more quality-oriented and so on. Their employees have to become more autonomous, more decisive, more creative and more entrepreneurial. They have to see change as the normal state of things. In order to accomplish this, they are expected to use all of their powers to the fullest. In fact they are asked to behave as if they were completely free beings, who are working for themselves, and not for an organisation owned by other persons. Older employees have problems with this kind of reasoning:

(1) They see the flaw in its logic. The adage 'be free and serve us' appears to remain more of a paradox to them than to younger employees.
(2) Older persons have learned to be suspicious of the intentions of those in charge of the company, and they have often learned it the hard way.
(3) During their work life they have effectively and efficiently unlearned how to act in the way that is now asked of them. As we saw in the last section, much of the organization of the production processes aims at exactly the opposite, namely to disempower the individual worker.

Conclusion

Most of the problems that older employees experience can be translated into a shortage or a loss of power of these employees. Their individual power becomes smaller: either because of the limiting of their decision latitude or because of the decay of their competence, skills, developmental potential and other resources. At the same time, a process of stereotyping confirms, reinforces and even legitimizes their loss of power. As a result they are, erroneously, seen as a more or less uniform group of people. Erroneously because, as older employees, having lived longer, diverging lives, they vary more than their younger colleagues.

A word of caution is appropriate: obviously, this complex amalgam of difficulties leads to a somewhat overly grim picture of trouble. After all, it is a summing-up of problems which leaves out the good things, and not all problems are as severe and insoluble as they look.

In order to be able to alleviate some of these problems, we must acquire a better understanding of the dynamics of the dialectical process of personal and organization-cultural development, and its effects on the functioning and well-being of older employees. Some serious research is needed here. Just as in Cicero's time, wisdom and virtue, especially in their original meanings of strength and courage, appear to remain the most important assets for older employees. As will be elaborated further in chapter 21, interventions should capitalize on the realization of individual opportunities to use these assets.

Notes

1. These changes can be described as consequences of the way in which we experience time. Progress, growth and our emphasis on being goal-directed and functionality are key words here, features which I have described elsewhere (Schabracq 1991).

References

Bourdieu, P. 1986. The forms of capital. In J. G. Richardson (ed). *Handbook of Theory and Research for the Sociology of Education* (Greenwood Press, New York).

Davies, K. E. 1988. *Power under the Microscope. Toward a Grounded Theory of Gender Relations in Medical Encounters* (Ph.D. thesis) (Foris Publications, Amsterdam).

Frijda, N. H. 1986. *The emotions* (Cambridge University Press, Cambridge).

Karasek, R. A. and Theorell, T. 1990. *Healthy Work. Stress, Productivity and the Reconstruction of Working Life* (Basic Books, New York).

Kohn, M. L. and Schooler, C. 1983. *Work and Personality: A Study. An Inquiry into the Impact of Social Stratification* (Ablex, Norwood, NJ).

Lee, J. S. 1991. *Abstraction and Aging. A Social Psychological Analysis* (Springer Verlag, New York).

Meyer, J. 1983. *Sekse als organisatieprincipe. Veranderende normen in de asymmetrische machtsrelatie tussen mannen en vrouwen. (Gender as an organizing principle. Changing norms in the asymmetrical power relation between men and women)* (Ph.D. thesis), Universiteit van Amsterdam, Amsterdam.

Schaie, K. W. and Schooler, C. (eds) 1989. *Social Structure and Aging* (Lawrence Erlbaum, Hillsdale, NJ).

Schabracq, M. J. 1991. *De inrichting van de werkelijkheid (The design of reality)* (Boom, Meppel/Amsterdam).

Schaufeli, W. (in press) *Burnout* (John Wiley, Chichester).

Thijssen, J. G. L. 1988. *Bedrijfsopleidingen als werkterrein (In company training as a field of work)* (Vuga, The Hague).

Vrugt, A. J. and Schabracq, M. J. 1991. *Vanzelfsprekend gedrag. Opstellen over nonverbale communicatie (Behaviour that goes without saying. Essays on nonverbal communication)* (Boom, Meppel/Amsterdam).

18

The neglect of growth and development of employees aged over 40: a managerial and training problem

J. G. Boerlijst

Summary. The consequences of growing older were investigated by comparing three age groups of junior, medior and senior employees aged over 40, who were working in middle and higher functions in 10 large-scale Dutch industries. Central themes were functional utility and learning value, the quality of function, mobility and participation in training and development programmes and the way in which the managerial activities could stimulate the over-40s in these fields. The data indicated that most of the over-40s are not equipped too well for other functions outside their own field of work. Main causes are an age-related lack of relevant training. This is especially true for training courses in other functional fields and personal development. The results show that between 16 and 20 per cent of junior and medior over-40s in a 5-year period did not follow any training programme in their own field at all. This figures doubles for senior over-40s. The necessity of participation in courses was emphasized in view of the benefits for the elderly employee himself as well for the organization. The drawbacks of non-participation in courses are described and since vertical mobility is, in general, not possible for over-40s in middle and higher functions, horizontal mobility and the creation of a continuous learning environment were discussed. Since motivational and other factors may hinder the development of such a stimulating environment, arguments were offered to redesign or restructure functions by participation in courses not only in the employee's own field of specialization but especially in other functional fields.

Keywords: over-40s, middle and higher functions, overspecialization, mobility, participation in courses, concentration of experience, utility value of function, learning value of function.

Introduction

In many European countries the greater proportion of workers aged 55 and over leave the work-force before reaching their 'official' pensionable age. One of the main roots of this phenomenon is the worsening competitive position of these countries compared with cheaper ones in terms of personnel costs. Many organizations feel an urgent need to reduce their manpower to improve their situation. Mostly they try to discharge their older personnel in the first place. The main motives are that these people are less productive and, in terms of salaries, more expensive than their younger colleagues. In other words, their early retirement or dismissal is more profitable. Sometimes it is argued that by sending away older people, room is being made for youngsters who otherwise would fail to find a job.

Since most older people don't seem to reject an earlier close of their work career in principle, and many private and public regulations have been set up that make it acceptable for people, both socially and financially, to quit or to retire voluntarily or to allow their employer to give them notice, there are good conditions for an exodus of aging workers. Figure 18.1 illustrates the example of that exodus in The Netherlands at the end of 1992.

From this figure we can see that only one out of every four people aged 55–64 years still belongs to the active work-force. This situation is completely different for men (40 per cent) and women (only 10 per cent). When we compare these percentages with those for the younger age group of 45–54 years, the differences are striking. There, 64 per cent (male: 87 per cent; female: 40 per cent) are still active on the labour market. The non-working

Total population of age group 55-64:   1.398.000 (end year 1992)

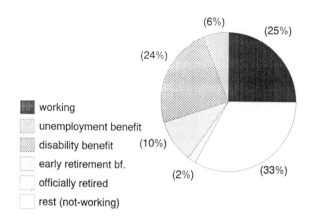

Figure 18.1. Participation and non-participation of older people (55–64 years) in the Dutch labour market.
(Source: Centraal Bureau voor de Statistiek, 1992).

remainder receives for the greater part some sort of benefit from governmental or private funds. In The Netherlands there exist not only many early retirement programmes, but also a network of social security laws and regulations, for instance for handicapped or otherwise unfit people, regardless of their age. Recently a parliamentary inquiry has made clear that this system has been more or less improperly used on a large scale, in order to phase out obsolete or superfluous personnel (older persons mostly).

The financial burden of these benefits rests on the shoulders of the working part of the population. Viewed in the light of demographic developments (aging of the working population), it seems that it will become unbearable rather soon. The way of financing the growing crowd of older non-workers has become an important political issue. To stem the tide one thinks of measures to be taken, for instance reducing allowances and social security payments, or abolishing early retirement provisions and restoring or even postponing the original pensionable age. The last mentioned 'solution' seems attractive, but the main problem is whether or not older people are able to stay economically active after a certain age, that is satisfactorily productive and easily employable. This is a valid problem, since up to now employers seem to have separated from their older employees rather easily, even willingly. Could this be a sign that because of their age they are no longer able to work effectively in the setting of an ever-changing organization? Or is this attitude of employers simply inspired by the above-mentioned economic motives or by irrational prejudices about elderly employees?

Working employees, aged 40 and over, as observed by their supervisors: an empirical study

In a study concerning individual employees of age 40 and over, working in 10 different Dutch work organizations (Shell, AKZO, PTT Telecom, IBM Unilever Research, DSM Research, DHV, Heidemij, De Nederlandsche Bank and the Directorate-General 'Management & Personnel Policy' of the Dutch Home Office), we tried to find out what the over-40s contribute to the organization, and what they will contribute in the future. Further, we asked questions like, 'What can be said about the resilience and versatility of the over-40s?', 'Do this resilience and versatility decline as employees grow older?' and 'Is management able to exert a positive influence on these qualities and, if so, what methods can it employ?'

These questions, in different forms and variations, were put not to the over-40s themselves but to their immediate superiors. In our study we were particularly interested in the standpoints and opinions of management on these matters. We consider the superior to be one of the best sources of information on individual personnel in a company. But, moreover, the supervisor is probably the person best acquainted with *management activities* intended to monitor the tasks and the functions of the individual over-40s in the organization and to optimize the quality of their functioning in his or her department. Ideally, the superior himself plays a key role in this.

As we are particularly interested in the consequences of growing older, we have made a comparison of three successive over-40 age groups, namely 'juniors', 'mediors' (a neologism, meaning something like 'more in the middle') and 'seniors', active at middle or higher levels of functioning within their respective organizations.

In every company or organization we tried to gather information about 100 employees, evenly spread over the age of 40 and over, active at least on a middle level of functioning or in a middle-management position. Besides these two conditions there were no further sampling restrictions, that is sampling could be done on a random basis. We were not in the position to control the randomization process at full scale. Most sampling preparations were done by personnel departments on sampling prescriptions. For this study we combined the available data of all 10 participant organizations. Since some of them didn't succeed in getting 100 protocols and there were no prescriptions made about the division between the two function levels, the available sample totalled 738, and it comprised fewer employees on the middle functioning level $n_{middle} = 308$) than on the higher one ($n_{higher} = 430$).

The reason we have restricted ourselves to the over-40s with functions at a middle and higher level is as follows. Looking for study data that can be generalized for application in the future, we made allowance for the possibility that the present over-40s, particularly the mediors and seniors, will be difficult to compare, on one point at least, with the over-40s who will be populating our companies in 20 years' time. Until 20 years ago simple functions and simple tasks were dominant in most work organizations. As a consequence, the bulk of over-40s in our existing work-force has a rather low level of education. As the complexity and level of difficulty of future functions will on average be higher than they are now, we have every reason to expect that the average educational level of the over-40s will likewise have undergone a sharp rise by the year 2010.

It is generally assumed that resilience in the face of changes, adaptability, versatility and so on depend, among other things, on a person's learning capacities. If this assumption is valid, then we may expect that the relatively

Table 18.1. Over-40s employees active in middle and higher functions.

		Middle-level function (n)	Higher-level function (n)
Juniors	40–46 years	100	161
Mediors	47–52 years	91	148
Seniors	53 years and older	117	121

better educated older employees of the future will be more resilient than those of the present time. The number of employees with a lower vocational education will, in any case, decrease relatively sharply and there will be an increase in those with an intermediate and higher vocational education as well as academics. To increase the generalizability of these study results for use in the future, it seemed to us preferable to restrict our observations to those bearing a closer resemblance, in terms of level and capacities, to the average over-40s of the future.

When analysing the study data, we made a distinction between employees at middle function levels and employees at higher levels. The contributions made by the over-40s with an intermediate vocational educational background are generally of a different nature than those who function at a higher vocational educational or an academic level. In this connection, we were particularly interested in whether management makes a significant distinction between the two levels, for example in the evaluation of their functions and their functioning. In addition to this, we investigated whether there is any difference in the management's approach to and treatment of these two categories.

The 10 organizations participating in the study occupy a prominent position in our country, but they do not, of course, offer a representative picture of the Dutch business community as a whole. Our impression is, however, that there are certain relationships within the personnel sphere of activities which are found to operate in virtually the same manner in every company, possibly to a greater extent in one case than in another. For this reason, we saw no objection to combining the observations on over-40s and their executive management in the 10 companies into one whole. This, then, allowed us to test, with a fairly large margin of certainty, certain consequences we had previously formulated on the over-40s aging process against their reality content. In so doing, we chiefly made use of an ANOVA design.

Central themes

The central themes in our study were:

— how does the value of the functions occupied by over-40s, at intermediate and higher levels in the organization, develop as they grow older?
— how do the quality of the functioning of these over-40s and the influence which they exert inside and outside the organization develop?
— what can be said about their mobility in the sense of transitions to other functions?
— what is their position with regard to their participation in social networks inside and outside the organization and what is their role in this?
— to what extent and in what field do they participate in training and development programmes?

As already stated, we were particularly interested in the extent to which certain management activities can stimulate the development of the over-40s in these fields.

Within the narrow confines of this paper, we can only deal very superficially with the findings of our study. (For details see Boerlijst et al. 1993). So we shall limit ourselves to a few highlights an dwell somewhat longer on the question of training and development and matters relating.

The utility and the learning value of the function

We have assigned two value aspects to functions:

(1) The value which the function has within the framework of the organization or, in a smaller context, the department in which the function is positioned. From now on we will designate this aspect as the 'utility value' of the function for the organization or department, respectively. By 'function' we mean the total package of tasks and activities, responsibilities and obligations, as well as the attendant results, with which an employee in the organization has been delegated, or which he/she has taken on him/herself with the knowledge of the management. (Some employees 'create' their own function.)

(2) The value which the function has as an ingredient for the employee's further development. This aspect will be termed the 'learning value' of the function for the employee.

Figure 18.2 contains an overview of the utility value of the functions exerted by over-40s. Of every function level and of every age group it gives the percentage of those whose function is indispensable or rather essential for their organization. The table underneath contains the ANOVA results.

In general we can conclude from these data that the utility value of the functions or positions held by over-40s is by no means that bad. Particularly on the higher level only a minority of these functions can be missed. With the increase in age, a slight decline in mean utility value is noticeable, but certainly not a dramatically large one.

What is more of a problem is the relatively low learning value of the functions for development of new expertise, especially at the middle level. In the next two figures, 18.3 and 18.4, we can demonstrate this.

Table 18.2. Over-40s whose function is indispensable or rather essential for their organization..

Source of variation	df	F-value	p-value
Age level	2	9.40	<0.01
Function level	1	5.80	<0.01
Age level × function level	2	0.23	ns

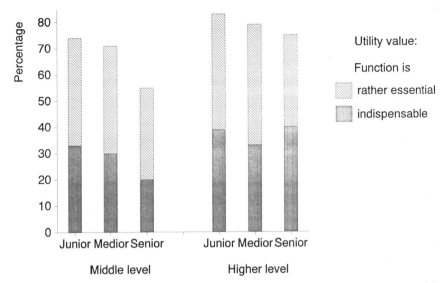

Figure 18.2. Percentage of over-40s whose function is indispensable or rather essential for their organization.

Figure 18.3 shows that the percentage of employees with a function offering too few opportunities for acquiring new learning experiences and, more specifically, for learning new expertise, is higher at the middle level than in the higher functions. At both levels the seniors are worse off than their younger colleagues. In a relatively large number of cases, the function does in principle offer opportunities for growth in expertise, but there are impediments or thwarting circumstances. Unfavourable working conditions and being allowed too little time to extend him/herself beyond the scope of the 'normal' work, tend to be to the disadvantage of many employees. In practice, nothing can then come of what might have been a development. Such 'impediments' occur more often at the middle level. This category also covers those cases of the employees themselves not taking advantage of the potentially present opportunities offered by their function. In their superior's opinion, they do not make any particular efforts themselves.

Table 18.3. Over-40s whose function is not a good ingredient for developing new expertise or for whom there are serious impediments or thwarting circumstances.

Source of variation	df	Function is not a good ingredient		df	Function is not a good ingredient or serious impediments or thwarting circumstances	
		F-value	*p*		*F*-value	*p*
Age level	2	6.63	<0.01	2	8.75	<0.01
Function level	1	58.01	<0.01	1	81.15	<0.01
Age level × function level	2	1.18	ns	2	2.96	ns

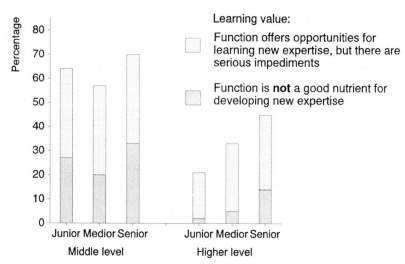

*Figure 18.3. Percentage of over-40s whose function is **not** a good ingredient for new expertise.*

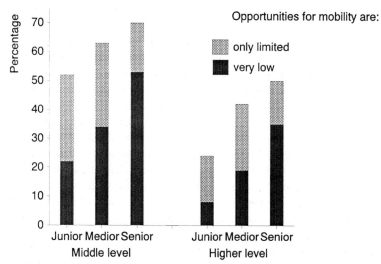

Figure 18.4. Percentage of over-40s whose function offers only limited or low opportunities for mobility to other function(s).

Table 18.4. Over-40s whose function offers very low or only limited opportunities for mobility to other functions.

Source of variation	df	F-value	p-value
Age level	2	14.17	<0.01
Function level	1	41.64	<0.01
Age level × function level	2	0.45	ns

Another important qualitative aspect is the question of the extent to which the function offers an opportunity for 'mobility' to other functions. In figure 18.4 we will show the negative side: the percentage of functions which offer only a limited opportunity for such an interchange, or no opportunity at all.

Figure 18.4 shows that if the opportunities for circulation are reasonable at the higher level, those at the middle functions are on average substantially smaller. In so far as middle-level functions provide opportunities, these mainly relate to functions within the employee's own company. There is far less likelihood of an interchange to other positions in neighbouring companies within the same concern or outside it at this level than in the case of higher functions.

As far as age differences are concerned, the picture varies. The functions of seniors often, though not on all points, offer fewer opportunities than those of mediors or juniors. The last are mostly the best off.

The quality of functioning

Within the circle of top management of companies there is a widespread stereotypical opinion that functioning of subordinates diminishes in quality as they grow older. On the other hand, findings from gerontological studies reveal that people who are not afflicted by any illness, handicap or stress, can continue to carry out their normal work for a long time without its undergoing any appreciable decline in quality (Warr, 1993). We therefore expected that the actual quality of the functioning perceived by superiors would not, or at most only partly correspond to the previously mentioned stereotype. As one can see from figure 18.5, this expectation is justified.

As a rule, the quality of the functioning of over-40s meets the norms set by the superior. This applies equally well to the effect or influence of the work performed by the over-40s inside and outside the organization. In so far as there are any age differences, these are not dramatically large. In some cases, the seniors, too, do just as well as their younger over-40 colleagues.

There are, however, systematic differences between both function levels. Employees at the middle level on average achieve a slightly lower score than those at the higher level. Here too, however, one can state: not much lower. It can further be said that the work performed by the over-40s is mainly oriented to the employee's own organization. In general, it has a greater effect within the organization than outside it.

Mobility

The mobility of over-40s, in the sense of transition to other functions or positions, is low. Most of the over-40s are 'stayers'. In figure 18.6, the development of mobility is portrayed succinctly. In this illustration, information on the length of time that the over-40s have occupied their present functions has been combined with their superiors' expectations as to their mobility over the next

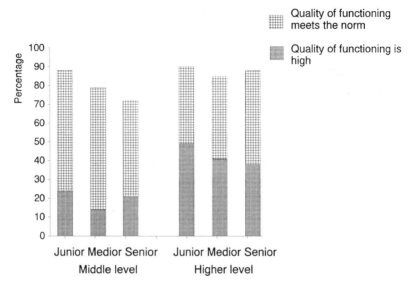

Figure 18.5. Percentage of over-40s whose quality of functioning is high or meets the norm.

5 years. Each circle represents an age category and is proportionally divided into six segments.

As one can see, in the case of juniors there is already a substantial degree of immobility, but seniors, in particular, tend to be marking time. Other data from this study, not shown here (Boerlijst, van der Heijden and Van Assen, 1993), demonstrate that in so far as there is any question of employability, this is restricted to a change of function in the person's own, already familiar territory and he/she stays as close to home as possible. Internal mobility outside the immediate circle of that employee's own company or company segment is scarcely present at all. The opportunities for external mobility to another employer's are estimated as being considerably lower.

'Hanging on' in a function does not always at first sight have to be a disadvantage. If a person has an interesting function, where new developments and new experiences continually present themselves, and if the organization continues to attach great value to the function itself as well as to the employee

Table 18.5. Over-40s whose quality of functioning is high or meets the norm.

Source of variation	df	F-value	p
Age level	2	3.40	<0.05
Function level	1	7.90	<0.01
Age level × function level	2	2.63	ns

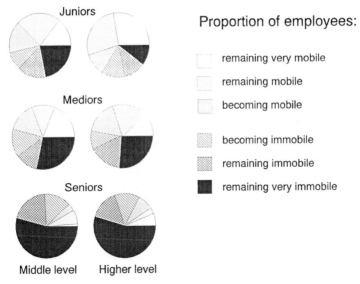

Proportion of employees:

▢ remaining very mobile

▢ remaining mobile

▢ becoming mobile

▨ becoming immobile

▨ remaining immobile

■ remaining very immobile

Middle level   Higher level

Figure 18.6. Mobility and immobility of over-40s.

Indication	Meaning
remaining very mobile	these employees have performed their present function for <3 years
remaining mobile	these employees have performed their present function for 3 to 6 years
becoming mobile	these employees have performed their present function for ≥7 years and ... will probably change functions in the next 5 years
becoming immobile	these employees have performed their present function for <3 years
remaining immobile	these employees have performed their present function for 3 to 6 years
remaining very immobile	these employees have performed their present function for ≥7 years and ... will probably *not* change functions in the next 5 years

who performs it, then one will not automatically start to think in terms of mobility. There are, however, certain risks attached to such a favourable situation. So-called 'growth in one's own function' often leads to over-specialization. A person's expertise in a particular area does, it is true, increase, but more depth-wise than breadth-wise (Thijssen 1993, refers to this as 'concentration of experience'). One shuts one's eyes to other areas—even adjacent ones—and opportunities for developing skills and expertise in other areas pass one by.

If the organization shifts its course and in consequence the function loses significance, the employee is all fingers and thumbs if he or she moves to another task or function. In other words: even in such a seemingly favourable case, immobility results in the disadvantage that from a certain moment switching over to another function is no longer a viable proposition. Tactics to catch up so as to be able once again to function well in a new function would take too much time.

Much more of a problem, however, is immobility in such cases where the employee's function has nothing more to offer. There are no new learning impulses, the function leads to monotony, and probably holds less interest for the organization as well. At such a time, an employee is at a dead end and the organization loses the opportunity to take advantage of his/her additional capacities and opportunities for development.

Mobility has important advantages. A transition to another function calls for considerable energy, motivation, fortitude, decisiveness and adaptability on the part of an employee, but on the other hand it provides fresh impulses. Circulation to other functions offers both organization and employees alike a number of rather obvious advantages. Experience and expertise acquired elsewhere in the organization are transferred to other functions and positions *en route*. Social information networks can be expanded and probably enriched. The employees concerned gain opportunities to acquire new experiences. They are given new learning opportunities which may contribute to their further growth and development. New tasks and responsibilities require an integration of relevant 'old' and 'new' skills and know-how. Sometimes they entail the necessity to call on present but not previously used capacities on the part of the employee.

Is it possible to curb 'immobility' and foster 'mobility'? Our study has produced leads which enable this question to be answered. Information (Boerlijst, Van der Heijden and Van Assen, 1993) shows that opportunities offered by the employee's present function for his/her further development form a major stimulating factor for mobility. On the contrary, functions forming an intensely closed-in-on-itself world of experience in which there is little to be learned that can be applied elsewhere, or in which the employee can discover nothing new, foster immobility. We also see that dynamic functions, which are highly susceptible to change, foster mobility too and that 'traditional' functions tend rather to raise impediments to this. Our study revealed that most over-40s in our companies have such a 'traditional' function, meaning that it has already existed for some time without striking transformations of content or aim.

The above-mentioned causal factors may seem obvious, but they would appear not to bring 'the whole truth' into the open. For, it would seem that, both in the case of the 'remaining mobile' and 'remaining immobile' there are a considerable number of examples of the opposite. For example, at least one in five functions of those 'remaining immobile' does in fact offer some opportunity for an interchange to another function. There would seem to be other factors involved impeding such a move. Two of these are directly employee-related: his/her capacities and his/her motivation or readiness to change. In figure 18.7 we demonstrate the differences between 'mobile' and 'immobile' employees. '(Im)mobiles' include both those who are already (im)mobile and those who will become (im)mobile within the next 5 years.

The difference between individuals 'remaining mobile' and those 'remaining immobile' lies not only in their abilities or capacities, but also, and even more,

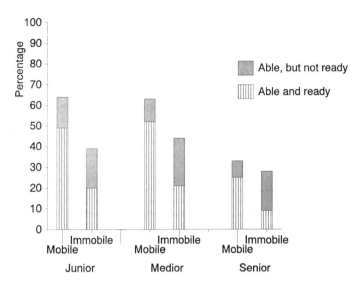

Figure 18.7. Percentages of mobile versus immobile over-40s who are judged as being able and ready (willing) or able, but not ready (not willing) to change to a function in a completely different field.

in their readiness to work in an entirely different territory. In the case of the 'immobiles' this is considerably less, but it is, however, not always present in the case of all the 'mobiles' either.

Now we can ask ourselves whether a lack of mobility and/or readiness to become mobile among the greater proportion of mediors and seniors and a big minority of juniors is as alarming. Immobility is perhaps no problem for employees who perform satisfactorily (as said before, this is the case for most over-40s) and for whom it can be foreseen that their present function or position will not alter very much and will not lose its significance or importance for the organization during their remaining career. In former times this was the rule, not the exception. Nowadays in most companies and organizations it is just the reverse. Reshuffling of targets and aims, reorganizations and transformations of departments, positions and functions are the order of the day,

*Table 18.6. Over-40s who are judged to be **able**, or **both able and ready** to change to a function in a completely different domain.*

Source of variation	**Able**			**Both able and ready**		
	df	F-value	p	df	F-value	p
Mobile/immobile	1	25.42	<0.01	1	65.17	<0.01
Age level	2	8.91	<0.01	2	8.01	<0.01
Function level	1	1.07	ns	1	1.76	ns
All interactions			ns			ns

because of changing markets, technology and so on. Quite a few 'traditional' functions lose their utility suddenly, and often unexpectedly. For the employees involved this can imply that they lose their present job and have to search for something different, within or outside their company. If their existing expertise and skills are not wanted or needed elsewhere, and if they lack the ability, buoyancy or motivation to face a more or less new functional field, with new work targets and demands of different skills and expertise, they are in serious trouble. Their immobility means that they cannot be placed in another job or position without serious losses of productivity. Learning the skills and expertise for a new, unknown job mostly asks for serious investments in time and money, at least when this job is no sinecure, and *qua* position and income comparable with the lost one. In the case of mediors and seniors the question arises whether these investments can be earned back with a certain profit, given the time that remains between now and their retirement age. Under the present circumstances, we presume the answer will be 'no' in most cases.

If we now turn back to the question of whether immobility can be counteracted or mobility can be promoted, respectively, then we should wish to utter a cautious 'yes' on the grounds of the above-mentioned findings. In our opinion it should in theory, be possible to make *every* function instructive and dynamic and to maintain its interest. This possibility could in any case be looked into. This will sometimes mean that new elements will have to be added, and the question of course remains as to whether this is feasible and can be 'technically' realized in the established business culture. In addition, the motivation (readiness) of 'unwilling' employees must in principle be able to be awakened. For this we would have to know why this readiness is often so slight. We suspect that in large part this can be ascribed to apprehension for the risks of having insufficient relevant expertise for a new function and probably not being able (quickly enough) to perform at an acceptable level in that new function. As a matter of fact, other factors can also play a role, for instance fear of not being able to learn, comprehend or memorize new things, lack of self-confidence and so on. Moreover, not only management has more or less negative stereotyped opinions about the abilities and capacities of older employees in general. Such stereotypes also dominate the thinking and behaviour of quite a few older persons themselves.

Training and development

In our study (Boerlijst et al. 1993) we have been able to ascertain that certain managerial activities help to promote the utility value and learning value of the functions as well as the quality and influence of the functioning of the over-40s. This is irrespective of the age and level of functioning of the over-40s concerned. On the other hand, we see that such activities are often not forthcoming, especially in the case of seniors. In addition, these activities are often much less frequently allocated to the middle level than to the higher level. This form of relative neglect appears on all kinds of fronts, including in the area of the stimulation of training and development.

Without training and development, one will not get very far. In the business community, they occupy an increasingly more important position. Schools and other external vocational training courses are increasingly less able to turn out professional practitioners who are, comparatively soon after their arrival and without too many extras, capable of being productive and effective in a working environment. Moreover, business dynamics require prompt adjustments to new developments and targets, and switching to new areas of know-how and experience. For employees in the organization this means that they cannot look forward to a more-or-less predictable working career from their school desks right up to retirement. Nowadays everyone should allow for the considerable likelihood of there being drastic changes of course in his/her own career. Companies should provide facilities for this, whether they like it or not. Retraining and in-service courses are both indispensable and unavoidable.

The costs of training and development programmes are high. It is entirely understandable that the management of work organizations feels little inclined to invest in these if there is no certainty of sufficient yield. There are very few training programmes for which the revenue begins to flow in, as it were, the day after their completion. More usually a great deal of time is involved; certainly in the case of training programmes in complex and new areas of knowledge.

The starting-point and the duration of a training and development course during a career are not neutral issues for management or for the employees in question either. The time still remaining for the recovery of costs of investments in individual training and the effort involved, are critical factors in the consideration as to whether to start or not. It is not likely that people will be overkeen to offer new training perspectives to employees with retirement in sight. Moreover, the seniors concerned will most likely not be highly motivated to start on a new and possibly tricky training course if no personal or commercial profit can be expected.

The current practice of early retirement schemes probably does nothing to improve the situation. It is obvious that such provisions will result in employees being 'written off' for further training sooner rather than later. In short, we may expect that seniors will be given less opportunity to follow training and development courses and that they themselves will deploy fewer initiatives than their younger colleagues.

There are also counterforces. If seniors occupy a function which is of vital importance for the company and which would soon devalue if they did not 'keep up-to-date' or 'keep pace with the times', then it is obvious that training and development courses will be embarked on after all and the argument of 'cost recovery' will be less relevant.

Let us see whether these expectations are confirmed by the study data. We will stay close to home by at first limiting ourselves to training and develop-

ment courses in the field in which the employee currently functions: his/her
'own' territory.

In figure 18.8 we have recorded the data relating to a recent 1-year period
(1990/91), assuming that even superiors who have known their subordinates
for a relatively short time have provided reliable information on this. At the
middle level the supervisors in question have a supervisory relationship of
only 2 years or less with approximately 40 per cent of the over-40s. At the
higher level, it is even higher: about 50 per cent.

There are no differences between the middle and the higher level for the
percentage of participation in training and development courses in *one's own
field*. The proportion of rather longer courses is greater than that for the
shorter ones. This, too, applies all the way across the board. There is, however,
a gap between juniors and mediors on the one hand, and seniors on the other:
the percentage of employees who have not followed any training or development course at all is significantly higher among the seniors and the share of
longer courses is substantially lower than in the younger age categories.

If we examine these data over a longer period, we see that non-participation
in courses is largely a phenomenon that is structurally person- (or function?-)
related. Approximately 20 per cent of the juniors and mediors and 25 per cent
of the seniors have never participated in any sort of training activity during
the past 5 years. At the higher level, these 'structural' percentages are somewhat lower among juniors and mediors (10 and 15 per cent, respectively), but
higher among the seniors (30 per cent). All in all, we can ascertain that the
older section of the over-40s are much less favoured with courses in their own
field than the younger section.

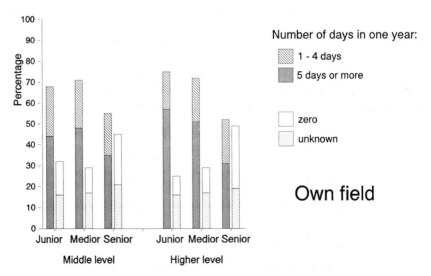

Figure 18.8. Training and development of over-40s in their own function field.

Table 18.7. Number of days of training in a recent-year period, own territory.

Source of variation	df	Own territory F-value	p
Age level	2	18.92	<0.01
Function level	1	2.13	ns
Age $\times$ function	2	2.51	ns

$\text{Mean}_{\text{junior}}$ (6.1 days) $>$ $\text{Mean}_{\text{senior}}$ (2.9 days)
$\text{Mean}_{\text{medior}}$ (6.6 days) $>$ $\text{Mean}_{\text{senior}}$ (2.9 days)

In the case of courses in a *different function field*, the number of days involved is much smaller. At least 65 per cent of the over-40s are not, or hardly ever, given (or take advantage of) the opportunity to do this.

Again, there is no systematic difference between both function levels, and also there are no significant differences between the three age levels. The mean number of days spent on courses or training in a different function area is much lower than the number spent in the 'home area'. In so far as participation in such training programmes in another function area occurs at all, the mean number of days spent and the proportion of longer training courses are much lower than with courses in their own territory. We see, then, that training and development courses for the over-40s concentrate primarily on their *own function area*. There, as one grows older, participation in courses and suchlike decreases, as was our previous assumption. Courses in other function domains are only seldom undertaken by over-40s in general.

The number of days per year that is devoted to further *personal development* of employees, thus not to function-oriented courses, is zero for more than 55 per cent of employees. Only at the higher level can a decline be traced in the higher age groups. In general, the training courses in question are of shorter duration.

From a comparison between the participation in training courses over the past year and the annual average for the past 5 years, we may conclude that whether or not to participate is to some degree person-related (or possibly

Table 18.8. Number of days of training in a recent-year period, different function field.

Source of variation	df	Different function field F-value	p
Age level	2	1.32	ns
Function level	1	0.00	ns
Age $\times$ function	2	1.81	ns

$\text{Mean}_{\text{overall}} = 0.9$ days

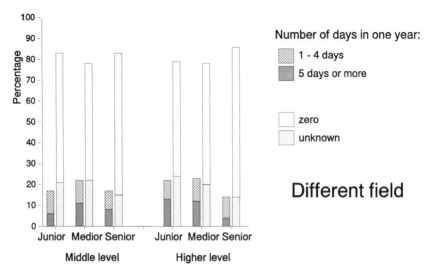

Figure 18.9. Training and development of over-40s in a different function field.

function-related). Anyone who acquires the taste stays in the running; anyone who makes no effort will persevere in this.

The annual number of days devoted to courses is also, to a certain extent, person-related. There are typical 'short' and 'long' course participants. The correlations between the number of course days in the past year (1990/91) reported by the supervisor and the yearly average for the past 5 years, both for training courses in one's own or in a different function area and for training

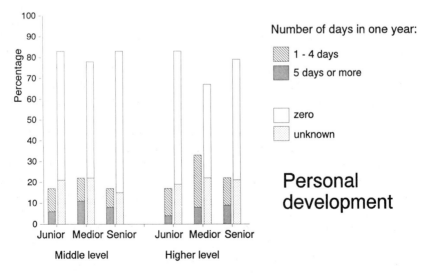

Figure 18.10. Personal training and development of over-40s (not function-oriented).

Table 18.9. Number of days of training in a recent-year period, personal development.

Source of variation	df	Personal development	
		F-value	p
Age level	2	2.49	ns
Function level	1	2.33	ns
Age × function	2	6.82	<0.01

Higher function level: Mean$_{junior}$ (1.6 days) > Mean$_{senior}$ (0.7 days)
Mean$_{medior}$ = 0.9 days
Lower function level: Mean$_{overall}$ = 0.8 days

courses geared to personal development are in the order of $r = 0.40–0.50$. Moreover, the fact is that anyone taking part in training courses in his/her own function area is also given the opportunity or opts more often for training courses in a different area (which, as stated earlier, occur much less frequently).

Conclusion

From our study we conclude that working over-40s don't seem to have serious problems with their present function or position, in so far and as long as this function stays indispensable or essential for their organization. As a matter of fact, most over-40s have a rather essential function and have already fulfilled the requirements of their job for a long time. Besides supervisors are of the opinion that most over-40s (seniors also) are performing quite satisfactorily. However, when an over-40s function becomes obsolete or superfluous, chances that the person in question is at a loss a.e quite high, since most over-40s, seniors more than juniors or mediors, are simply not equipped at all for other functions outside their own well-known territory. In other words, they lack the skills and expertise needed for other fields outside the immediate scope of their present job. They can't be put in other positions without serious losses of time and money, and in that sense they are evaluated by their management as being and unemployable.

In this chapter, we focused on one of the main causes of this unemployability: a lack of relevant training and development during the second part of the career. Measured over a 5-year period, one in five to six juniors and mediors never take part in any form of training programme or course in his/her own field. Among the seniors, this proportion rises to one in three to four, and seniors also follow longer-length courses somewhat less frequently.

These rather high percentages merit attention because training and development are indispensable for the 'maintenance' of functions and functioning. Training and development courses are predominantly geared to the employee's own function area. In the case of courses in a different area or more geared to personal development, more than 55 per cent of the over-40s are excluded. This is an alarming fact. Development in areas other than the familiar one is virtually the only way of avoiding one-sided 'over-specialization'. Where switching to another area necessitated by company-related circumstances, reorganizations or market changes, all those who have developed too one-sidedly, with all their acquired specific expertise in a single area, will be left by the wayside.

Whether or not to participate in training and development courses is person-related, in two ways. Anyone following courses now has a considerable chance of continuing to do so. Anyone following courses in a different field will usually do that in his own field as well. Vice versa, the non-participation in courses and suchlike is almost structurally person-related (or possibly also related to specific functions). This is not a healthy state of affairs, either. In every organization every person should count and it is not wise to leave anyone, whoever that may be, to fend for him/herself. If people are not given the opportunity or do not take advantage of the opportunity to continue their development, this will lead sooner or later to destruction of capital invested in people, to loss of work motivation and productivity, to decline in utility and learning value of functions and of informal networks, to lack of creativity and capacity to adapt and innovate.

Fifty-five per cent of the juniors and mediors at middle level and 40 per cent of those at the higher level are considered by the superior to be at 'the top of their ability'. In the case of seniors, this percentage is around 90. Transition to a higher function is deemed no longer to be feasible for these persons. We wish to point out, however, that horizontal mobility would also be refreshing for these people and profitable for the organization. On the one hand, their expertise and competence would end up elsewhere in the organization, while at the same time they would be able to expand, supplement and renew this expertise and competence. In this connection, it is something of a problem that large numbers of over-40s are deemed no longer able and/or no longer willing to work in a completely different territory at a later stage. The management is faced with the task of investigating what can be done about this. Accepting other tasks naturally involves risk, both for the organization and for the employee him/herself. Some risks may possibly be reduced or mitigated. In any case it is important to explore why certain employees who are *able* to start doing other things are not *willing* to do so.

It is important that what is passed on in courses can be tested and refined in practice. The qualities of the function as *learning environment* should never be ignored. Experience is the best teacher. Our study shows that the application of know-how obtained during training courses does not in most cases present any problem, not even in the case of courses in different fields. None the less, there is still a high percentage (mostly among seniors) of employees who find application of know-how within the scope of their function impossible or difficult. Moreover, the percentage of over-40s who follow courses in a different field is small.

Possibly many—if not too many—functions are so restricted that the employees concerned are not able to apply anything that is in any way outside the scope of their function. In such cases, the possibility of learning something new is absent and undergoing training and development in a different field has no point. In functions where overspecialization and rigidity prevail, development towards other functional fields and horizons becomes virtually impossible. Here lies a challenging task for both management and employees: designing or restructuring functions in the organization to improve their 'continuous learning quality' for the function-holders in question.

In the whole issue of training and development, it is wise not to confine oneself to courses in strictly function-bound fields, but to include courses aimed at the personal and social development of employees. The quality of an organization as an effective information-processing vehicle depends in large part on its social infrastructure. In its turn this infrastructure depends above all on the people taking part, on their qualities as partners and co-workers involved in social interaction with others both inside and outside the organization.

References

Boerlijst, J. G., Van der Heijden, B. I. J. M. and Van Assen, A. 1993. *Veertig-plussers in de Onderneming (Over-forties in the Organization)* (Van Gorcum, Assen).

Centraal Bureau voor de Statistiek 1992. *Enquête Beroepsbevolking 1992 (Questionnaire concerning the Occupational Population in Holland, 1992)* (Ministry of Economical Affairs, Gravenhage).

Thijssen, J. G. L. 1993. Ervaringsconcentratie: drempel voor kwalificatievernieuwing in de tweede loopbaanhelft (Concentration of Experience: a barrier for the renewal of job qualifications in the second part of the career). In: Boerlijst, J. G., Van der Flier, H. & Van Vianen, A. E. M. (eds) 1993. *Werk maken van loopbanen (To go in for Careers), 121–141.* (Lemma, Utrecht).

Warr, P. 1993. Age and employment. In: M. Dunnette, L. Hough, & H. Triandis (eds), *Handbook of Industrial and Organizational Psychology, vol. 4, 485–550.* (Palo Alto: Consulting Psychologists Press).

19

Matching vocational training programmes to age-related mental change— a social policy objective

R. Cremer

Summary. Vocational education and training are an important part of an adequate age-related social policy in work organizations. Continuous training and education extend and improve a worker's career and his or her motivation. Moreover, this aspect of career planning will also be beneficial for the work organization by creating well-motivated and competent workers. In order to develop educational programmes, a clear understanding of the mental capacities of older workers should be obtained and myths about reduced capacities should be refuted. This chapter will therefore discuss how an individual's learning potential and development of experience may change over a lifetime and how an optimal training environment can be created. To this purpose some suggestions are given in a set of guidelines which, in particular for older workers, may be helpful in learning.

Keywords: guidelines for training programmes, acquisition of knowledge, mental work ability, information processing.

Introduction

Worldwide economics affect the position of elderly workers, particularly the age cohort preparing for retirement who may easily lose their jobs as a result of reorganization programmes. On the other hand, demographic and also emancipatory forces dictate an increasing emphasis on work participation of the older age group. Several parties in society have participated in plans to reconsider the position of the aging worker. Initiatives for work integration are expected from groups in society at large, politicians, work organizations or individual workers. For example, political legislation on social policy issues is needed to provide directions for age-related matters at work, work organizations should develop and implement age-related social policies and groups of workers should become aware of their potential work ability in relation to the demands of work.

An important question for the management of work organizations is why they should be concerned with having to retain older workers. Some obvious reasons are that older workers show less absenteeism, that they are less likely to transfer to another organization and that they report greater satisfaction and commitment with their work. Also, replacing existing staff with new employees may be more troublesome and expensive than providing a continuous training programme which is adapted to an individual's career. Another argument for providing a structured education programme is the technological knowledge which is required for rapidly changing industrial technologies. For example, information technology and software packages are renewed frequently, sometimes within 2 or 3 years. These developments in production or processing technology require permanent education and training facilities. As these services are effective for the broad age range of employees, changes in mental development across the life-span should be accounted for.

Age-related policies

In a field study Cremer and Meijman (1993) examined age-related issues in social policies in a number of large work organizations. To this purpose, personnel or occupational health managers were interviewed. The results showed that the majority of officials stated that their organization had only reached the stage where 'the seriousness of the aging of the working population is acknowledged'. Generally, however, in their organization no internal evaluations were performed, solutions considered or implemented.

Furthermore, questions were posed with respect to possible measures that would be effective in age-related social policy. In the interviews it was repeatedly mentioned that the worker's experience should be better utilized, and education and training should be on a permanent and regular basis. It was often reported that such programmes were stopped at a certain age. Additionally, it was established that old adults are reluctant to participate in vocational training programmes because of age-related intellectual and motivational limitations. Common-sense opinions about age-related changes in performance or productivity are mental slowing and reduced learning or memory capacity. Moreover, learning new skills and procedures was not always felt to be useful at a later age. For reasons of disuse of learning and a limited number of years to work, it was thought to be less effective to put effort in education. It was argued by Cremer and Meijman that alternative views should be developed in order to make vocational education and training an integral part of social policy. Finally, in their study, the need to focus objectively on issues concerning education and training and age-related changes in learning abilities was emphasized.

Functional changes with age

Chronological age is often taken as an index for the aging process. Some researchers have attempted to find an index for functional age as opposed to chronological age (Dirken 1972). However, the usefulness of such an index has been debated (Welford 1980). Welford stated that 'within one individual, different functional capacities change at different rates with age, and the rage for any one functional capacity differs between one individual and another. Moreover, changes of capacity that limit performance on a task, making one set of demands may not do so for another task making different demands'. So, the aging process is just one more factor contributing to individual differences.

Aging has its effects on both physical and mental functions. Generally, studies on physical capacities in relation to work demands demonstrate a gradual decline with age (see Ilmarinen 1991, part II in this book). In the present chapter effects of cognitive aging in relation to changes of abilities over the active life-span are discussed. In comprehending age-related change in learning abilities, it is important to distinguish between structural and strategic effects. Structural effects (the hardware analogy) relate to changes in brain tissue, for example brain damage or effects of biological aging leading to, for example, mental slowing, reduced processing efficiency and mental capacities (Welford 1980). Strategic effects (the software analogy) are reflected in emotional or motivational changes affecting the general state of a person (arousal, activation). Modifying forces such as strategic or compensatory methods may also be useful to improve learning abilities. The use of strategies affects work motivation directly and is important in the ability to learn.

Many psychological mechanisms may play a part in the maintenance of abilities and capacities. A clear cause for reduced learning capacity may be 'disuse' (not using) of certain mental abilities on a regular basis. This psychological effect relates to the 'freshness' of knowledge of facts or procedures. Regular learning and the maintaining and updating of knowledge and skills are essential requirements for a successful career development. In this respect, the saying 'use it or lose it' is applicable. Older workers who have not been able or did not have the opportunity to use their intellectual abilities to deepen and widen their knowledge base will reduce their chances of employment.

Work motivation may vary owing to variables such as autonomy at work, the availability of feedback or the opportunity to learn. Such factors give direction to performance. Positive emotions are likely to reinforce, whereas negative emotions usually interfere. Negative emotions such as fear of failure in a training programme may be reversed by choosing an alternative learning strategy. Structural deficits in intellectual functioning are more definite and irreversible. Strategic or compensatory methods may also be useful to improve learning abilities. Cognitive abilities required for education and training and their sensitivity to the aging process will be considered in some detail in the next sections. The discussion is thought to be relevant for the design or adjustment of education programmes.

In sum, the process of aging can be considered as a dynamic process with continuous biological, structural and psychological change. This view is also expressed in the life-span perspective and ideally fits in an approach in which a worker's life career frequently adapts to an individual's work ability. In the next section, changes in the ability to learn will be discussed.

Mental work ability

A person's mental work ability embraces the availability of various capacities at a certain time. Availability of capacities is determined by a person's willingness to put effort into actions. In table 19.1 some variables are presented which play a part in motivation at work, variables that are also important in making an education programme successful for both student and teacher.

In this section, the relation between age-related cognitive change (structural or strategic, see previous section) and the concept of mental work ability will be discussed. Strategic mechanisms are in control of the application of abilities—they may be of a static or dynamic nature. Static aspects of work ability may be seen as fixed beliefs about society, personal opinions and values—they are rather stable over a lifetime (e.g. 'I always try to give the best of my abilities'); whereas dynamic aspects of work ability are shaped by changing emotions or motives (e.g. 'In this stage of my career I feel more at ease with office work' or 'I should not be learning any more but should retire and allow a young colleague to take my job').

Generally, work can be explained in terms of information processes, usually proceeding from mental to physical activities. In assessing an individual's capacities, one may apply the common information-processing model, which is based on assumptions regarding the flow of information through a system

Table 19.1. *Work ability: work capacity, and willingness to work, with static and dynamic variables*

| Ability | Work ability | |
nature	Work capacity	Willingness to work
static	structural capacities	values, beliefs
dynamic	ability to learn, solve problems	emotions, motivation, flexibility, creativity

Table 19.2. Information processes and suggestions for adaptations in work conditions

Information processes	Practical effects	Implementations in the work situation
sensory	resolution, (signal-to-noise ratio) S-N strength of signal/background noise, relevant/irrelevant signals	improve resolution, increase signal-to-noise ratio more light, stronger signals, reduce irrelevant signals
sensory/perceptual	perceptual capacity, short-term memory, concentration	increase simplicity and uniformity of signals, present information in small 'chunks', reduce distraction
cognition	working-memory capacity, problem solving, abstract reasoning, controlled processing, attention, memory	reduce complexity of information and attentional demands, apply expert knowledge, use transfer of knowledge, use well-known concepts
central/motor	stimulus—response (S–R) relation	present direct S–R relations, make choices compatible, logical and simple
motor	actions to be executed	response reinforcements, mechanical aids

(Sternberg 1969). Working may be considered to be a continuous flow of information processing. At the input stage, the sensory modalities (vision, hearing or feeling) have to be sufficiently sensitive to detect the 'signal among the noise'. Then a perceptual process has to be completed by (usually) temporary storage of impressions before some decision process can follow. Perceptual processes may be vulnerable to dual input, for example distracting visual or auditory stimuli (Cremer 1993) or deficiencies in attention which become evident when capacities are reduced (Rybash et al. 1986) by internal factors such as aging or external factors such as high processing demands or time pressure.

In many work situations, it is necessary to be attentive to relevant signals (and ignore irrelevant signals) and to make decisions on the preparation of actions. Typical in such situations is that, with increasing complexity of a task, old adults are proportionally more slowed than young adults. For example, it is known that old adults are less accurate and require more processing time than young adults when stimulus–response relations are indirect or not logical (Cremer 1993). Finally, overt behaviour or a motor response is likely to be delayed in old adults because of general cognitive slowing in the nervous system (Welford 1984). In short, information processing involves sensory stimulation, perceptual coding, memory storage, goal setting, problem solving, preparation of motor actions and the execution of actions. All processes are more or less affected by cognitive aging because of general mental slowing and specific age-related processes. Decomposition of work into broad categories of information processes makes the assessment of mental work abilities more efficient. Subsequently, when age-related limitations in an individual's work ability are known, it is easier to redesign work or compose a vocational training programme. In table 19.2 some basic information processes are summarized and some examples of the process in the work situation are given. Additionally, possible measures are suggested to improve the quality of work when, for example, age-related limitations in specific processes occur.

The meaning of age-related change in learning

Cognitive aging has important consequences for the ability to learn. Learning skills require perception and comprehension of data; a transient short-term memory (STM) store, limited in capacity and vulnerable to interference; consolidation of the STM trace into long-term memory (LTM), requiring some active mental work, for example, rehearsal, organization or coding; an enduring LTM store of large capacity; and active search and retrieval, which are thought to be involved in recall memory and to be absent in recognition memory (which merely requires matching of information in storage and in the environment) (Schludermann and Schludermann 1983). Age-related changes in memory are expected to deteriorate any of the above mechanisms, leading to comprehension difficulties, lower capacity or greater vulnerability of STM storage, consolidation problems (e.g. coding deficits), changes in LTM storage, or search and retrieval problems. So, it may be that all mental processes are affected by age, or that primarily some specific processes are affected. A definite answer to this question cannot yet be given. Horn (1982), however, distinguished between so-called 'fluid' mental abilities, which are suspected to decline with age, and 'crystallized' abilities, which are more resistant to the process of aging (see also chapter 12 by Kok). Fluid abilities refer to cognitive functions such as speed of decision making, short-term memory and attention which play a part mainly during the acquisition and transformation of information. Crystallized abilities are associated with verbal, acquired or specialized skill. In other words, there are abilities that are well preserved, even at an older age, and that are vulnerable and may deteriorate with aging. These abilities are characterized in table 19.3.

Usually, in cognitive functioning the two categories of functions are used in combination. In many daily-life activities, however, the passive use of acquired knowledge, learned skills or procedures can be observed. They comprise the actions which are more or less effortless and fully automatic (crystallized abilities). It should be noted that behaviour based on automatic processing may mask limitations of fluid functions which usually occur in aging individuals (Uttal and Perlmutter 1989). Therefore, in learning new skills (the acquisition of knowledge), it is more likely that age-related limitations are revealed. Aging can be considered to be a dynamic process with continuous biological, structural and psychological change. This view is expressed in the life-span

Table 19.3. *Characteristics of cognition divided in abilities which are well preserved and which are vulnerable and may deteriorate during the aging process*

| Chracteristic | Cognitive abilities | |
	Fluid	Crystallized
intellectual processing	effortful processing requires attention and is mentally demanding	passive use of available knowledge, learned skills
knowledge handling	acquisition (learning) problem solving	application (rules, procedures)
nature, nurture	innate (analogue hardware)	acquired (analogue software)
accessibility	requires much effort	effortless
mental load	high	low
relation with age	negative	positive

perspective and ideally fits in an approach in which a worker's life career is frequently adapted to the work ability of the individual.

Adapting learning programmes to age-related change

'No one is too old to learn.' This saying implies that age is not the most important factor in an individual's potential to learn. However, it is likely that more effort has to be invested or specific learning strategies have to be developed for the older individual. For this purpose, it is useful to know how training programmes can be designed so that older workers can benefit from their (crystallized) experience. Then, it is important to understand how crystallized knowledge or gained (work) experience plays a part in the capacity to learn. For a novice young worker who learns a trade, the acquisition of skills requires much effort in the learning process. Abilities and skills do not develop spontaneously. New actions require the conscious investment of effort: one has to pay full attention all the time. Over the years, when activities become more familiar, the nature of the processes change: actions are of an automatic nature more often. Working skills are increasingly based on experience, and the processing of information becomes less demanding. The integration of information becomes faster and the uptake of related knowledge is facilitated, leading to a work strategy which characterizes an experienced worker. A training programme that is aimed at the integration of related knowledge with existing knowledge is fast and efficient, whereas training aimed at unfamiliar abilities will require more time and takes more effort. This principle is important for the design and development of training and education programmes. In conclusion, generally, but particularly for older adults, training is most effective when existing knowledge is extended.

During a career, experience is developed by a diversity of actions which usually have been repeated many times and which were more or less successful. The accumulation of experience during a career is also known as concentration of experience (Thijssen 1991). This means that a knowledge domain can deepen as opposed to widen. Workers become experts in a limited domain of activities. They stay in a certain function for much of their career while further development would have been better for their personal development. Such 'experts' become deprived of new incentives and personal development. Their work situation it is not too interesting or challenging and does not encourage motivation at work. An important tool in personnel management is a worker's lifetime opportunity to learn and master new skills.

Principles and guidelines of learning

On condition that age-related change is taken into account, a well-adapted education or training programme can be designed. For this purpose working from a set of guidelines will be useful. From existing literature, some basic principles about learning in adults were isolated (Thijssen 1991). For example:

(1) Active processing: learning by doing, having the opportunity to practise skills as they are taught, and by evoking interest and engagement to develop an active learning strategy. Old adults appear to be able to reach similar levels of knowledge and skill as young adults if an active, structured approach is applied.

(2) Systematic feedback and support: frequent information about the quality of study work facilitates adjustments and helps acceptance of corrections. It is often reported that older learners ask for more support, suggesting that expert support is particularly useful to older learners.

(3) Reward systems for new learning: promotion or recognition may serve as positive incentives.

(4) Recognition of concepts: well-known concepts offer structure in extending a knowledge base; 'transfer of knowledge' may be facilitated. In designing a training programme, the distinction between fluid and crystallized abilities could prove to be of importance. For example, building on existing knowledge is a good starting-point for the extension of related knowledge. Or, in other words, transfer of knowledge is facilitated when learning materials are built on existing knowledge.

(5) Direct applicability: insight in practical use and applicability of study materials increase motivation.

(6) Adapted social context: a learning situation should not become a confrontation with incapability, but a capitalization on available capacities. Competition as a negative group component should be avoided, and a positive attitude should be fostered, directed at continuation and progression. Social support from fellow-learners is important at all ages. Learning in small groups or pairs is more effective than learning singly.

(7) An adapted logistic context: a flexible study planning tuned to individual capacities, study content and available time is necessary. Age-related slowing of intellectual processes and changes in memory capacity justify a self-paced learning technique. Ordering educational material into small, meaningful units lowers the demands for memorizing material.

Conclusion

In private situations and at work, individuals differ to a large extent in capacities, motivation to work and style of coping with problems. With advancing age individual differences may increase further, so that, in practice, chronological age is a bad predictor of changes in performance. In the previous section it was discussed that chronological age is only a relative concept with respect to the individual's capacity, flexibility and motivation. Therefore it is essential to base conclusions about the work ability of an elderly worker on objective evaluations. The work ability of the individual comprising a changing set of functional abilities tends to diminish with advancing age. It should, however, be realized that in human development both growth and loss of abilities occur; some abilities are gained but others are lost over the years. With increasing age, the loss of fluid abilities especially is evident, such as the conscious use of intellectual functions which are needed in learning. On the other hand, consolidated, crystallized knowledge and abilities are less vulnerable.

In an effort to maintain the ability to learn it is necessary to offer educational opportunities regularly. It is important to realize that there is more to cognitive aging than simply decay of functions, and to acknowledge that an accumulation of knowledge is present in the aging adult. It may be expected that individually matched training programmes may extend the career of elderly workers positively. This reasoning is likely to motivate workers of all ages to follow training programmes on a regular and continuous basis.

Finally, preventive measures in work situations should be taken when effects of work on health are still reversible. When the health status of a worker has become irreversible (e.g. back disorders), special corrective measures can help—for example, adapting working conditions to match individual capacities or creating alternative job opportunities. In terms of social policy, vocational education and training, starting at the onset of a work career and lasting a life career, should be a normal part of proactive policy.

References

Cremer, R. and Meijman, T. F. 1993. Social policies for the aging worker in some work organizations. In J. Ilmarinen (ed.) *Aging and Work*, International Scientific Symposium on Aging and Work, Proceedings 4 (Institute of Occupational Health, Helsinki), 195–201.

Cremer, R. 1993. Cognitive aging, visual task demands and noise. Ph. D. thesis. University of Amsterdam, Department of Psychonomics.

Dirken, J. M. 1972. *Functional Age of Industrial Workers: A Transversal Survey of Aging Capacities and a Method for Assessing Functional Age.* (Wolters-Noordhoff, Groningen).

Horn, J. L. 1982. The theory of fluid and crystallized intelligence in relation to concepts of cognitive psychology and aging in adulthood. In S. E. Craik and S. Trehub (eds) *New Directions in Cognitive Science* (Ablex, New York), 69–87.

Ilmarinen, J. (ed.). 1991. The aging worker, *Scand. J. of Work Environment and Health*, 17(suppl. 1): 1–141.

Rybash, J. M., Hoyer, W. J. and Roodin, P. A. 1986. *Adult Cognition and Aging* (Permagon Press, New York).

Schludermann, E. H. and Schludermann, S. M. 1983. Halstead's studies in the neuropsychology of aging, *Archives of Gerontology and Geriatry*, 2: 49–172.

Salthouse, T. A. 1982. *Adult Cognition: An Experimental Psychology of Human Aging* (Springer-Verlag, New York) Sternberg, S. 1969. The discovery of processing stages: extensions of Donders method, *Acta Paychologica*, 30: 276–315.

Thijssen J. G. L. 1991. Een model voor het leren van volwassenen in flexibele organisaties (A model of adult learning in flexible work organizations). In H. P. Stroomberg, J. G. L. Thijssen, L. Simonis-Tabbers and H. W. A. M. Coonen (eds) *Didactiek en volwassen educatie (Didactics and adult education)* (Dekker & van der Vegt, Assen).

Uttal, D. H. and Perlmutter, M. 1989. Towards a broader conceptualization of development: the role of gains and losses across the life span, *Developmental Review*, 9: 101–132.

Welford, A. T. 1980. Where do we go from here? In W. Poon (ed.) *Aging in the eighties* (APA, Washington, DC) 615–621.

Welford, A. T. 1984. Between bodily changes and performances: some possible reasons for slowing with age, *Experimental Aging Research*, 10: 73–88.

20

Older employees' participation in organizational and technological changes —experience from a company undergoing changes

G. Holm

Summary. Older employees are often marginalized if they are not given a chance to take an active part in organizational changes and technology training. Experience from a Danish company shows, however, that older employees can satisfy the demands that arise with changes on equal terms with their younger colleagues. In connection with the introduction of an advanced production control system, the company carried out a 12-week training course for 120 operators, of whom 20 were over 50 years of age. The operators were divided into 18 production groups with extended autonomy. The older operators were afraid of participating in the change process. Positive expectations from the surroundings helped eliminate their worries and liberated their resources for active participation in the changes. The older operators stabilize the production group's function though they are not the fiery souls of the development. They are equal members of the production groups and, therefore, have to participate in the training on how to use the new machines, take on the coordination of the group's work, participate in the employment and introduction of new operators, man the machines and develop the group's work-related and social life.

Keywords: new technology, production groups, training, experience, course participation.

Introduction

When organizational changes are being implemented within the private and public sector, older employees are often marginalized or forced to leave the labour market. Experience from a Danish company shows, however, that older employees can adjust to new technology, and they can take an active and constructive part in organizational changes. Older employees possess a long working life's resources and experience that are valuable for production and collaboration in connection with the performance of the work processes. Older employees are, indeed, able to cope with the challenges of the introduction of new technology and organizational changes. It is, however, a precondition that they are given a fair chance to adjust and that they are trained in the new work functions.

Older employees and the labour market

The age-related drop in business participation has been pronounced since the end of the 1960s. The reason why so many individuals over 60 years old are leaving, or have chosen to leave, the labour market is that high unemployment pushes elderly people out of the labour market. Other important reasons are the attitude of the older work-force, technological readjustment and the physical and psychological working environment.

The consequences of unemployment in the labour market means that elderly people often leave a job to make room for young, unemployed persons. Moreover, the general feeling is that the older generation is less productive, mobile and innovative than the younger workforce. They are therefore willing to leave their job, or special severance settlements make them choose to leave the labour market (Friis 1990).

New demands on operators

Technological development has increased demands on the work-force's readjustment abilities, qualifications, attention, speed and the ability to stay. The introduction of new technology will often be followed up by organizational readjustment, job development and changes in the working environment. The consequences may be that many jobs will become one-sided, repetitive, monotonous and characterized by a high work rate. Older employees are vulnerable to this development, because a long working life has left its marks in the form of chronic fatigue, muscle joint aches and psychological stress.

Another developing tendency with introduction of new technology is that new demands are being made on employees, who are being required to readjust from routine jobs to more autonomous jobs with greater demands on attention, learning and action. Older employees may experience these changes with anxiety because the new challenges are in sharp contrast to their earlier working experience (Lorentzen and Bendix 1989).

The introduction of advanced production technologies often leads to active involvement of younger employees in the development process. Older employees are transferred to jobs with no built-in possibilities of development and no means of influence. Many have not participated in training activities or development of their job for many years. Experience from a Danish company, a producer of cables, however, proves that older employees can adjust to new technology and group organization without being marginalized in relation to other employees.

Reorganization in a company

In the middle of the 1980s, the company decided to invest DKK 325 million (£32.5 million) in new technology. During 1989 and 1990, a new production plant was constructed with the latest CIM (computer-integrated manufacturing) technology and had 40 employees. At the same time, a plant with 70 employees was reorganized to handle the new and more advanced technology. Of the 70 operators, 20 had passed the age of 50. At the same time, the company's management decided that all people employed in production should participate in extensive training and that organizational changes would be implemented after the training. The Danish Technological Institute developed the organizational concept and implemented it in the company.

The purpose of the changes

The company's decision to invest in education and development of the organization, in parallel with the introduction of advanced production technology, was made on the following grounds:

(1) The operators are to have more responsibility and obtain more competence. Furthermore, the job content and the job quality must be improved if, the new technology is to be used optimally.
(2) Orders are no longer communicated through a foreman, but appear on a screen. New technologies demand an increased independence and commitment from the operators to be able to take in and initiate production following the orders on the screen.
(3) New technology requires the operators' comprehensive view and understanding of production planning, and they must be able to man the machines independently.
(4) The foreman no longer plans or controls work procedures, but has more superior planning tasks.

Furthermore, the company's management and shop stewards were aware of the fact that it might be a problem in the future to maintain the labour force, if the industrial work was not improved, that is made more interesting and challenging.

The traditional Taylorist management strategy was changed into a strategy where the human resources of the operators were strengthened and where the decision making was delegated to the operators. The organization was changed so that the operators were given a new position, a new status within the company.

Production groups

At the same time, in cooperation with the shop stewards the management decided to introduce autonomous production groups of seven to eight operators. Each group was asked to participate in the planning, execution and control of its own work. The groups were made responsible for reaching the production targets and for ensuring that the products were in line with ISO 9001 standards. The operators were also to participate in the hiring of new operators to the groups and the laying down of the groups' development targets, including the need for education. The reorganization meant that the foremen got new job functions and that some of them were transferred to other tasks.

Today, 18 groups function as autonomous production groups. All operators have undergone a training course during which they have been working with team building, group function and targets for group development.

Training of operators

Each operator participated in a 12-week training course focused on the development of qualifications such as collaboration, communication and training in computer technology. The company invested 60 000 hours in training and 10 000 hours in connection with the introduction of group organization. The company's costs for employee development amounted to DKK 12 million (£1.2 million), inclusive of wages for operators during their training. The operators themselves did not have any expenses in connection with course participation.

Right from the beginning, everybody agreed that all operators should participate in the education programme-this meant that older operators were forced to participate. Only very few older operators preferred not to participate; they were the ones who had decided to retire within the next 1 to 2 years. During the entire development process, no special considerations where shown for the older operators. They participated in the project on the same terms as everybody else.

The interesting thing is to know how the older operators experienced this new challenge and how they functioned during the process of transition.

Methods

No research evaluation has been made of the implementation of the new technology. This means that no measurements of the effect of the intervention in the form of comparing results before and after implementation have been carried out. Assessments of older employees' functions and ability to participate in organizational changes have been collected through interviews and observations made during the training of 120 operators. Further, during all of the 2-year process, interviews were made with managers, shop stewards and training consultants.

A more systematic collection of the older operators' experiences was made by means of qualitative interviews with six older operators. They were interviewed on the basis of a questionnaire comprising the following main themes: technology; training and adjustment; changes in and functioning of the group organization; participatory influence; training; psychological and physical working conditions.

Results

The majority of the older employees have been with the company for as long as 20 to 27 years; a few have been with the company for only 2 years. Some of them have had serious illness such as heart attacks, but all had resumed their jobs after long periods of sick leave. The six oldest employees were men between 54 and 60 years of age. One 54-year-old operator had been with the company for 2 years. Most elderly employees had attended elementary school for 7 years and were all, in essence, formally seen as unskilled workers. By and large, they have all had the same tasks at the same machines, and their position/status within the company was low.

Technology changes the work

The older employees noticed that their work changed with the introduction of new technology. Tasks formerly carried out manually are now performed by machines. At the same time the number of employees has been reduced and the foremen no longer give orders, since orders appear on a screen. The introduction of computer-controlled production systems has made the work more monotonous. For a major part of the working day, the operators control by monitor the correct running of processes. In purely physical terms, the distance between operators has become larger, because large machines are spread over the production area.

One of the consequences for the operators is reduced social contact. It has become more difficult to leave the workstation to talk with colleagues, and now the foreman does not pass by very often. The collaboration and the daily contacts within the groups compensate a bit for earlier work-related possibilities of contact.

A 55-year-old operator expressed the changes in the following way:

The daily chat by the machines has disappeared—the work has become more monotonous.

New technology makes us more lonely and very solitary.

Older employees experience changes in different ways. Some feel anxious about working with new technology, and there is a tendency that older employees are given jobs that require less knowledge or no knowledge at all about computers, whereas other older employees work with complex knowledge-demanding computers systems. They experience the job as a challenge and they are very much committed to the new technology. Yet, all of them agree that it is easier for younger operators to acquire knowledge of EDP (electronic data processing)-not necessarily because they are young, but because they have an interest in EDP and computers, and most often they were taught EDP in school.

The operators who took part in the introduction of the new technology participated in meetings with engineers and in the purchase of equipment, for instance, in France, Germany and Sweden. In addition, the operators played an important role in the implementation of the technology into the production. This active participation had a positive effect on the operators' experience of the new technology.

The introductory EDP training, which all workers received during the training course took away the initial fundamental scepticism and anxiety towards the new technology. Several workers became eager to broaden their knowledge of EDP by attending more courses, and some of the older employees have become so motivated that they are eager to learn more whenever an opportunity occurs during the working day.

A 56-year-old operator puts it this way:

> I am beginning to learn to push the buttons again. When you have been away from the screen for a period, you forget what to do, but the younger ones just know it and they are willing to help. I may be a little slower than them, but then, on the other hand, they have been brought up in front of a screen.

Both management and colleagues expect the older operators to participate in the reorganization and process of changes on the same terms as everybody else. This means that they participate in education and training on the operation of the new machines. Each production group has set targets for colleagues' training, which means that, in principle, everyone must be able to perform all tasks with the different machines of the group's area. Older operators are expected to participate in colleagues' training and, through that, in training in the use of new technology.

Group organization

Organizational changes do not affect older employees very much, and only a small number of elderly employees are frightened of these changes. There are several explanations why older employees do not have as great expectations of work organization in production groups as their younger colleagues:

(1) Older employees handle their work with greater independence thanks to many years in the job and their long life experiences.

(2) Older employees control their expectations, and take things and changes as they come with a philosophical attitude in order not to be disappointed.

(3) Older employees are not willing to invest their efforts in bringing about innovations themselves, they leave that to their younger colleagues.

Older employees can participate in processes of change, and they will adapt to their new functions. They stabilize the process, of change among other things by slowing down or holding back the younger ones a bit. However, older employees do not impede or stop processes of change.

Older employees are not allowed to stand still with younger colleagues around. They contribute to the group in a constructive manner with suggestions and they take on new tasks in relation to the group's function. As a rule, however, they are not the fiery souls of the group. Working in groups means that all group members collaborate more efficiently and are more target-oriented on production. During recent years the relationship between management and operators has changed considerably. Today, management appreciates the production operators a lot more than it used to do. The distance between the two has been reduced and the tone between them is different.

Training

The majority of older operators were afraid of participating in a 12-week course; others took it calmly. This insecurity eventually disappeared when positive experience with the training course was 'handed down' to these operators.

A 60-year-old operator experienced the process like this:

> I got nervous when I was told to participate, but after 2 weeks I had settled in. Indeed, I am very glad I participated in the course. It gave my life new content, both professionally and personally.

Another operator of 56 years of age put it this way:

> We got to know each other a lot better, so our unity is very good. The young were really good at helping the older employees with computers, arithmetics: and when we came across English expressions we did not understand, they were helpful too. In return we could help them in more practical matters.

Owing to word-blindness, troubles occurred quite early in the courses, in connection with the theoretical subjects. It turned out that 30 per cent of the participants were word-blind, or had some difficulties in reading and writing, and extra courses for this group of operators had to be arranged. For the first time in their working life some older employees had to recognize that reading and writing were a problem to them. This recognition was painful, but it eventually had a positive effect. The recognition meant that they no longer had to hide their handicap. Everybody showed consideration for the fact that some of them needed more time than others for reading and writing and that more needed help or assistance.

In general, the older operators felt that learning was difficult because so many years had passed by since attending education. Maybe age does affect the ability to learn, but age is not decisive for the capacity to learn.

Physical reactions

Older operators did not experience aging as troublesome. Several of them felt that many changes were taking place, but said that life has always been like that. They felt that they had control of the production and that they could keep up with the changes at work. Their bodies may have become a little stiff, and they could not run as fast as before, and from time to time some of them need colleagues' assistance when they have to carry or move heavy burdens, but they definitely did not consider age to be an obstacle to carrying out their work. Generally, they all felt fine physically, though some suffered from chronic diseases. Their good physical condition was confirmed by the fact that they had fewer days lost through sickness than their younger colleagues.

Psychological reactions

Elderly operators do not think they have changed psychologically because of age, but life experience has improved their ability to react more appropriately to the influences they are exposed to.

Operators worded their experiences as follows:

I have always been quiet and calm, that has not changed.

Years have made me more tolerant and patient, which only makes the collaboration with colleagues run more smoothly.

One's temper gets more quiet, and one gets less aggressive.

I have learned to keep out of discussions that may excite me, so I have become more calm than I used to be.

A few have experienced stress-like reactions to the changes, especially in connection with the training course. These reactions were limited, however, to short periods just before the courses and at the beginning of the courses.

The older operators left in the company feel well, both physically and psychologically, and they have been able to adjust to the new demands.

Conclusion

The general experience from Danish companies is that older employees are marginalized when new technologies are introduced, or when major organizational changes occur. Often, older employees are forced to leave companies before they reach the age of retirement.

The experience from the studied company, a producer of cables, shows, however, that by involving the older employees in long-term development projects older employees can be kept in the company. By putting older and younger employees together, the company obtained a series of positive results. The two groups of employees can learn from each other, which is resource-efficient: older employees can transfer their work and life experience to younger employees and the company's stability is maintained.

Older employees can learn to work with new technology if they are given proper opportunities to do so. They can adjust, just as younger ones do, although they cannot be expected to be the linchpin of the process of change.

No doubt, positive expectations of older employees are important in terms of their ability to participate in the reorganization. Positive expectations and equality in relation to younger colleagues mean that older operators keep self-confidence; it invalidates the general attitude that older employees do not have sufficient resources for change. Usually, the expectations of the elderly worker are negative, leading to a lack of confidence in their ability to meet new challenges.

Older employees can confront challenges if they are engaged actively in the reorganization process and only then when they can make substantial contributions to this process and are involved actively with their working life and life experiences.

References

Friis, H. 1990. *Nye tider - nye æ ldre: Arbejde og Afgang* (*New Times—New Senior Workers: Work and Retiring*) (Senior Forlaget, Copenhagen), 13–40.

Lorentzen, B. and Bendix, K. 1989. *FMS, indflydelse og arbejdsforhold*, Dansk Teknologisk Institut (*FMS, influence and working conditions*) (Danish Technological Institute), 53–90.

21

Training and education programmes for older employees

M. J. Schabracq

Summary. The general idea of the human resource management (HRM) approach advocated in this chapter is the empowering of the individual. Attention is focused especially on the elderly employee, since for several reasons he or she suffers from greater powerlessness than younger workers. Improvements in this situation can be made by the creation of training opportunities, greater transparency of organizations and function assessment. A possible intervention method to mobilize self-empowered resources are age-conscious personnel policies. Part of such a policy might be the HRM approach. For optimal HRM, employees should themselves be given responsibility, supplemented by adequate training opportunities. The aimed-for higher autonomy and freedom of the individual worker can be attained by taking measures like job redesign, task enrichment, higher mobility and so on. The success of such measures depends on adequate training of skill, implementation of age-conscious personnel policies and self-management training. Possibilities and techniques to attain this success are discussed. The gains of the HRM approach are more freedom for the employee, to arrange his or her own work and to choose challenges. Direct benefits are a strong identification with and more interest in the job and improvement of health and well-being; indirect benefits are less sick leave, higher productivity and so on. There are some contra-indications, however, namely its dependence on the right organizational culture, a basic trust in the sincerity of the organizational intentions and the intrinsic motivating qualities of the work. Other factors are union support and the cost of implementing the approach. In sum, the emphasis of the HRM approach lies in the decrease of power differences by empowering the less powerful. It seems an attractive way to cope adequately with the continuously changing environment.

Keywords: human resource management, age-conscious personnel policy, freedom, responsibility, organizational culture, training, self-management, autonomy, individualization.

Introduction

The presence of many older employees in an organization is often considered to be a problem by such an organization (see also chapters 17 and 20). The demographic developments in many western countries (an increasing proportion of older employees and a diminishing proportion of younger employees to replace them) have made this problem urgent. So, new ways of solving it have become necessary. An initial impetus is given.

Most of the problems that older employees experience can be traced back to a shortage or loss of power in the older employees. Using this statement as a point of departure, it is only logical that solutions of these problems should focus on enlarging individual power, that is giving it back to the employees involved. So, the general idea of the approach is one of empowerment of the individual employee.

The relative powerlessness of older employees is, however, reinforced by a process of stereotyping of the elderly as people with diminished potential for change and adaptation, which is foremost a counter-productive way of conceiving reality. A stereotype like this one shows a tendency to develop self-fulfilling characteristics and it can also be considered as a case of blaming the victim, which leads attention away from solutions in the direct work environment and the organization.

Interventions have to deal with the negative effects of these stereotypes. Moreover, as older employees vary greatly, every designer of such intervention in this respect has to reckon with this variability. In fact, the great amount of individual variation pre-empts the possibility of success of general measures.

The general approach advocated here, besides alleviating the stereotype problem to a workable level, is to capitalize on the potential of individual employees to adapt and cope, the older as well as the younger ones.

Giving employees sufficient freedom and responsibility to develop their own further career is considered to be the best guarantee to commit them to the success of this undertaking. After all, they are the best experts when it comes to their own predicaments and, in this respect, such an approach is a simple matter of respect for the personal dignity of older employees. However, in order to mobilize the adaptive potential of individual employees effectively, some changes are necessary: changes of the organization as a whole, as well as changes of the individual employees involved.

A main objective here consists of creating enough opportunities for individual adaptive changes. This requires, apart from improvements in the quality of jobs (especially in the direction of a greater job decision latitude), that organizations have to provide their (older) employees with the room to change jobs in ways that are rewarding, both to these employees and to the organization.

To attain this objective the organization has to become more transparent which can be achieved by different forms of periodical assessment of opportunities and limitations of change in combination with an explicit communication of results in this area to all parties.

Training also can play an important role here. Training of older employees does, however, ask for a substantial revision of training practices. Training should result in more responsibility of older employees for their own development. On the other hand, additional training should become more accessible to them. Apart from training in occupational skills and self-management techniques (for the employees), programmes on stereotypes and leadership skills (for lower and middle management) are important here.

In order to solve problems of and with older employees, interventions must result in a well-integrated, widely accepted and supported 'age-conscious' career policy. Such a policy should not focus exclusively on elderly employees, but should address employees of all age groups. It is obvious that such an approach demands substantial change of the organizational culture, a process that may require a substantial amount of time. Possible elements of the approach sketched are outlined below.

Human resource management

In order to deal successfully, and in the long term, with problems involving elderly personnel, policies in this direction have to be embedded in some form of 'human resource management' (HRM). HRM here is defined as an integrated policy to optimize the development of the employees' potential. The idea behind it consists of an optimal integration of organizational objectives and objectives of individual employees. To this end, the organization invests in career planning and guidance, as well as in training and education. A secondary effect is that the employees, to a greater extent, experience the organization as a part of themselves, a part which they do not give up easily (Schabracq 1993).

An age-conscious personnel policy should be a more or less self-evident and central characteristic of a consistent HRM approach. Over the years an HRM approach leads to an accumulation of substantial financial investments in older employees. Seen from this perspective, negligence of problems concerning older employees, which affect their functioning, would be a foolish destruction of capital. So, from an HRM perspective, it is only a matter of sound financial reasoning to prevent such problems.

Organization philosophy

As a central characteristic of HRM, an age-conscious personnel policy should be made part of the organizational philosophy. As such it should be explicated in a kind of mission statement, which should contain some passages on the equal rights in this respect of specific groups, such as women and foreigners. Within such a statement intended objectives in this area can be formulated.

Subsequently, these objectives can serve as points of departure for the elaboration of policies to realize these objectives and as touchstones to test other policies for their effects on elderly employees and aging personnel.

By making an age-conscious personnel policy a central item of the organization philosophy, it becomes possible to lay the responsibility for the health and well-being of (older) personnel on the employees themselves and their direct managers. In this way, the organizational system can become more self-supporting. When insoluble problems arise, specialists of the personnel and the medical department can be called in. However, the central task of these departments may only consist of helping to create conditions in which the direct participants can solve their own problems. As a matter of fact, these direct participants, the elderly and their direct managers, are the best equipped to do so, since they know most about the problems (Doeglas and Schabracq 1992).

Transparency and the necessity of adequate assessment

In order to be self-supporting the organization has to be very transparent for all its members. Organizations have to communicate in clear and unambiguous terms what they want from all employees. This implies that organizations have to spell out the opportunities and limitations of the organization, to specify the rights of the employees, to explicate the specific means and facilities which employees may receive to help them in their enterprise and to explain the consequences of different lines of action.

To these purposes, management itself must have a clear picture of the ecology of the organization, the state it is in and the niches it provides. Besides, those who are responsible for this policy must have some idea of the possibilities and talents of the employees involved. To get such information several forms of periodical assessment are necessary:

(1) Those in charge must have clear figures about the workers' age distribution in the different divisions and departments to get a general picture (Schlüter 1993).

(2) It is useful to relate these figures to statistics on phenomena like sick leave, turnover and quality of production (Van der Vlist and Pot 1993). In this way an indication is gathered of the distribution of possible problems over different age groups.

(3) Indications should then be followed up by more specific forms of assessment: interviews or questionnaires can be used to get a more comprehensive picture. Such an approach focuses on information concerning the employees, about the quality of functioning, social networks and—very important—their job quality. To get usable information about job quality, job classification in terms of the personal level needed for the execution of the different job elements can be helpful (Dresens 1991). An outcome of such a survey could exist of

an overview of so-called 'killer jobs', jobs that lead to permanent health damage if employees hold on to them too long. Another important outcome could be an inventory of jobs that are suitable for different categories of older employees.

(4) Both for those in charge and for the employees themselves it is useful to have valid data about the employees' capacities, values, motives and objectives. Occasionally this can mean a form of psychometric testing, but most of the time a series of periodical job and career interviews is more appropriate. To be effective such a procedure requires cooperation and trust from the employees involved, which is not always granted.

As a result, both parties should ideally have the same information about the employee's actual and potential specific abilities and capability. Also, it should be known what training needs each individual employees have, what they wish to do, and their motives and objectives. Also, it should become clear what individual employees are allowed to do, especially in terms of organizational constraints and, to a lesser degree, of the limits set in this respect by their own personal values.

Apart from information about the employee's personal characteristics the interview also surveys the job as it was done during the last period and as it will be performed in the next one. Consequently, the interview has to lead to a mutually shared evaluation of the past work done, as well as to some explicit mutual understanding about targets, in the area of task performance and training results, to be met in the next period.

Measures to be taken within an HRM context

As was stated before, an age-conscious personnel policy within an HRM context should capitalize on the potential of adaptation and coping of the employees involved. General solutions, as we have seen, being out of the question, the individual employees are considered to be the best experts at their own possibilities. Also, they are the ones who gain most by making things better and who are most strongly motivated to do so. So, the most promising line of action is to push down responsibilities in this respect to those immediately involved: the employees themselves. The employees have to have the possibility to make their own choices in a way which also is of benefit to the organization at large. Consequently, measures to be taken should, on the one hand, be aiming at the creation of an optimal degree of freedom or autonomy for the employees to individualize their jobs. On the other hand, the organization should provide its employees with adequate training opportunities.

Within the organizational constraints, this freedom, aimed at optimal decision latitude, should enable employees to arrange tasks and functions according to their own preferences, capacities and limitations. At the same time, interventions should also enable them to change jobs—when they want to—in order to prevent plateauing effects (chapter 17). It should, however, be remembered that an approach like the one suggested is only possible within a highly transparent organization with a serious and consistent HRM policy.

Of course, an approach like this demands sufficient trust towards the organization, as well as sufficient willingness to take the responsibility from the employees involved. A substantial lack of trust can be regarded as a contra-indication to use the approach advocated here. In less serious cases, however, some specific additional training in self-management will do.

To attain a greater freedom and individuality, the following measures can be taken:

(1) Job redesign in order to ensue a greater job decision latitude. A more optimal sequence of different task elements and breaks may result in a greater personal autonomy to arrange one's own job (Karasek 1979).

(2) Task enrichment, for instance, the instalment of task groups. Cooperation patterns can be redesigned, as in the so-called socio-technical approach. This approach aims at the joint optimization of the technical system and the social system at the level of work groups by diminishing the division of labour (Kompier and Pot 1993).

(3) Tuning jobs to the potential and capacity of the worker. For instance, the most taxing task elements can be diminished or eliminated, such as working night shifts and variable hours, or individualization of performance norms.

(4) Ergonomic improvement of work conditions. Physical work circumstances can be improved and attuned to the needs of the individual employee. Especially for employees with physical impairments, technology-based systems can make the job more practicable (Brzkoupil, chapter 26).

(5) 'Demotion', a voluntary transfer to an already existing and less taxing job often with lower responsibilities.

(6) Transfer to other jobs can take yet another form. When an operational job becomes too burdensome for an older employee, it is not unusual that such a person is transferred to a more supervisory or instructional kind of job. Especially older employees do have a lot of experience and knowledge, and it is only logical to use these assets. Apart from the obvious advantages for the individual older employee, the organization at large can gain a lot from such an approach too.

(7) Institutionalized job rotation to ameliorate qualitative underload owing to permanently doing the same job: either to a completely or partially different job or to the same job but in other locations.

(8) Creation of jobs in other organizations, for example, consultancy jobs in smaller companies or exchange of jobs between related

organizations. In The Netherlands, at the moment, we are witnessing the first attempts to exchange civil servants between municipalities and ministries. Furthermore, in Japan it is not uncommon for greater companies to establish daughter companies, mainly for the purpose of replacing older employees.

(9) Instigating special temporary projects. In such a project employees can participate and take in their particular skills and knowledge and in this way widen their competence and expand their social network.

(10) Changing working hours schedules. People can work full time, part time, with fixed or variable work hours. They can take longer, shorter, more or fewer breaks. They can take study leaves and, lastly, they can opt for a complete sabbatical year (Schlüter 1993).

(11) Different retirement schedules: earlier or later, partially or completely.

(12) Employee-adapted benefit systems (Lawler 1981). Within such a system employees can choose those benefits which suit them best. For example, they can choose among more holiday days or other benefits, such as financial benefits or opportunities of career advancement.

A number of these measures can supply older, and younger, employees with sufficient decision latitude and time to realize useful adaptations. In addition, however, the organization, that is the management and the specialists from the personnel department, have to pay close attention to their employees. Managers should be available to give information, advice, guidance and support, to the older as well to the younger employees, at least when it is asked for (Doeglas and Schabracq 1992). Besides, it would be sensible to provide employees, especially elderly ones, with periodical medical check-ups. Lastly, organizational health and fitness programmes, if on a voluntary basis, are also options (Fielding 1982, Falkenburg 1987).

Training

As stated before, training plays an important role within HRM. Training can act as a preventive as well as a curative measure, in particular for the aging worker. Three kinds of training will be mentioned:

(1) Skill training necessary for good job fulfilment, especially aimed at aging employees.

(2) Training for managers to implement age-conscious personnel policies.

(3) Self-management training for individual employees.

Training of skills

Elderly employees often get insufficient additional training. In order to improve this state of affairs the organization has to offer relevant and accessible courses and programmes, from which employees, by mutual arrangement with managers and personnel officers, can choose. However, such a menu should be tailored to the needs of older employees also. Often this is not the case: most training courses do not focus on and are not sufficiently adapted to the needs of older personnel.

A problem with some training courses is that they obviously conflict with the personal values of many older—and younger—employees. Obvious examples are many of the so-called humanistic approaches of about two decades ago and the recent 'new age' variants. In this kind of course, constant factors are the indiscriminate use of techniques guaranteed to evoke deeply felt emotions, images and meanings, and the presence of a leader who is fond of power. These courses still appear to be popular in organizations. As such, the whole management of the lubricants division of a petrochemical multinational has gone through a 'guided fantasy' programme and the regular management development programme of a solid international retail company includes 'encounter groups' including a variety of Gestalt techniques. Though many employees seem to go along with these programmes, these practices do influence the image of training courses in a negative way.

As stated before, criticisms of many training courses include: insufficiently tailored to the needs of the trainees, they go too fast for older employees or they are not geared to the specifics of the organizational culture. For instance, it is not uncommon that complete departments get the same training, notwithstanding the diverging needs of different employees. This may turn some courses into compulsory exercises in absurdity, which obviously reinforces negative opinions and prejudices about additional training.

A last problem with training is not inherant in the courses themselves but has to do with the fact that trainees often do not get the opportunity to practise and by that to reap the fruits. For this reason, it is necessary to enforce an active policy of job rotation, project management and job redesign, as described in the former section. If this does not happen, the newly learned skills and knowledge will vanish and the training effort becomes useless and meaningless. Because older employees, more than younger ones, attach considerable value to usefulness and meaning, sufficient and adequate room for practice is of essential importance.

In conclusion, a rather radical revision of training policies is called for.

Training in implementing age-conscious personnel policy

Middle and lower management are key to the introduction of the new policy, since they deal directly with aging personnel. Some additional training for this group is often necessary, and should be aimed at two areas: learning to neutralize the stereotyping process and managing the change inherent in the implementation of the new policies. Though it is common knowledge that stereotyping processes do operate when it comes to older employees, there exists a awkward lack of relevant research data to support this. However, notwithstanding the robust empirical finding that older employees show a relatively strong variation on all relevant variables, line managers[1] often regard them as a homogeneous group, characterized by low flexibility, energy, health and production, and as conservative and old-fashioned.

This stereotyping is very counterproductive when it comes to solving problems. Two processes are important here:

(1) Stereotyping can be considered as a case of blaming the victim, a strategy which exclusively attributes the cause of a problem to the problematic party, that is the elderly. This leads attention away from solutions outside the person, for example in the immediate work environment or the organizational culture. At the same time a favourable image of the organization and its management can be kept intact, although this way of thinking hinders the solution of the problem within the organizational context. Consequently, expelling the elderly from the organization seems the only feasible and reasonable 'solution'. The feasibility of this solution has, at least in most western countries and in Japan, been diminished considerably by the aging and dejuvenilization of the working population.

(2) Stereotyping has a tendency to act like a self-fulfilling prophecy. It implies an identification with and, in a way of speaking, an incorporation by the elderly of the status of being a problematic person. It determines and legitimizes the elderly individual's position as a problematic person within the organization, and considerably adds to the persistence of the problem. It is a well-documented fact that people who are treated as inferior develop ways of functioning that reinforce the ways in which they are treated (Vrugt and Schabracq 1991). They learn to live with their ascribed identity and find some compensation in it (less responsibility, lower standards, fewer demands, claims for support on other parties).

Courses about stereotyping can to a certain extent neutralize these processes by providing relevant knowledge and help to become more sensitive in this respect. One is made aware of the way in which stereotypes operate and learns to recognize these processes in the functioning of all parties involved. Though such forms of awareness are not always sufficient to completely prevent all stereotypical functioning, they often do serve as effective mental signals in this respect. Role-playing exercises, including exaggeration of role behaviour and different forms of role reversal, can be effective techniques here (see, for instance, French and Bell 1984).

Since the managing of the change inherent in the implementation of the new policies is a difficult task, some additional training in this respect is often needed, such as training in goal-setting techniques and general leadership skills (conducting different kinds of interviews, judging performances, setting clear targets acceptable to all parties involved, giving support and guidance) (Yukl 1990).

The main difficulty is the negative attitude to change among many (older) employees (Holm, chapter 20). Training in this area can profit from the managers' experiences of their own transitions as a start. Such an experience may give some understanding of resistance to change from an inside perspective. Subsequently, techniques can be taught to use obvious cases of resistance to change in a constructive way. Besides, using one's own experiences of change, sometimes made more clear in role playing, often can provide some interesting ideas about what, at the occasion, would have been good advice, support and help, as well as some insight into the courses of action a manager can better abstain from (Doeglas and Schabracq 1992).

Self-management

Obviously, our advocated approach requires a fundamental change in attitude and orientation from many individual (older) employees. He or she has to accept more responsibility for his or her own job and career. In a sense, they even have to take more responsibility for their own lives.

Such a change can be brought about by, for example, a 'self-management' programme (for details, see Doeglas and Schabracq 1992, Schabracq 1993). By emphasizing and reinforcing some values and motives, such a course can instruct employees to make better use of the opportunities for further development; improve their apprehension of their own position, possibilities and limitations; learn to set their own goals in this respect, learn to make plans in this direction and to structure their own jobs in such a way that their decision latitude is increased.

The emphasis on values and motives concerns the following issues:

—A general attitude of sufficient perceived self-efficacy: a solid belief in one's own potential and competence, and willingness to learn from one's own faults.

—The conviction that one can develop one's own competence and skills (by exposing oneself to new experiences and by seeking additional schooling), and that such enterprises are decisive in the steering of one's own life.

—Willingness to go into one's own strong and weak points (what one can do, wants to do and allows oneself to do). Often this includes the exploration of one's own so-called 'irrational thoughts' (Ellis and Harper 1975), inadequate assumptions about the nature of reality which steer one's perceptions and actions in an undesired way. Often these irrational thoughts can be traced backed to stereotypes operating in the organisation.

—The conviction that a good social network within the organization is an important asset and that one has to be actively involved in building and maintaining such a network.

—The conviction that one is, to a certain degree, responsible for one's

own well-being and health, and, accordingly, should arrange one's work so that it is the least damaging in this respect.
—The conviction that it is necessary to have a meaningful life, with significant relationships and challenging activities.
—Readiness to screen the organization and one's own job on the points mentioned, including opportunities for change, in order to come to personal goal-setting.

When there is sufficient insight into and compliance with these motives, employees are challenged to set personal goals in different areas (job, career, personal development, social network and leisure activities). In addition, the (older) employees learn how to make effective plans in this direction. They learn how to:

—translate long-term goals into a series of subgoals,
—establish priorities,
—phase and group activities,
—combine goals,
—delegate,
—prevent interferences with the realization of their plans,
—minimize effects of such interferences, and
—guard the ongoing process (Yukl 1990).

Essentially, the purpose of such a course is one of empowerment: employees learn to use the power that is offered to them—if it is offered to them—in a constructive way.

Conclusion

Gains and contraindications

When the human resource management approach described above comes off well, nine out of 10 of the categories of difficulties described by Schabracq (chapter 17), Cremer (chapter 19) and Holm (chapter 20) are directly or indirectly met, completely or to some degree.

On the individual level, employees gain some control: they get opportunities and power to arrange their own work and to choose their own challenges and safeties. By optimizing the balance between these two, they can become more centred in their work and experience something akin to a 'flow' of experience (Csikszentmihalyi 1988), the experience of being completely one with a challenging and exciting task. As a result, they may develop a stronger identification with their work, and maybe somewhat paradoxically, experience this as a kind of freedom. Generally speaking, this may imply a significant gain in individual well-being and, indirectly, health.

On the organizational level, those in charge have to let go of some of the control structures with which they used to rule the organization. This is not necessarily a real loss of power, as the loss is made up by a gain in employee trust and commitment. Also, the organization may make a more optimal use of its human resource potential, its organizational flexibility increases, as does the quality of the internal communication. Turnover and sick leave regress, and the risk of an acute shortage of key employees decreases.

Though the suggested approach can yield significant gains it by no means is a panacea; contraindications can be stated:

(1) The approach is only possible within a certain type of organizational culture. For instance, motives and values have to be compatible, at least in principle, with those operating within the organizational culture. Clearly this is not the case in most 'machine-bureaucracies' (Mintzberg 1979), or in organizations where the culture is thoroughly authoritarian. What is needed is a more democratically oriented organization with some experience with a form of systematic HRM, which needs to operate in a flexible way and which sincerely wants to make better use of the potential of its human resources.

(2) Among the employees involved, there must be some 'fundamental trust' in the intentions of the management. This implies that deeply rooted conflicts or the presence of many demotivated employees form serious contraindications. The latter often occurs in organizations with high sick leave and/or turnover (Van der Vlist and Pot 1993). Fundamental trust is related to another limiting condition, namely it needs rather quiet circumstances. That is, the chances of success are considerably smaller in times of acute crisis, since the concomitant emotional climate often is not an optimal condition for open-mindedness and voluntary change (Doeglas and Schabracq 1992).

(3) The work to be done has to provide sufficient intrinsic motivation; it must enable employees to attain work goals of sufficient importance (agreeable work, personal contacts, status, prestige, autonomy, meaningful products and so on; Schabracq 1993). Organizations where production processes are based on a rigid division of labour—resulting in fragmentary, short-cycled, machine-paced work tasks, with a small decision latitude and little opportunity for social contacts—are unsuitable for this approach.

Examples of organizations which do meet the demands of the advocated approach are mostly found among professional organizations: medical, advertising and consultancy companies, legal offices, research institutes, schools and training bureaux.

A condition of prime importance is that the approach requires a substantial change of the organizational culture. Acceptance and support in the organization for such an approach of all parties involved must be high. Additionally, without the unconditional support from top management the approach cannot succeed. The top management has to be convinced of the necessity and benefits of the proposed approach.

Union support is another critical condition. Unions may regard the approach as an instrument to discipline employees into complying with the organization goals. In this respect it is important to emphasize that the most important goal is one of increased freedom of choice for individual employees. No one is forced to change in a certain way. Also, it can be argued that the goals of the organization do not necessarily interfere with the goals of individual employees, since significant synergetic consequences are just as plausible.

A main difficulty, however, is that the advocated approach is expensive. Though some of the proposed measures, foremost in the area of terms of employment and variants of early retirement schemes, can be compensated for by some insurance arrangements, there are considerable costs which the organization needs to meet. Not taking these measures may, however, cost considerably more. In fact, inadequate use of the human resource factor may be one of the most costly forms of waste in terms of individual misery and missed opportunities, bad publicity, low flexibility and low morale, low product quality, substandard performances, vacancies which cannot be filled, high turnover and high social-security contributions.

On a national level this kind of training approach can help to reduce the costs of social security substantially. It might thus be a wise decision to spend a part of the social-security money on approaches like this.

A meta-issue

Although the chosen approach is one of reinforcing and empowering personal individuality, this may seem a strange approach as the growth of individualism is often designated as one of the main causes of crisis in our culture. In that view, inflated individualism is a destructive force, corrupting the structure of our society and leading to ecological disaster.

Besides, the psychoanalytical concept of 'narcissism' has been put forward to discredit individualism even more. As a semi-clinical term it locates the origins of this kind of functioning in a pathologically disturbed personality development. It recalls a frightening image by its mythological origin: Narcissus, literally drowning in his own reflection.

The foregoing line of thought has some validity, but only under certain circumstances. Individuality becomes inflated only when it is allowed to; when there is no solid structure of rules to function in; when there are too few limits to what one may do. Because these limits are social by nature—as they are set by the rules governing social interaction—inflation of individuality typically occurs when one is essentially alone, either physically alone or socially isolated. Through this moderating effect of loneliness and isolation, inflation of individuality is closely linked with power.

There are on the one hand, individuals who try to exercise as little power as possible. They avoid other people as much as they can: they isolate themselves, and live mainly in the private domain of their imagination and fantasies. As they can be in full control here—no limits whatsoever!—they can inflate their individuality as much as they wish: they have found a way to escape from the operation of social rules and the limits which they set. A few of them come up with something, a product of their fantasies, that also works in the outside world and that takes away their isolation. Most of them stay in their fantasy world, however, a way of functioning which is not unusual among people who get stuck in their jobs.

On the other hand, inflation of individuality also can occur when a person has so much power over other people that he or she is in a position to act out his fantasies in reality, because no one is able to stop him. Obviously, such a person too has withdrawn himself from the system of social rules and the limits inherent in it. However, this withdrawal implies that he or she, too, is in a very isolated spot: it is lonely at the top.

In conclusion, it can be said that withdrawal from the limits set by the social system, mediated by isolation and/or great differences in power, is an essential moderating factor for the development of inflated individuality. It makes it easy to understand that the approach sketched in this chapter does not represent much danger in this respect, since its main objectives consist of decreasing isolation and power differences. So, individual power here is not considered to be pernicious in itself, as long as it is moderated by an adequate structure of social rules. After all, this configuration of individuality and rule structure is, and always has been, one of the defining characteristics of western civilization. What is different here is the emphasis on the decreasing of power differences by empowering the less powerful. It may well be that, at least at the level of organizations, such an approach turns out to be a viable way to cope with ongoing change, which has become a defining characteristic of western culture as well.

Notes

1. It is interesting that many personnel managers, at least in the non-profit area, for example in Dutch municipalities (Friele 1993), hold a diametrically opposite position, even denying the existence of serious problems.

References

Csikszentmihalyi, M. 1988. Introduction. In M. Csikszentmihalyi and I. S. Csikszentmihalyi (eds) *Optimal Experiences* (Cambridge University Press, Cambridge).

Doeglas, J. D. A. and Schabracq, M. J. 1992. Transitie-management ('Transition management'), *Gedrag en Organisatie*, **5**: 448–466.

Dresens, C. S. S. H. 1991. *Hebben zij mij niet meer nodig? Facetten van een preventief leeftijdsbewust personeelbeleid (Don't they need me any more? Facets of a preventive age-conscious personnel policy)* (Koninklijke Vermande BV, Lelystad).

Ellis, A. and Harper, R. A. 1975. *A New Guide to Rational Living* (Prentice Hall, New York).

Falkenburg, L. E. 1987. Employee fitness programs, *Academy of Management Review*, **12**: 511–522.

Fielding, J. E. 1982. Effectiveness of employee health improvement programs, *Journal of Occupational Medicine*, **24**: 907–916.

French, W. L. and Bell, C. H. 1984. *Organization Development* (Prentice Hall, Englewood Cliffs, NJ).

Friele, J. 1993. *Leeftijdsbewust personeelsbeleid bij de gemeentelijke overheid* (*Age-conscious personnel policies in municipal administration*) (University of Amsterdam, Amsterdam).

Karasek, R. A. 1979. Job demands, job decision latitude, and mental strain: implications for job redesign, *Administrative Quarterly*, **24**: 285–308.

Kompier, M. A. J. and Pot, F. D. 1993. Diagnose van stress en welzijn ('Diagnosis of stress and well-being'). In M. J. Schabracq and J. A. M. Winnubst (eds) *Handbook 'Arbeid and Gezondheid'-Psychologie. Deel II: Toepassingen* (*Handboek Health and Work Psychology. Part II: Applications*) (Lemma, Utrecht).

Lawler, E. E. 1981. *Pay and Organization Development* (Prentice Hall, Englewood Cliffs, NJ).

Mintzberg, H. 1979. *The Structuring of Organizations* (Prentice Hall, Englewood Cliffs, NJ).

Schabracq, M. J. 1993. *Stress* (Kluwer Bedrijfswetenschappen, Deventer).

Schlüter, I. 1993. *Mogelijkheden voor ouderenbeleid en leeftijdsbewust personeelsbeleid* (*Possibilities for policies for the elderly and age-conscious personnel policies*) (University of Amsterdam, Amsterdam).

Van der Vlist, R. and Pot, F. D. 1993. Diagnose van inadequaat management ('Diagnosis of inadequate management'). In M. J. Schabracq and J. A. M. Winnubst (eds) *Handboek 'Arbeid and Gezondheid'-Psychologie. Deel II: Toepassingen* (*Handbook Health and Work Psychology. Part II: Applications*) (Lemma, Utrecht).

Vrugt, A. J. and Schabracq, M. J. 1991. *Vanzelfsprekend gedrag* (*Behaviour that goes without saying. Essays on nonverbal communication*) (Boom, Amsterdam/Meppel).

Yukl, G. 1990. *Skills for Managers and Leaders* (Prentice Hall, Englewood Cliffs, NJ).

22

Age and job performance

P. Warr

Summary. This chapter examines what is known about employee age and the effectiveness of job behaviour, concentrating on studies carried out in work settings and linking those with laboratory investigations. Rated job performance is shown to be in general unrelated to age; overall absenteeism tends to be greater among younger employees; accidents are more common at lower ages; and staff turnover declines with age. A framework is proposed in terms of three main components of job behaviour: physical ability, adaptability and general work effectiveness. The major practical need is seen to be the implementation of procedures which enhance among older employees the second of those components. An outline model is suggested, with the objective of identifying those jobs in which greater age is either a benefit or a limitation. That model indicates that, in research as well as in practice, the key issues now concern the nature and enhancement of adaptability.

Keywords: adaptability, effectiveness, expertise, job competence, learning, stereotypes, training, work performance.

Introduction

The overall finding from more than 100 research investigations is that there is no significant difference between the job performance of older and younger workers. The average correlation coefficient is about $+0.06$, but separate correlations range from -0.44 to $+0.66$ (McEvoy and Cascio 1989, Warr 1994). Research results demonstrate differences between jobs and between different dimensions of performance, but the general pattern is clear: there is no overall difference between the performance observed for older and younger staff in the same job. Incidentally, in almost every case variations within an age group far exceed the average difference between age groups.

The large majority of studies have assessed performance solely through supervisors' ratings of specific competences or of overall effectiveness. Furthermore, in most cases results are presented only as simple correlation coefficients, so that we cannot identify the pattern of scores (linear or non-linear) across different ages. However, some published results about actual job performance (rather than ratings) at different ages have been collated in table 22.1.

Standardizing the scores around those for workers aged between 35 and 44, it can be seen that in these studies there is generally an increase in output up to that age and then either a continuing increase, a plateau, or a small decline. In respect of the equipment service engineers, there is a more rapid decline after the middle range. It is important to stress, however, that the decline at older ages in that study was found *only* for employees who had not recently been trained; I will come back to that issue later.

(It is often found that older workers are more consistent in their performance than younger ones.) For example, Walker (1964) observed in a study of mail sorters a steady increase in consistency, with less variation in output from week to week, across groups from under 25 years to 60 and over; the oldest group was 60 per cent more consistent than the youngest group. British shoe-leather cutters were found by De la Mare and Shepherd (1958) to work more slowly at older ages, but older employees produced work of a higher quality, at least up to age 60. Bowers (1952) reported more positive appraisals of conscientiousness and attendance for older workers, in a sample where ages extended into the late 60s.

The absence of an overall age effect in the level of job performance does, of course, contrast sharply with laboratory evidence of very clear age decrements in many forms of information processing. And it is inconsistent with a wide-

Table 22.1. *Age and job performance: Examples of research findings through measures of output.*

	under 25	25–34	35–44	45–54	55 and above
1. Equipment service engineers (UK)		99	100	94	
2. Skilled manufacturing operators (USA)	77	85	100	106	106
3. Semi-skilled assembly workers (USA)	89	87	100	105	101
4. Mail sorters (USA)	101	102	100	101	99
5. Office workers (USA)	92	99	100	99	98
6. Manufacturing machine operators (USA)	96	100	100	97	94

Sources: 1. Sparrow and Davies 1988; 2. Giniger et al. 1983; 3. Schwab and Heneman 1977; 4. Walker 1964; 5. Mark 1957; 6. Kutscher and Walker 1960.

spread negative stereotype about older workers. We clearly need to learn about the factors differentiating between age patterns in different situations and different behaviours. Looking at individual research reports is not very helpful in that respect, because they usually give very little information about the nature of the jobs and the individuals concerned. This point will be developed later.

In terms of other job behaviours, the research evidence is as follows. Although sickness absence is often found to be greater at older ages, the opposite is the case for 'voluntary' absence, when people take time off work without medical or organizational approval (Martocchio 1989, Hackett 1990). The overall age pattern of absenteeism thus depends on the mix of those effects in a particular organization, and in many cases there is either no overall age difference or younger employees are absent to a greater extent than older ones. Second, accidents are more common at younger ages, especially among inexperienced workers; that significant negative association with age is stronger for men than for women (Dillingham 1981). And, third, older staff are less likely to leave their employer voluntarily (e.g. Doering et al. 1983), partly of course since they tend to be relatively unattractive on the labour market. One implication of the lower staff turnover at older ages is that the financial payback from training older employees can be greater than in the case of younger ones, since younger staff are more likely to move away to another employer.

Views held by personnel managers

In order to learn more about the structure of perceptions in this area, Warr and Pennington (1993) asked members of the (British) Institute of Personnel Management to compare older and younger non-managerial workers at their site on a range of different performance dimensions. 'Older workers' were for that study defined as those aged 40 or above.

Analysis of variance with varimax rotation revealed that there were two clear factors in their judgements. Details are shown in table 22.2. In respect of the first factor, older workers were perceived to be *more* effective than younger ones, but in respect of the other factor (labelled 'adaptability'), older workers were instead seen as the *less* effective group.

In relation to the second component, it is not surprising that older employees were seen as less adaptable and slower at acquiring new skills and knowledge. We know from other research that that is the case (e.g. Sterns and Doverspike 1989, Warr 1994). For example, American air-traffic-control trainees were found by Trites and Cobb (1964) to have progressively greater difficulty in acquiring job skills between the ages of 21 and 51. However, it must be stressed that there are wide differences in adaptability at all ages.

The larger factor was what we have labelled general work effectiveness. That covered a wide variety of behaviours, and in every case it was the older group

Table 22.2. Factor analysis of perceived differences between older and younger workers at a respondent's site, with rotated item loadings in brackets.

Factor 1: General work effectiveness	
are conscientious	(.73)
are reliable	(.69)
work hard	(.67)
are effective in their job	(.62)
think before they act	(.60)
are loyal to the organization	(.55)
have interpersonal skills	(.55)
do not take things easy	(.53)
work well in teams	(.50)
are receptive to direction	(.38)
Factor 2: Adaptability	
are able to grasp new ideas	(.77)
adapt to change	(.76)
accept the introduction of new technology	(.72)
learn quickly	(.88)
are interested in being trained	(.46)

who were seen as significantly *more* effective: in terms of reliability, conscientiousness, work effectiveness, thinking before they act, working well in teams, and so on.

That pattern of views held by managers appears to reflect the findings from research into individual employees summarized above. In many respects, older employees are on average more effective than younger ones, but in certain key respects they present a problem. It is helpful to think in terms of table 22.3.

We can view job activities from three perspectives. Physical abilities are discussed in other chapters, and some declines clearly do occur as workers grow older. Of the other two factors, adaptability tends to decline (again with wide individual variation), but the many aspects labelled as general work effectiveness remain stable or increase with age. Older workers thus have a great deal in their favour, but they have certain *specific* limitations, in respect of the first two factors.

'The problem' of the older worker is thus a restricted one. Despite a common stereotype of generalized decline with increasing age, in many or most respects age brings definite and recognized benefits: those identified as General Work Effectiveness. However, cognitive difficulties particularly arise in terms of Adaptability, and those are especially significant in the present

Table 22.3. The older worker: Three components of job behaviour.

1. Physical abilities:	Tend to decline
2. Adaptability:	Tends to decline
3. General work effectiveness	Remains stable or tends to increase

period of rapid technological change. It follows that a sound practical approach is for organizations not to reject older workers on the basis of an excessively pessimistic overall negative stereotype. Instead, they should concentrate on enhancing Adaptability, in order to obtain the widespread benefits of experience which older employees provide. I will return to that issue later.

A Four-Category Framework

A general theme in this chapter is that we should not expect to find a decline in all aspects of job performance with increasing age. There are many work activities in which age is a definite advantage or at least neutral. That positive perspective contrasts sharply with the findings from many laboratory studies of cognitive processes. Those investigations focus upon *maximum* intellectual performance in relatively complex activities, often in circumstances where speed of response is specially important. Work behaviour is not usually of that kind. A person rarely has to operate at maximum levels for more than a short period, and success in most jobs is a function of motivational factors and interpersonal behaviour, as well as merely aspects of cognitive functioning.

A related point concerns the effectiveness of people's overall work strategies. Irrespective of *specific* cognitive competences, as assessed in laboratory experiments, people differ in their overall approach to a job, its planning, how they combine different activities together or in sequence, and how they choose which tasks to prioritize and which to treat as less important. Those higher-level strategic processes are sometimes referred to as 'metacognitive' activities, and there is no reason to expect them all to deteriorate with age in the way that many specific maximum performances undoubtedly do decline. In fact, the opposite pattern is often to be expected: with increasing experience, a person can develop particularly effective strategies and broad-based styles of working.

In seeking to identify aspects of jobs that are likely to remain stable or to change across the working years, I have found a four-category framework quite useful. A version of that is presented in Table 4.

The four categories of course overlap rather than being completely separate as in the table. And the key problem is to determine which job activities are likely to fall into each category. However, the table clearly makes the point (in the fourth column) that a negative relationship with age is expected in only one out of four sets of conditions.

Two features have been combined into a 2 × 2 table. The first feature is in terms of basic capacities of a physiological or fundamental psychological kind. For instance, it is well established that response speed, working memory and selective attention decline with age, especially in difficult tasks (e.g. Warr 1994). In those activities (task categories 1 and 2 in table 22.4), basic capacities are more exceeded with increasing age. Incidentally, most declines of that kind are likely to be greatest after the age of retirement from the labour force.

Table 22.4. Four categories of job activity and expected relationship of performance with age.

Task category	Basic capacities and more exceeded with increasing age	Performance can be enhanced by experience	Expected relationship with age	Illustrative job content
1	Yes	No	Negative	Continuous paced data-processing; rapid learning; heavy lifting
2	Yes	Yes	Zero	Skilled manual work
3	No	No	Zero	Relatively undemanding activities
4	No	Yes	Positive	Knowledge-based judgements with no time pressure

However, there are also circumstances in which basic capacities are not further exceeded as people grow older; those are shown in task categories 3 and 4. Working within capacity usually occurs in settings where response speed is of no great consequence, where established routines are being followed, or where the task is quite simple. We should remember that many working activities are of those kinds.

The second factor in this framework (in column three) concerns the gains from experience which come with increasing age. People acquire helpful knowledge and skills, routines of behaviour, effort-saving procedures and strategies and wider perspectives on problems than may be the case at younger ages. However, those gains are of benefit in only some activities, so it is important to consider both possibilities: where performance either is or is not enhanced by experience ('yes' or 'no' in column three).

Let us look briefly at the four categories in the framework, recognizing that they are not in fact completely separate.

Task category 1: Age-impaired activities

Within task category 1, basic capacities are exceeded to a greater extent for older people and experience cannot help. Tasks of that kind include continuous rapid information processing and some forms of strenuous physical activity. For example, reaction time in a wide range of tasks is known to be longer as people become older. Salthouse (1985) illustrated this by presenting more than 50 correlations between age and speed of performance on laboratory tasks (single and choice reaction time, card-sorting, digit–symbol substitution and so on). The median correlation was −0.45, with a range across the studies from −0.15 to −0.64. Most values were derived from samples extending beyond working age, but even within the 15 groups aged below 65, the median correlation with age was −0.43.

Older people's slower handling of information may give rise to poorer performance on any task that requires rapid cognitive processing. That is obvious in the case of reaction time and similar activities, but can also be expected in situations where active mental rehearsal, comparison between alternatives and temporary storage in memory are needed. Cognitive capacities are limited and there is a need to pass information through the system as rapidly as possible before it is lost or overtaken by other material. Particularly complex tasks, requiring a large number or processing steps, are especially likely to be susceptible to cognitive slowing (e.g. Myerson et al. 1990).

Specially important in category 1 are processes of 'working memory'. In those cases, a person has to carry out mental operations on one set of material, while temporarily retaining in storage material that has to be brought into active processing later in the task. Items in working memory are more likely to become forgotten or inaccessible as a person grows older. Since working memory is centrally involved in many complex mental processes, that deterioration can have negative effects on a range of different activities (e.g. Salthouse 1991).

Other circumstances in which a negative age gradient is expected include jobs whose content is changing rapidly, so that previous knowledge and skills can become obsolete. For example, Dalton and Thompson (1971) reported a marked (cross-sectional) decline in the performance of professional engineers after the age of about 40, as their technical knowledge became increasingly out of date. This process was viewed in part as a negative spiral: an older engineer who lacks current knowledge receives poorer appraisals and less challenging assignments, so that he or she comes to feel discouraged and less willing to make an effort for retraining; this discouragement in turn leads to more negative appraisals and further discouragement. Such a process has been analysed in more formal theoretical terms by Fossum et al. (1986); the importance of training at older ages is emphasized again later in this chapter.

Task category 4: Age-enhanced activities

At the opposite extreme (in task category 4), however, work remains within basic capacities despite advancing years and performance is improved as a result of experience. Note, incidentally, that this effect of experience may not be linear, continuing equally across all years in a job; there are grounds for expecting the benefits from experience to level off after an initial period of enhancement.

Examples of task category 4 often involve substantial knowledge relevant to the job. For instance, in a study of an American company's sales staff older employees were rated much more positively than younger ones in almost every respect. In terms of an employee's product knowledge, the correlation with age was as high as $+0.67$ (Maher 1955). That positive association is surprisingly strong, until you read that 'since the company carries over 1000 items in its catalogue, a man may require many years to know the technical characteristics of the majority of them'. Increased age and time in the job are clearly helpful in this respect. The value of experience in enhancing social knowledge and interpersonal skills was particularly emphasized by Perlmutter et al. (1990). In their study of food-service employees between 20 and 69, age was correlated $+0.36$ with performance effectiveness.

Increased experience in a particular area of work (which is strongly associated with age) brings about cognitive and affective changes often identified as 'expertise'. In several different domains, comparisons between the performance of novices and experts have distinguished between 'declarative' and 'procedural' knowledge (e.g. Anderson 1982). Expertise is accompanied by greater knowledge of both kinds. In the first case, a person acquires information about individual facts and their relationships in different situations. During learning, this declarative knowledge is gradually converted into a set of behavioural procedures through which it is applied in dealing with the environment. These procedures become collapsed into increasingly long strings of action, which the person can execute as a whole and which he or she can readily assess for appropriateness and generalizability.

Expertise thus includes greater 'automatization' of behaviour, as people move from controlled, effortful cognition to execute fast strings of action, which are not under direct control once initiated but which permit simultaneous processing of new information. For example, only about one-third of the speed-up that Charness and Campbell (1988) observed in a two-digit mental squaring task was a result of faster execution of elementary arithmetic operations; most of the speed-up was a consequence of learning how to chain together the sub-goals quickly and efficiently. Other aspects of expertise have been reviewed by Charness (1989), Glaser (1988) and Warr and Conner (1992). For instance, experts perceive and recall large meaningful patterns in their domain, made possible by their superior and more organized knowledge base. However, research has emphasized that expertise is very much domain-specific; an expert in one area may be relatively ineffective elsewhere.

Task category 3: Age-neutral activities

Activities in this third category are those in which work is unproblematic and fairly routine, such that no noticeable age gradient in either direction is expected. Many work tasks are of course of that kind, especially those that are fairly easy. For instance, 'primary' memory is apparently unaffected by age;

older people are as able as their younger counterparts to hold in memory small amounts of information (within their span of primary memory) that are being used in uncomplicated cognitive activities (e.g. Poon 1985). Furthermore, while increased age is associated with slower learning of new material (see above), it is unrelated to the speed of forgetting of material after it has been acquired (Rabbitt and Maylor 1991). Age differences are also absent in what Miles (1933) described as 'spontaneous imagination', generating ideas based on ambiguous stimuli.

Task category 2: Age-counteracted activities

This task category is particularly interesting. In these cases, older people may have increasing difficulties because of a decline in information processing or physical capacities, but they are able to compensate for that decline in some way. Additional knowledge or helpful habits may be available, employees may take more written notes, or behaviours may be modified to capitalize on personal strengths or to avoid situations known to present difficulties. For instance, Birren (1969) has described how middle-aged managers may have learned to conserve their energy and time by operating through day-to-day tactics which reduce cognitive and affective load.

Category 2 in the framework of table 22.4 includes activities to which the models of Baltes and others are most applicable, in terms of 'selective optimization with compensation' (e.g. Baltes and Baltes 1990). Those models are based on the fact that, as people age, they are able to increase their effectiveness in areas of specialization. Continued interest and practice in a limited number of areas permit the growth of knowledge-based competence; and individuals are sometimes able to learn how to compensate as necessary for limitations arising from deteriorating basic capacities. Outside their areas of specialization, however, older people may find it more difficult to acquire new skills. That difficulty may be a result of greater capacity limitations with age and/or through reduced confidence in some kinds of new situations.

Older people's successful cognitive compensation for decline in information-processing ability has been documented in relation to chess-playing and transcription-typing. In the first case, Charness (e.g. 1989) points out that chess depends heavily on the ability to think ahead, which in turn makes continuing demands on working memory. He has shown that, among equivalently skilled players, older people have poorer memory for individual chess positions than younger ones. On the other hand, older players were found to be equally good at choosing the best moves from given positions. They searched less widely (although they looked ahead to the same extent) and from that more economical (and more rapid) procedure they were as effective as younger players, despite some memory deficits with age.

A second example of the combined effect of capacity decrement and a gain from experience is the study by Salthouse (1984) of typists aged between 19 and 72. Although older typists were clearly impaired in separate measurements of response speed, they were able to type as fast as the younger ones. It turned out that older typists achieved this by means of looking further ahead along the line to be typed, so that they were processing at any one time longer chunks of material than were younger typists. That greater anticipation permitted older people to compensate for declining perceptual-motor speed.

Published research into job performance is not easily fitted into the four-part framework of table 22.4, since very little information about task activities and sampling procedure is included in most reports. Much of the available evidence is also rather old and we now need some new and more complex investigations. My colleagues and I are working with several organizations to try to create a firmer research base about performance and age; we would be delighted to hear from anyone who might like to collaborate in that project.

Some key issues

Three key issues need attention at the present time. First, we must look much more carefully at those features of jobs and career progression which might underlie an observed age pattern. There is at present too great a willingness to take an observed correlation with age at its face value.

For instance, a *negative* age gradient, where older job-holders are seen as less effective than younger ones, can arise for a number of different reasons. Although laboratory psychologists tend to emphasize the importance of cognitive declines with age, those are not relevant in many jobs and they will often be of little practical significance.

More important factors underlying research results may arise from a company's policies and procedures for promotion. In some cases, the best workers have already been promoted out of the position. Comparisons between older and younger people in such a situation are thus not comparing like with like. In cases of selective promotion out of a job, we cannot infer from observed lower scores at older ages that older job-holders are necessarily less effective. We can conclude only that the current older incumbents are relatively poor performers, perhaps because of differences in the kinds of people involved at the different ages.

In other cases, a *positive* age gradient might also arise artefactually, because an investigator has combined within a single study several levels of job within a career progression. Older people tend to be in the higher grades and higher grades tend to obtain more positive ratings. If a study has included staff from several hierarchical grades, the statistical tendency will be toward a positive age gradient: higher scores for older people, who are usually in the higher grade. In that case, the pattern may be very different from the findings one would obtain within merely a single grade of job.

For instance, in a study of bank employees, we found that for the complete sample the correlation between average rating of work effectiveness and age was around zero. When *separate* analyses were carried out for the two grade-levels making up the overall sample, however, there was a strong negative association with age among the older grade. If we had published only the overall pattern (which is what most investigators do), the results would have been quite misleading.

Another important issue concerns the bases of between-person variations. The magnitude of any age gradient is not the same for all kinds of people and we need to learn more about the reasons for differences between people in their pace of aging. If we knew more about that, we could perhaps act to reduce decrements or potential decrements in particular cases.

For example, what about the decline in basic capacities: how might we explain between-person variations in the steepness of those declines? In respect of cognitive decrements, there is accumulating evidence that differences in cardiovascular fitness may be implicated, associated with oxygen transport to the brain. Inter-individual differences in that form of fitness, assessed in terms of lung function, have been examined among male British white-collar workers by Bunce et al. (1993) The frequency of brief mental blocks in a choice reaction-time task was found to increase between the ages of 17 and 63, but that increase was significantly greater for unfit individuals than for men who were relatively more fit. Future research is likely to identify more precisely the physiological bases of different kinds of age-dependent cognitive processes.

Another possible influence on the pace of aging is the degree of exposure to an intellectually challenging environment. It is possible that workers who are required to operate at a high level of cognitive complexity throughout their adult life may show less intellectual decline than those who have little cognitive challenge in their jobs. For example, Kohn and Schooler (1983) summarized an extensive programme in which several forms of 'intellectual flexibility' were found to be influenced by the complexity of a person's current job and that held 10 years previously. Related research has indicated that age is less important in the performance of people who are expert at a particular task. Rabbitt (1991) reported that, in the completion of crossword puzzles, experts' performance was not associated with their age (between 55 and 75), whereas the (poorer) performance of novices declined cross-sectionally across those years. Similarly, Allen et al. (1992) illustrated how highly practised aspects of mental multiplication are unaffected by age.

A third set of issues concerns the second feature in the framework of table 22.4: performance enhancement by experience. Relevant experience can often be provided by an employing organization, so that in many cases a specific negative age gradient can in fact be prevented or reduced. To put that another way, if we find that older employees are at present less effective than younger ones in a particular job, that difference may not be fixed and unchanging. It could be that the older workers mainly need more opportunities to learn.

In most organizations, training is directed especially at younger employees. This is partly because employers think they are likely to gain a better financial return from younger people, since they have more years ahead of them.) But the emphasis on younger workers' training is also sometimes a consequence of what is in effect collusion between management and older workers themselves. Many older people are nervous about being trained, they feel that they are less able to learn than previously and they are not sure that they will subsequently have opportunities to use many of the new skills they might acquire.

So there is sometimes a self-fulfilling prophecy: older workers in a particular company may indeed be less adaptable, but usually they receive no assistance to be otherwise. As a result, they remain as they are and the negative stereotype is reinforced. A major organizational requirement is to create a culture in which learning and development are given a high priority among older as well as younger employees. Companies should assist older workers to move laterally into novel and challenging jobs and provide training that is specially tailored to their requirements. Despite widely held negative beliefs, older employees can learn extremely well; but the training techniques devised for younger learners are often inappropriate at older ages (e.g. Belbin and Belbin 1972, Sterns and Doverspike 1989).

Issues of training are thus at the heart of 'the problem' of older workers. In relation to the job performance issues discussed here, a key requirement is to learn more about the specific nature of experience which gives rise to greater expertise. A negative age gradient in job performance may indeed sometimes be observed, but in many cases that is likely to be at least in part reversible through appropriate training and learning activities.

At present, too many people are liable to overgeneralize. Recognizing that older employees are less adaptable than younger ones, they conclude that this limitation applies in all respects. But it does not. Older employees score particularly well on those other important features, which extend widely across what has been labelled as general work effectiveness. It is important to emphasize those positive features, while also recognizing the need to enhance adaptability.

References

Allen, P. A., Ashcraft, M. H. and Weber, T. A. 1992. On mental multiplication and age, *Psychology and Aging*, 7: 536–545.

Anderson, J. R. 1982. Acquisition of cognitive skill, *Psychological Review*, **89**: 369–406.

Baltes, P. M. and Baltes, M. M. (eds) 1990. *Successful aging* (Cambridge University Press, Cambridge).

Belbin, E. and Belbin, R. M. 1972. *Problems in Adult Retraining* (Heinemann, London).

Birren, J. E. 1969. Age and decision strategies, *Interdisciplinary Topics in Gerontology*, **4**: 23–36.

Bowers, W. H. 1952. An appraisal of worker characteristics as related to age, *Journal of Applied Psychology*, **36**: 296–300.

Bunce, D. J., Warr, P. B. and Cochrane, T. 1993. Blocks in choice responding as a function of age and physical fitness, *Psychology and Aging*, **8**: 26–33.

Charness, N. 1989. Age and expertise: responding to Talland's challenge. In L. W. Poon, D. C. Rubin and B. A. Wilson (eds) *Everyday Cognition in Adulthood and Late Life* (Cambridge University Press, Cambridge), 437–456.

Charness, N. and Campbell, J. I. D. 1988. Acquiring skill at mental calculation in adulthood: a task decomposition, *Journal of Experimental Psychology: General*, **117**: 115–129.

Dalton, G. W. and Thompson, P. H. 1971. Accelerating obsolescence of older engineers, *Harvard Business Review*, **49**(5): 57–67.

De la Mare, G. C. and Shepherd, R. D. 1958. Ageing: changes in speed and quality of work among leather cutters, *Occupational Psychology*, **32**: 204–209.

Dillingham, A. E. 1981. Age and workplace injuries, *Aging and Work*, **4**: 1–10.

Doering, M., Rhodes, S. R. and Schuster, M. 1983. *The Aging Worker: Research and Recommendations* (Sage, Beverly Hills).

Fossum, J. A., Arvey, R. D., Paradise, C. A. and Robbins, N. E. 1986. Modeling the skills obsolescence process: a psychological/economic integration, *Academy of Management Review*, **11**: 362–374.

Giniger, S., Dispenzieri, A. and Eisenberg, J. 1983. Age, experience and performance on speed and skill jobs in an applied setting, *Journal of Applied Psychology*, **68**: 469–475.

Glaser, R. 1988. Thoughts on expertise. In C. Schooler and W. Schaie (eds) *Cognitive Functioning and Social Structure over the Life Course* (Ablex, Norwood, NJ), 81–94.

Hackett, R. D. 1990. Age, tenure and employee absenteeism, *Human Relations*, **43**: 601–619.

Kohn, M. L. and Schooler, C. 1993. *Work and Personality* (Ablex, Norwood, NJ).

Kutscher, R,E. and Walker, J. F. 1960. Comparative job performance of office workers by age, *Monthly Labor Review*, **83**(1): 39–43.

Maher, H. 1955. Age and performance of two work groups, *Journal of Gerontology*, **10**: 448–451.

Mark, J. A. 1957. Comparative job performance by age, *Monthly Labor Review*, 80, 1467–1471.

Martocchio, J. J. 1989. Age-related differences in employee absenteeism: a meta-analysis, *Psychology and Aging*, **4**: 409–414.

McEvoy, G. M. and Cascio, W. F. 1989. Cumulative evidence of the relationship between employee age and job performance, *Journal of Applied Psychology*, **74**: 11–17.

Miles, W. R. 1933. Age and human ability, *Psychological Review*, **40**: 99–123.

Myerson, J., Hale, S., Wagstaff, D., Poon, L. W. and Smith, G. A. 1990. The information-loss model: a mathematical model of age-related cognitive slowing, *Psychological Review*, **97**: 475–487.

Perlmutter, M., Kaplan, M. and Nyquist, L. 1990. Development of adaptive competence in adulthood, *Human Development*, **33**: 185–197.

Poon, L. W. 1985. Differences in human memory with aging: nature, causes and clinical implications. In J. E. Birren and K. W. Schaie (eds) *Handbook of the Psychology of Aging* (2nd edn) (Van Nostrand Reinhold, New York), 427–462.

Rabbitt, P. M. A. 1991. Management of the working population. *Ergonomics*, **34**, 775–790.

Rabbitt, P. M. A. and Maylor, E. A. 1991. Investigating models of human performance, *British Journal of Psychology*, **82**: 259–290.

Salthouse, T. A. 1984. Effects of age and skill in typing, *Journal of Experimental Psychology: General*, **113**: 345–371.

Salthouse, T. A. 1985. Speed of behavior and its implications for cognition. In J. E. Birren and K. W. Schaie (eds) *Handbook of the Psychology of Aging* (2nd edn) (Van Nostrand Reinhold, New York), 400–426.

Salthouse, T. A. 1991. Mediation of adult age differences in cognition by reductions in working memory and speed of processing, *Psychological Science*, **2**: 179–183.

Schwab, D. P. and Heneman, H. G. 1977. Effects of age and experience on productivity, *Industrial Gerontology*, **4**: 113–117.

Sparrow, P. R. and Davies, D. R. 1988. Effects of age, tenure, training and job complexity on technical performance, *Psychology and Aging*, **3**: 307–314.

Sterns, H. L. and Doverspike, D. 1989. Aging and the training and learning process. In I. L. Goldstein (ed.) *Training and Development in Organizations* (Jossey-Bass, San Francisco), 299–332.

Trites, D. K. and Cobb, B. B. 1964. Problems in air traffic management III. Implications of training-entry age for training and job performance of air traffic control specialists, *Aerospace Medicine*, **35**: 336–340.

Walker, J. F. 1964. The job performance of federal mail sorters by age, *Monthly Labor Review*, **87**(3): 296–301.

Warr, P. B. 1994. Age and employment. In M. Dunnette, L. Hough and H. Triandis (eds) *Handbook of Industrial and Organizational Psychology* (2nd edn), Vol. 4 (Consulting Psychologists Press, Palo Alto), 485–550.

Warr, P. B. and Conner, M. T. 1992. Job competence and cognition. In B. M. Staw and L. L. Cummings (eds) *Research in Organizational Behavior*, Vol. 14 (JAI Press, Greenwich, Conn.), 91–127.

Warr, P. B. and Pennington, J. 1993. Views about age discrimination and older workers. In *Age and Employment: Policies, Attitudes and Practices* (Institute of Personnel Management, London), 75–106.

Conclusion

In this summarizing conclusion of part III, the disadvantages of an exclusively cross-sectional approach are commented on and a possible line of future research into the problems with which the elderly are confronted within organizations is elaborated.

The infringements on the integrity of individual functioning, and the experienced loss of power and identity which result from it, form the main source of problems that many older employees experience. The causes of these infringements are, however, so deeply ingrained in our culture and our ways of thinking and acting that we are liable to see them as necessary parts of our reality. Consequently, these causes are not easy to influence. In order to be able to change organizations effectively, a better understanding is needed of the dynamics of the dialectical process of personal and organizational development, and its effects on the functioning and well-being of older employees.

Until now, most studies in this respect, including the one described by Boerlijst (chapter 18), have been survey studies of a cross-sectional nature. However, this kind of research has some serious limitations. They are, necessarily, limited to employees who are still working, leaving aside those who have left the organization. Though useful as a way of taking stock of that current situation by supplying us with a survey of the values of different variables, it has little to offer when it comes to explanations.

For example, data from transversal studies show rather consistently that elderly employees, on average, see themselves as relatively healthy (Broersen et al. 1993) and satisfied with their work (Haasnoot 1993). To make sense of these data, several explanations have been offered. So, there is the so-called 'healthy worker effect' hypothesis (Ilmarinen 1991), which states that the cause must be sought in the one-sided 'natural' selection of subjects. In addition, however, there are explanations which state that some kind of 'response shift' is the relevant variable (the 'grinding down' and 'accommodation' hypotheses).

Still another explanation points to the overall positive change in the functions of elderly employees as the variable responsible for this phenomenon and, lastly, there are explanations based on cohort differences (see Mottaz 1987, in Haasnoot 1993). The point is that the relative validity and importance of these explanations cannot be examined by an exclusively cross-sectional approach.

A longitudinally oriented approach seems to be more promising. The case-study by Holm (chapter 20) is a good example of such an approach. However, an obvious problem with longitudinal studies often is that they are awfully time-consuming. A feasible alternative, though, consists of a retrospective approach, using the memories of those involved as research material. In this way, the development of individual careers, organizational cultures and their interactions can become serious research objects. Though this kind of study has its own known limitations, such studies can be useful instruments to enrich our understanding of the development of problems of older employees in fast-changing organizations.

For example, in a study of 30 disabled supermarket employees from a large Dutch retail company (Krijnen 1993) retrospective interviews were used in order to study the history of their disability. A rather unexpected, but consistent, finding was that subjects considered their period working for the retail company generally as the most agreeable and happy one of their lives (compared with elementary school, high school and first job). Also most of them described their current work as a time where they felt more popular than in former periods of their lives. Though the purport of these data is far from clear-cut, it does throw some doubt on the apparently self-evident validity of the 'healthy worker' hypothesis.

Apart from supplying us with a better understanding of individual and organizational functioning, and their potential drawbacks for well-being, health and productivity, such studies can also point out possibilities for feasible interventions. Obviously, evaluation studies with respect to these interventions are another instance of such a longitudinal approach.

An interesting research area for this kind of approach would be the relation between work motivation and variables such as workload, fatigue and health. For example, studies into the motivating qualities of different kinds of work, as well as studies into the role played by the individual application of present cultural meanings and ideas in this respect, may give some answers to the following questions:

—What makes people keep up work which looks, at least to external observers, completely unattractive?
—What makes people try to keep up high standards of performance?
—How are these factors interfered with: what are critical incidents here?
—How are critical incidents dealt with in terms of coping and power tactics?

This approach implies a modest kind of research—actually the same as was used by Holm in her case-study. The emphasis is on the use of observation of the actual work process and of interviewing informants. This means that this kind of research, in addition to the traditional psychometrical, psycho-physiological and statistical methods and measures, can make use of the methods developed in the fields of ethology and cultural anthropology.

Finally, it is obvious that this line of research is not the only possible one and that other kinds of research would also be viable here.

References

Broersen, J. P. J., de Zwart, B. C. H., Meijman, T. F., van Dijk, F. J. H., van Veldhoven, M. and Schabracq, M. J. 1993. *Atlas Veroudering, Werk en Gezondheid* (*Atlas of Aging, Work and Health*) (University of Amsterdam, Amsterdam).

Haasnoot, M. 1993. *Leeftijd en arbeidstevredenheid* (*Age and work satisfaction*) (University of Amsterdam, Amsterdam).

Ilmarinen, J. 1991. The aging worker, *Scandinavian Journal of Work, Environment and Health*, **17**(suppl. 1): 1–141.

Krijnen, M. A. 1993. Onderzoek WAO-instroom 1991 AH-operations (Study WAO Influx 1991 AH Operations). Internal Report, Faculty of Psychology, University of Amsterdam.

Mottaz, C. J. (1987). Age and work satisfaction, *Work and Occupation*, **14**: 387–489.

Part IV
Social policy and
perspectives

Introduction

In part IV some reasons as to why elderly people are less and less involved with work are discussed.

Lately, work and aging do not seem to be very compatible. As has been evident from the proceeding chapters, getting older nowadays can be characterized as a stage in which the individual 'suffering' from it slowly but surely loses some of his or her most enviable abilities. Thus, the elderly, and more specifically the elderly worker in the production process, run the risk of becoming a burden rather than a contributor to society. The elderly worker, also subject to this way of perceiving getting older, more or less aligns to this way of thinking and feeling: they tend to see themselves as less worthwhile to society as well. Thus the stereotype is confirmed by both sides, younger people as well as the elderly themselves: getting older and becoming less of a complete person—in one way or another—is by now a truism. In a way, the publication of this book testifies to this development of values. In short, aging at this point in time could be described as a disability we all have to deal with sooner or later in our lives. However, it should be mentioned that the way the elderly are seen nowadays, or rather the current perception of young adulthood as the most favourable age, has not always been the same (Ariès 1962).

One might ask oneself what reasons lie behind this development. Though there might be many, one that seems very important is economic. Economies that have been able to support most citizens in western society with a previously unrivalled standard of living had to pay a price. In sustaining the level of prosperity, efforts in western economies were directed at keeping production at the highest possible level, both quantitatively and qualitatively. One prerequisite to this is to employ workers who are able to perform at the highest possible level. This means that achievements are evaluated mainly by (high) standards of productivity and—for instance—not by achievements in the areas of sociability, experience in handling conflicts, stability and possible other issues not directly connected to the production process itself. With the advent of competitive economic powers from a different culture background (e.g. those in the far east), the stress on efficiency has become stronger than ever before. It does not seem unreasonable to suspect that these economic forces help in shaping the conception of aging as a disability, a disadvantage, if one sees aging as a process that stands in the way of economic efficiency only. Moreover, there is probably some truth in seeing the elderly as such if one judges them only from the viewpoint of economic value. And, if the elderly worker wants to try to stay at work, this means that he or she has to put more and more effort in keeping up to the high standards required. A logical result of this is that elderly workers will sooner or later not be able to keep it up and fall ill, or become unhappy with work and eventually try to escape from work one way or another.

It has been known for a long time now that work can be rewarding or, inversely, that people out of work may start losing their 'edge'. Being unemployed can make one miss out on social contacts and engender in unemployed individuals the idea that they are not being useful: they can lose their sense of identity if they cannot find alternative ways to test (social) reality. For such reasons work can be thought of as structuring human experience, and even more so if having work is perceived as a value in itself. In short, work can be seen as a means of identification. Losing it can have serious consequences mentally, socially and, of course, financially.

Over the past decades, while western society has witnessed an economic high, attempts have been made to give aging a new place in society. One of the policies used to tackle the problem has been simply to take the elderly out of the production process. In most western countries, retirement, in itself commendable as a period of life in which one doesn't have to work any more, has been followed by early retirement. Laws have been passed that enable workers to stop working because of chronic disease or disablement. Because of the growing pressure, more and more elderly workers have taken these opportunities to leave work. It was made relatively easy to do so too. Employers as well as employees were happy that these possibilities existed and made good, sometimes excessive, use of them. However, regulations such as these cost a lot of money and eventually have to be paid for one way or another. Also, one tended to forget that the elderly workers themselves might want to keep on doing their job for the reasons referred to earlier. Again, work in itself can be a rewarding activity that can render a sense of self-fulfilment. Thus, it was forgotten that people who were laid off before their time could suffer from it.

It is because of all this that a re-evaluation of these practices is seen as necessary. This re-evaluation suggests that the elderly worker should be given the opportunity to stay at work for as long as possible.

In the following chapters an attempt is made to define the problems elderly workers face and then to suggest what could be done to ameliorate this situation. In chapter 23 the recent history of the diminishing participating of the elderly worker on the labour market is described by van Dijk within the framework of a disturbed equilibrium between elderly workers' capacities and current working conditions and demands. Reasons are given as to why the elderly still leave the workplace in such numbers. Also, some policy changes are proposed that might help to keep the elderly worker at work.

In chapter 24 Bruyn-Hundt focuses on the total workload (the amount of paid and unpaid work) men and women have during different phases of their working life. It analyses the influx of women on the labour market over the past 20 years, and the changing division of paid and unpaid work between the sexes. Differences in total workload along lines of age and sex and its possible influence on eventual burn-out of employees, specifically the elderly, are brought forward. Proposals are made that aim to reduce the total workload for both men and women in order to keep them on the labour market as long as possible.

Chapter 25 by van der Velden discusses the need to implement an 'age-conscious social policy' within companies. Preconditions that have to be fulfilled by the government and other institutions to make the policy into a success are addressed. What is actually meant by an 'age-conscious social policy' is explained next, and two short examples are given.

In chapter 26 Brzokoupil gives an example in which the already-mentioned connection between getting older and disability is explored. That is, if the elderly worker can be compared to the disabled worker to a certain extent (justifiable or not), then it may be revealing to see how a company that aims at employing the disabled is organized and manages to operate on the market as a real competitor. Special attention is given to the kind of management that makes it possible to run a company that has to deal with employees in which the balance between ability and demand is precarious.

In chapter 27 Gerardu and Schabracq present the personnel policies of a company that actively tries to contribute to the continuing well-being of its (elderly) employees. Although there are no 'hard data' available to estimate the results of the implemented polices so far, it gives an inside view of the problems companies have to deal with in order to reorganize their personnel policy.

Reference

Ariès, P. 1962. *Centuries of Childhood. A Social History of Family Life* (Vintage Books, New York).

23

Health, aging, social policy and perspectives

F. J. H. van Dijk

Summary. In various cross-sectional studies elderly employees still active in paid work surprisingly are relatively healthy as a result of powerful selection processes. As a result of these processes, half of the potential working population of 50–65 years in The Netherlands receive a disability pension. Age-related diminished capacities, although present, cannot explain why the participation rate of elderly employees has declined dramatically from 1960 to date. A disturbed balance between capacities, broadly defined, and current working conditions is proposed as a useful scheme to explain recent history. To meet future labour-market demands we need a dramatic change in our companies towards an age-dynamic policy. In addition, we need new incentives for companies and employees in favour of the participation of the elderly.

Keywords: elderly employees, age-dynamic policy, working population, disability, participation rate, working conditions, working capacity.

Introduction: the problem

When we look at the health state of elderly workers still active in Dutch enterprises, at first glance there are no clear problems. Especially in physically demanding jobs but not only there, we see the phenomenon of healthy elderly workers, in the epidemiology known as the 'healthy worker efffect'. They are sometimes as healthy as or even somewhat healthier than those in younger age groups (see Figure 23.1) or cohorts in the recent past (Smulders and Bloemhoff 1991, Broersen et al. 1992).

In figure 23.1 we see a stabilized percentage of workers with backache in the construction industry as a whole (mainly blue-collar workers), above the age class of 35–44 years of age. This pattern has been found in various cross-sectional studies, for example for mean blood pressure (Dijk 1984, pp. 86, 162), health impairments in general (Kompier 1988) and hearing losses (Hees 1991).

back ache

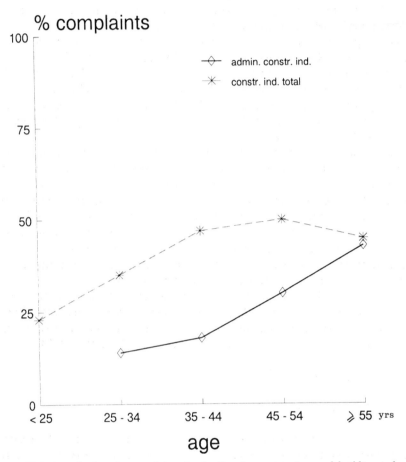

Figure 23.1. Low-back pain complaints in the Dutch construction and building industry. Administrative staff personnel in the construction/building industry and 'total construction/building industry' (blue-collar workers mainly), divided in five age categories, 1989/90 (Broersen et al. 1992).

These cross-sectional figures in fact show a survivor population as the product of a powerful selection process. In figure 23.1 we can also observe impressive work-related differences between the administrative staff personnel and the sector as a whole. Prevalence rates in administrative staff personnel increase with age, presumably reflecting mainly age-related processes not influenced by heavy physical labour and associated selection processes.

However, cross-sectional figures provide little insight into the dynamic course of events during the working career. For instance, in the Dutch construction industry the turnover is high: 3 years after the first entrance in the sector only 46 per cent of the workers has been continuously active in the construction industry, 5 years after entrance it is only 34 per cent. For unskilled workers the figures are 37 per cent and 26 per cent, respectively (Corten 1993). Many transitions are seen from one occupation to another. Other employees are leaving paid work using unemployment, pension or disability schemes. Elderly employees especially run a relatively high risk of disablement. In 1991, of the employees in the age classes 50–54, 55–59 and 60–64 years, respectively 3.5 per cent, 4.0 per cent and 3.6 per cent received a disability pension for the first time (GMD 1992). The overall incidence rate was 1.7 per cent. In the early 1980s the figures for elderly employees were even higher (Willems et al. 1990).

At first sight the impact of this selection process may not look very impressive. When 3 per cent of the 45–54-year-old working population enter a work disability pension scheme each year, however, this means (in a simplified model) that after 10 years about 25 per cent of the original population of 45–54-year-olds received a disability pension at any time in this period.

These kinds of processes were not uncommon in our country. As a consequence, at present in The Netherlands about half of the potential working population of 50–65 years have a disability pension. Moreover, many elderly employees use different forms of early retirement schemes, resulting in a participation rate of 40 per cent for 55–64-year-old men and 15 per cent for 55–64-year-old women in 1992. For 55–64-year-old men especially a dramatic change took place in comparison with the situation in 1960 when about 90 per cent of them were still active at work (Smulders and Bloemhoff 1991).

Disturbed equilibrium

In general, those that drop out of work are less healthy than the survivors. They represent those who have diminished functional capacities and those with age-related diseases. We know that various functional capacities decrease with age, but with a considerable variation in onset and rate of change. Examples are the hearing threshold, cardiorespiratory capacity and reaction time. Moreover, various diseases such as coronary heart disease are age-related. The percentage of the population with health-based diminished capacities increases with age, especially for women (see figures 23.2a and 23.2b).

In a longitudinal study health problems in general and strains owing to work and working conditions predicted the risk of disablement, especially for elderly workers (Winter 1991).

There is, however, no reason to assume that the health of our working population has considerably changed for the worse since 1960. Even above the age of 65, more than 80 per cent of the population perceive their state of health as (very) good or as fair (Vermeulen and Bosma 1992). Moreover,

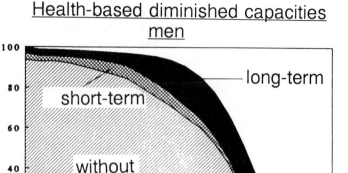

Health-based diminished capacities men

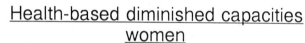

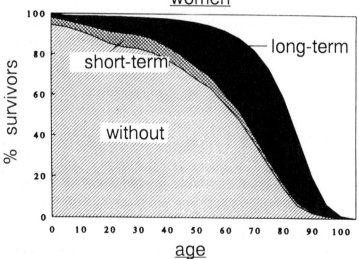

Health-based diminished capacities women

Figure 23.2a and 23.2b. Survivor figures and health-based diminished capacities in men and women, dependent on age. A division was made in men and women 'without diminished capacities', respectively with 'short-term' and 'long-term' diminished capacities (Maas 1989).

diminished functional capacities are not relevant for every job. Coping capacities and decision latitude are often decisive intervening factors (Karasek 1979, Dijk et al. 1990).

We have to conclude that the problem is not primarily the age-related impairment of health and functional capacities as such, but the lack of balance between the capacities of elderly employees and current working conditions. Working conditions in this concept include not only the task and environment but also social relationships, working schedules and decision latitudes. The capacities include not only physical characteristics but also knowledge, skills, attitudes and motivation (Dijk et al. 1990). This concept of a disturbed equilibrium is supported by the opinion of disabled workers themselves. One-third of them mention that timely adaptation of working conditions could have prevented work disablement (Ministerie van Sociale Zaken en Werkgelegenheid 1991).

For instance, working conditions that cause disequilibrium can be physically demanding work, shiftwork, and high work pressure. Another determinant is the growing gap between educational background and rapidly changing working environments. A third factor is the widespread lack of motivation of elderly employees to continue with paid work.

Thus, we have a substantial problem. At present, 21 per cent of our potential occupational population is aged 50 years old or over, in the year 2010 this percentage will be 30 per cent (FNV 1992). In the year 2020 about 40 per cent of the workforce in OECD countries will be between 45 and 65 years (Editorial 1993). Therefore, in the near future we need a much higher participation rate of elderly employees.

Future policy

To restore the balance between capacities including educational background and motivation on the one hand, and work demands and social policy on the other, we need a dramatic change of policy in our companies. Personnel and working conditions policy as well as occupational medical support have to become an integral part of the company policy that might be characterized as age-dynamic. Thus we need an adequate career policy that includes necessary changes in wage and pension structure, programmes for permanent education with special attention for the participation of the middle-aged and the elderly, a preventive working conditions policy especially for ergonomic and psychosocial aspects, occupational medical support of employees with age-related problems and diseases and a good social climate.

To date only a few companies have a comprehensive policy for elderly employees. In only 10 per cent of the collective agreements in The Netherlands is there a 'senior paragraph'. As far as is clear from collective agreements, shiftwork mostly is an obligation until the age of 55, sometimes even until the age of 60 (FNV 1992).

Thus, we have to address the willingness of the companies for a substantial change, both in The Netherlands and in the European Union, distinguishing words from deeds. In many companies including those in the public sector such as hospitals, ministries and schools, the competition and the company's struggle to survive are put forward as impediments to the preservation of jobs for elderly personnel. We are obliged to study these arguments very seriously. On the other hand, more optimal care for, and use of, all human resources in a company or country will also provide substantial profits in comparison with a pension scheme that does not offer any labour in return. In this context, we have to stimulate and study the implementation of new incentives, including financial measures, to encourage the participation of elderly employees. From another perspective it is important to study carefully the needs and demands of elderly male and female employees themselves, in order to find a better balance between capacities and work. More attractive and safe work will enhance motivation to continue with paid work. Finally, a discussion is needed on what role the government and the organizations of the employers and employees must play in the promotion of an age-dynamic policy in companies.

References

Broersen, J. P. J., Bloemhoff, A., Duivenbooden, J. C. van, Weel, A. N. H. and Dijk, F. J. H. van 1992. *Atlas gezondheid en werkbeleving in de bouw (Atlas of work and health in the construction industry)* (Coronel Laboratorium/Studiecentrum Arbeid en Gezondheid, University of Amsterdam).

Corten, I. W. 1993. Naast veel korte ook aardig wat lange bedrijfstakbindingen in de bouw (Short and long employment durations in the construction industry), *Arbouw Journaal* 3(1): 21–22.

Dijk, F. J. H. van 1984. Effecten van lawaai op gezondheid en welzijn in de industrie (Effects of noise in industry on health and well-being) Ph. D. thesis, University of Amsterdam.

Dijk, F. J. H. van, Dormolen, M. van, Kompier, M. A. J. and Meijman, T. F. 1990. Herwaardering model belasting-belastbaarheid (Re-evaluation of the model of work load and capacity), *T. Sociale Gezondheidsz.*, **68**: 3–10.

Editorial. 1993. Aging at work: consequences for industry and individual, *Lancet*, **340**: 87–88.

FNV (Dutch Trade Unions Confederation) 1992. *Werk en ouder worden, 50 plus en arbeidsvoorwaarden. Onderzoeksrapport (Work and aging, 50 plus and labour conditions)* (FNV, Amsterdam).

GMD (Gemeenschappelijke Medische Dienst) 1992. *Statistische Informatie 1991 (Statistical Information 1991)* (GMD, Amsterdam).

Hees, O. S. van 1991. Lawaaidoofheid bij orkest musici. (Noise induced hearing impairment in orchestral musicians), Ph. D. thesis, University of Amsterdam.

Karasek, R. A. 1979. Job demands, job decision latitude and mental strain: implications for job redesign, *Admin. Science Quarterly*, **24**: 285–308.

Kompier, M. A. J. 1988. Arbeid en gezondheid van stadsbuschauffeurs. (Work and health of city bus drivers), thesis (Eburon, Delft).

Maas, P. J. van der 1989. *Lang zullen we leven?* (*We shall live a long time?*). (Erasmus University, Rotterdam).

Ministerie van Sociale Zaken en Werkgelegenheid 1991. *Integraal beleidsplan arbeidsomstandigheden* (*Intergrated working environment policy planning*) (Ministerie van Sociale Zaken en Werkgelegenheid, The Hague).

Smulders, P. G. W. and Bloemhoff, A. 1991. *Arbeid, gezondheid en welzijn in de toekomst. Toekomstscenario's arbeid en gezondheid 1990–2010* (*Work, health and social well-being in the future. Future scenarios work and health 1990–2010*) (Bohn Stafleu van Loghum, Houten/Antwerp).

Vermeulen, C. A. and Bosma, A. M. M. 1992. De gezondheid van ouderen; epidemiologie en beleid (Health of elderly people; epidemiology and policy), thesis, University of Amsterdam).

Willems, J. H. B. M., Smulders, P. B. W. and Pot, P. D. 1990. De oudere werknemer: (g)een medisch problem? (The elderly worker: a medical problem?), *Ned. T. Geneesk.*, **134**: 1937–1943.

Winter, C. R. de. 1991. *Arbeid, gezondheid en verzuim also voorspellers van uitval uit het werk* (*Work, health and absenteeism to predict job drop-out*) (NIPG-TNO, Leiden).

24

Distribution of paid and unpaid work of people of 50 plus in The Netherlands

M. Bruyn-Hundt

Summary. Labour market participation of Dutch men younger than 50 remained the same in the period 1975–1985, but the participation rate of men over 50 is rapidly diminishing. Labour market participation of Dutch women younger than 50 years of age has rapidly grown. In an analysis of time budget surveys over the period 1975–1985 special attention is paid to the total workload of paid and unpaid work because the division of paid and unpaid labour between men and women is changing: women enter the labour market, men increase their share of unpaid work. To prevent younger people from burning out when they reach the age of 50 and over because of the double burden of paid and unpaid work, policy measures should be taken to make a combination of paid and unpaid work easier for both sexes. Some measures are proposed that could diminish this total workload.

Keywords: labour market participation, gender, age group, workload, policy, burn-out.

Introduction

Over the past 15 years there has been an accelerating trend towards early labour force exit among older workers. This trend is very strong in The Netherlands, as illustrated in figure 24.1.

Figure 24.1 clearly shows that labour market participation of men younger than 50 years of age remains the same over the period 1975–1985, but that the participation rate of men over 50 is rapidly decreasing. Similar developments can be seen in other European countries. This holds for the development of women's labour market participation as well. A publication of the European

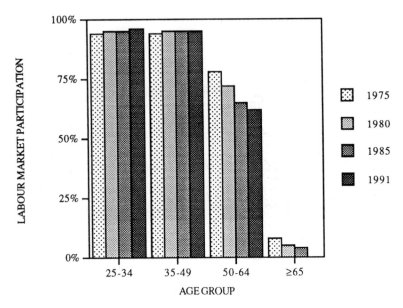

Figure 24.1. Labour market participation of Dutch men per age group from 1975 to 1991 as a percentage of all men of this age.
Source: Central Bureau of Statistics.

Community on this subject remarks that horizontal data about the activity of women is difficult to decipher, because the early exit from work among women is masked by the massive entry of younger women (European Community Observatory 1991: 17). This is particularly clear in figure 24.2, a graph of the labour market participation of Dutch women.

Figure 24.2 shows that the participation rate of Dutch women younger than 50 years has rapidly grown. Labour market participation of women over 50 years of age is growing only slightly: women over 50 do not re-enter the labour market easily because their professional abilities have become obsolete and their husbands are retiring.

The low participation rate for men aged over 50 is a problem in several EU countries. The EC (European Community Observatory 1991: 17) remarks that this development has been encouraged by public policy in response to rising unemployment in the 1980s. The early exit from activity is not the result simply of an advancement of the date of retirement. Other subsystems of income maintenance than those of retirement regulate the frontiers between activity and inactivity. The EC mentions invalidity pension, unemployment benefits and other programmes of pre-retirement (European Community Observatory 1991: 23). Now some countries are facing the problem of how to reverse this trend in order to compensate for a fall in the supply of young labour market entrants, or in order to reduce the cost of early retirement (European Community Observatory 1991: 3). Public authorities are now

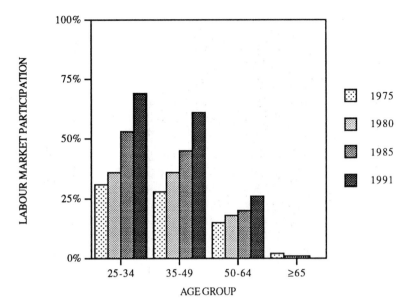

Figure 24.2. Labour market participation of Dutch women per age group from 1975 to 1991 as a percentage of all women of this age.
Source: Central Bureau of Statistics.

attempting to slow down or reverse the shift to early retirement by encouraging more flexible, gradual and progressive entry into retirement. For managements the elimination of older workers is no longer regarded as a panacea. As in other countries, the Dutch government wishes to increase labour market participation of older men and women of all ages.

But early cessation of activity has profoundly modified the practice and the conditions of labour market exit. Age discrimination against older workers often takes place: they are the first to be sacked and the last to be hired if they are hired at all. Unions, employers' organizations, governments and research organizations have recommended policies to increase labour market participation of older people in the future (Wiǫǫets et al) 1990, Ministerie van Sociale Zaken en Werkgelegenheid 1991, FNV 1992, OSA 1992). Some of these recommendations are as follows:

—in case of illness the firm should try to keep in contact with the employee and pay close attention to the re-entry of the employee, perhaps in another function; the administration must have a good registration of all illnesses;
—older employees should have the same chances of professional training and retraining as younger employees;
—the experience of older employees can be used by making them mentors for younger, less experienced employees;

—personnel management should pay attention to the age of personnel and adapt their functions if necessary; shiftwork should be abolished for older workers;

—extra free days dependent on age, such as shortening of the working week, or part-time retirement;

—no age discrimination in hiring and firing; a registration of the age structure of the personnel in the annual social report should be kept (FNV 1992, OSA 1992).

I would like to add a new element to this list. If a policy to increase labour market participation for people older than 50 is to be developed, attention should be paid to the whole workload of paid and unpaid labour because the division of labour between men and women is changing. I will clarify this point. I will take The Netherlands as an example for most European countries. Before the second world war a rather strict division of labour existed in The Netherlands between the sexes: men worked for pay on the labour market, women worked at home without pay. In the 1970s Dutch women began to re-enter the labour market as part-time employees after they had reared their children. The data were presented earlier. Few women over 50 participate in the labour market. For this reason, men of 50 plus who left the labour market in the 1980s either had a wife who had never participated in the labour market or a wife who left her part-time job in the labour market at the same time as her husband. These men never felt the double burden of paid and unpaid work. After they retired from paid work they started doing quite a lot of unpaid work. Lately, younger men and women are changing their habits regarding paid and unpaid work earlier and bear the double burden of paid and unpaid work before they reach the age of 50 years. If official policy is to keep older men and women in the labour market, more attention should be paid to this double burden. The double burden of paid and unpaid work can affect people's health, well-being and productivity. What do we know of this double burden?

Time-budget research

Time-budget research is the basis on which to answer this question. The time-budgets data I use in this section were collected by a commercial bureau, Intomart, on the initiative of the Dutch radio and TV organization, the Social and Cultural Planning Bureau and others. Intomart gathered data by questioning representative samples of the Dutch population in 1975, 1980 and 1985 (Knulst and van Beek 1990). By following the same method, the data of these three years can be compared. The number of respondents was enlarged from 1309 in 1975 to 3262 in 1985 to make analyses of subpopulations possible. Respondents were asked to write down how they used their time by keeping a diary for a whole week. The same questions were asked and the same codes used for all activities. The samples consisted of all members of households of

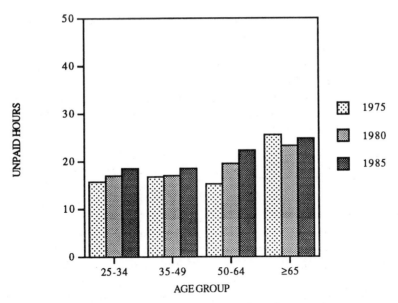

Figure 24.3. Average number of hours spent on unpaid work per week per age group for men, 1975–1985.
Source: Social and Cultural Planning Bureau.

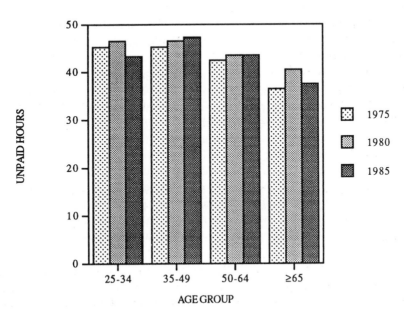

Figure 24.4. Average number of hours spent on unpaid work per week per age group for women, 1975–1985.
Source: Social and Cultural Planning Bureau.

12 years and older. Only the results for people of 25 years and over will be presented here.

In the time-budget study of 1980, the Social and Cultural Planning Bureau classified these activities in five subgroups: work, household work, personal care, education and leisure activities (Knulst and Schoonderwoerd 1983: 268, 269). In these Dutch studies, paid work time is time for which one receives a wage, salary or other remeneration. Time spent in transit to and from work is considered as work time. The same applies to study or leisure during work time. Unpaid household work is: household work, caring for members of the household, purchases, participation in unions, political organizations and other social organization and do-it-yourself activities. The other activities are considered to belong to the categories paid work, personal care, education and leisure.

Unpaid work by age group and sex

How is unpaid work divided over the sexes and how much time is used for unpaid work by both sexes? Figures 24.3 and 24.4 tell us the average number of hours all respondents of a certain age group spent on unpaid work in one week.

Figures 24.3 and 24.4 show that:

(1) The average number of unpaid hours increased over the period 1975–1985 for men of all age groups.

(2) One would expect the average number of unpaid hours for women to decrease over this period because men started doing more unpaid work and because women's participation on the labour market increased. This turned out not to be the case: only young women of 25–34 years of age spent fewer hours on unpaid work. The number of unpaid hours for women in all other age groups increased. On the other hand, women's share of unpaid work decreased slightly because men increased their hours of unpaid work. This development was most pronounced in the 50–65 age group. Nevertheless women are still doing the bulk of unpaid work—some 58–74 per cent.

(3) This becomes clear if one compares both figures: in 1985, 60 to 72 per cent of all unpaid work was done by women.

(4) The total workload of women of 65 and over is much higher than that of men.

Paid and unpaid work by age group and sex

Labour market participation of women under 50 years of age increased enormously, while labour market participation of men in the age groups 25–50

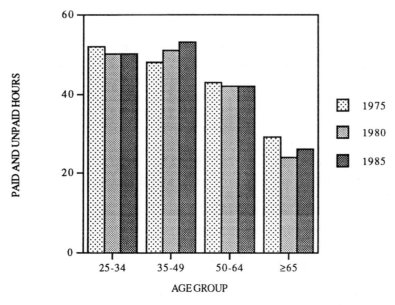

Figure 24.5. verage number of hours spent on paid and unpaid work per week per age group for men, 1975–1985.
Source: Social and Cultural Planning Bureau.

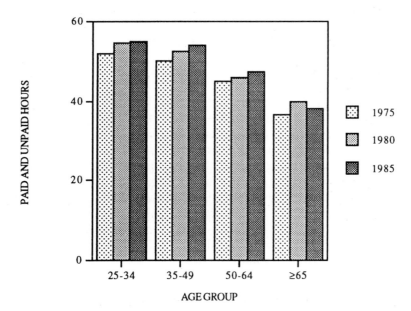

Figure 24.6. Average number of hours spent on paid and unpaid work per week per age group for women, 1975–1985.
Source: Social and Cultural Planning Bureau.

remained rather stable in the period 1975–1985. Labour market participation of men over 50 years of age has diminished. What has been the influence of the increasing labour market participation of women, the decreasing labour market participation of (older) men and the increasing number of hours of unpaid work done by men of all ages on the total workload of paid and unpaid work?

Figures 24.5 and 24.6 show that:

(1) The total workload for men aged 35–49 has increased, while the total workload for men in all other age groups was stable or diminished slightly.
(2) The workload for women in most age groups increased, especially for women younger than 50.
(3) If one compares both figures it becomes clear that women are working longer hours than men, especially in the age groups of 25–34 and 50 years and over.
(4) The total workload of women of 65 and over is much higher than that of men.

In all age groups women's average working week is longer than men's; the difference is smallest in age group 35–49: men work 53.5 hours, women 54. This age group has the highest paid and unpaid workload of all age groups. The division of paid and unpaid work between the sexes is different for the age groups: the older generations of women have a smaller share of paid work and a larger share of unpaid work than the younger generations.

What kind of unpaid work is done?

For policy makers it is interesting to know what kind of unpaid work is done, by whom, and how much time it takes. It figures 24.7 and 24.8 unpaid activities are differentiated in the five categories mentioned earlier.

Household work comprises activities such as preparing meals, cleaning and dusting, laundry and ironing. Care of the family is time spent on looking after children and other members of the family. Shopping, among others things, comprises daily shopping, and shopping for clothes and furniture. Do-it-yourself activities are gardening, looking after pets, jobs in- and outside the house, knitting and so on. Volunteer work comprises managerial activities for political and other social organizations and family help for other households.

Figures 24.7 and 24.8 show that:

(1) Household work is the most time-consuming activity for both men and women.
(2) For men shopping and do-it-yourself tasks are next in time use, although care of the family is also important for men aged between 25 and 35.

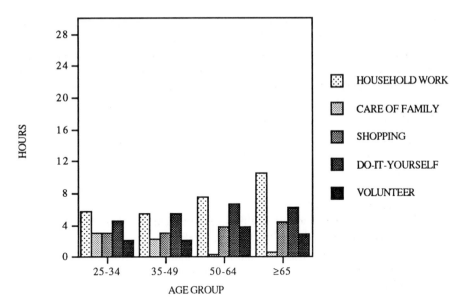

Figure 24.7. Average number of hours spent on unpaid work, by kind of unpaid work per week per age group for men, 1985.
Source: Social and Cultural Planning Bureau.

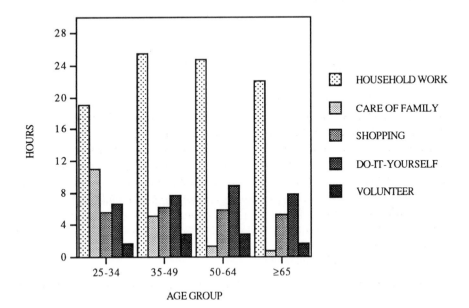

Figure 24.8. Average number of hours spent on unpaid work, by kind of unpaid work per week per age group for women, 1985.
Source: Social and Cultural Planning Bureau.

(3) For women aged between 25 and 35 care of the family takes up most time after household work.

(4) If one compares both figures it turns out that men of 50 years and over increase their hours of unpaid work, while women continue doing approximately the same amount of unpaid work.

In all age groups men are taking over a growing share in household work. At the same time, their hours of do-it-yourself activities have grown in the period 1975–1985.

Policy conclusions

Policies designed to increase the labour market participation of older people should take the amount of unpaid work into account.

(1) The unpaid workload of men and women over 50 cannot be lowered or altered significantly, their paid working week should be shortened. If not, especially women in this age group will not be able to stay in the labour market.

(2) To prevent the age group of 35–49 from becoming burned out once they reach the age of 50 and over, measures should be taken to diminish the double burden of paid and unpaid work for this age group. The following measures could be proposed:

—paid leave for calamities, such as illness of children or other members of the family;

—more household services should be available for reasonable prices, for instance by lowering value-added tax for these services;

—the paid working week should be shortened and/or more part-time jobs should be available.

(3) The workload of the age group 24–35 is heavy because they spend many hours on care of the family apart from their other unpaid tasks. Childcare facilities of good quality for reasonable prices could help this age group to lower the double burden and prevent them from being burned out once they reach 50. The same policies as those proposed for the 35–49 age group could be applied here as well.

(4) These measures should be taken together with measures to improve the situation in paid labour as suggested by others.

References

European Community Observatory 1991. *Social and Economic Policies and Older People* (Commission of the European Community, Brussells).

Federatie Nederlandse Vakbeweging 1992. *Ouderenbeleid FNV, het werkt!* (*FNV policies for the elderly: they work!*) (FNV, Amsterdam).

FNV 1992. *Werkend ouder worden (Getting older while at work)* (FNV, Amsterdam).

Kamstra, E. and Craats, W. van der 1991. *Eenrichtingsverkeer op de arbeidsmarkt (One-way traffic on the labour market)* OSA-Werkdocument.

Knulst, W. P., and Schoonderwoerd, L. P. H. 1983. *Waar blijft de tijd? (Where did the time go?)* (Sociaal en Cultureel Planbureau, Rijswijk).

Knulst, W. P. and Van Beek, P. 1990. Tnd Komt met de jaren (Time arrives with age) (Sociaal en Cultureel Planbureou, Rijswijk).

OSA-Rapport 1992. *Arbeidsmarktperspectieven (Labour market perspectives)* (OSA, The Hague).

Ministerie van Sociale Zaken en Werkgelegenheid, Loontechnische Dienst 1991. *Ouderenbeleid in arbeidsorganisaties (Policies for the elderly in labour organizations)* (Ministerie van Sociale Zaken en Werkgelegenheid, The Hague).

Ministerie van Sociale Zaken en Werkgelegenheid 1991. *Discussienota Leeftijdscriteria en Arbeid (Directives for the discussion on age criteria and labour)* (Ministerie van Sociale Zaken en Werkgelegenheid, The Hague).

Tweede Kamer, Zitting 1990–1991, stuk nr. 21814 nrs. 1 en 2. Ouderen in tel, Beeld en Beleid rond Ouderen 1990–1994 (The elderly do count, image and policy for the elderly 1990–1994).

Wiggers, J. A., Langeveld, H. M. et al. 1990. *Ouderen en Arbeidsmarkt, een onderzoek naar de arbeidsmarktparticipatie van ouderen (The elderly and the labour market, research on the labour market participation of the elderly)* (Ministerie van Wellijn, Volkshuisvesting en Cultuur).

25

Age-conscious social policy

J. van der Velden

*As I approve a youth with
something of the old man,
so I am no less pleased
with an old man with
something of the youth*

Cicero

Summary. In view of several developments it is argued that an 'age-conscious social policy' within companies is necessary. In order to make a concrete implementation of such a new social policy possible, however, certain pre-conditions ought to be set by, among others, the government. The greater part of this chapter will consist of a description of what an age-conscious social policy actually looks like. In the final part two examples of the implementation of such a policy are given.

Keywords: Age-conscious social policy, elderly workers, quality of labour, prevention.

Introduction: why an age-conscious social policy?

Many elderly workers of 55 years old and over have left employment early—either voluntarily or not—the past few decades. This 'exclusion' of older workers from employment is partly caused by high productivity demands that many older workers can no longer meet. Paid labour resembles top sports in several respects: participation is linked to performing maximally at work (Wetenschappelijke Raad voor het Regeringsbeleid 1990). If they can no longer succeed, the employees mostly cannot and/or do not want to make the change to a position which is adjusted better to their present-day capacities, or a less demanding job elsewhere. They are referred or have themselves referred to social security, which usually does not require a lot of effort. Early retirement has been considered a favourable option by most people until recently, and refusing it was remarkable. A change in this practice is considered necessary, for the following reasons:

(1) Demographic developments.
(2) The ratio of active to non-active people and its financial consequences.
(3) The social basis.
(4) The reduction of economic profit from public investments.
(5) The shortage on the labour market.
(6) The elderly worker's importance for the labour organization.

I will now elaborate on these aspects separately.

Demographic developments. Because the number of young people is decreasing (fewer births) and the number of older people is growing (aging), the percentage of older people in the potential work-force (all 15 to 65 years olds) will rise.

Table 25.1 shows that in all EU countries the number of older workers increases during the period 1990–2020. Germany and Italy are the countries which are expected to have a very large number of older workers in the year 2020.

The ratio of active to non-active people and its financial consequences. The increasing number of non-active people leads to financial problems within the public sector and in trade and industry. In the public sector, the number of people potentially entitled to social security will greatly increase in the coming decades owing to demographic developments. As a result, fewer and fewer workers will have to pay the premiums for an increasing number of non-active people, and premium increases will be necessary. This in its turn will lead to higher wage costs for companies. On the other hand, higher labour participation leads to fewer benefit claims, and thus results in lower premiums. The lower wage costs resulting from this can generate new jobs, especially in labour-intensive sectors.

Another financial consequence of older workers leaving the production process is the huge destruction of human capital. For instance, an employee for whom the employer pays fl. 50 000 annually, leaves employment when he or she becomes 60. It can be calculated that, in this case, the lost capital owing to early retirement approximates fl. 208 000 (Nederlands Studie Centrum 1991).

Social basis. A third development that leads to an increasing interest in the elderly worker is the question of how long a small group of workers will want

Table 25.1. *The number of older workers, per age group in 1990 and 2020 as a percentage of the total population in the European Union member states.*

Year	1990		2020	
	50	60	50	60
Age	plus	plus	plus	plus
Country				
Belgium	32.0	20.4	43.5	28.2
Denmark	30.8	20.3	40.4	25.8
Germany	34.1	20.8	47.1	29.7
Greece	32.8	19.4	40.2	26.0
Spain	29.6	18.5	41.1	24.9
France	29.7	19.1	39.5	26.1
Ireland	23.6	15.1	36.5	22.9
Italy	32.4	20.2	46.8	29.9
Luxembourg	31.1	19.1	42.0	26.8
Netherlands	27.4	17.2	40.5	25.2
Portugal	29.2	18.2	41.0	25.5
England	31.2	20.7	38.0	36.4
EC total	31.3	19.7	42.2	26.7

Source: From EC-statistics.

to provide the increasing financial means for the relatively large group of non-workers.

The reduction of economic profit from public investments. The costs for education, training and work experience are made assuming that the person involved, as well as society and the company, will profit economically in the long run. But these costs increase yearly as a result of a more restricted number of working years.

The shortage on the labour market. When people stop working earlier, this will lead to an increasing shortage of experienced employees.

The elderly worker's importance for the labour organization. Finally, the recognition that older workers have an added value for the labour organization is of major importance. More and more people are becoming aware that values of the 1980s—the desire for change, innovation, creativity and efficiency—are necessary to survive fierce competition, but should be aligned with values such as meditation, persuasion, diplomacy, coherence and transfer of culture. Elderly workers can play an important role in transferring such values.

Who has what responsibilities to make an age-conscious social policy possible?

Figure 25.1 depicts the process of why an age-conscious social policy is necessary (the top three boxes). Going further down there are two branches. One deals with the contents of a policy change: what does a change of policy entail? The other branch deals with the responsibilities or influences that various institutions or groups of individuals have in realizing such a policy. The following section will highlight some of the 'participants' involved in this process. They represent influential forces that can enable pre-conditions to be met if any age-conscious social policy is going to work.

Social basis

A major condition for change to another policy is that there is a social basis. Judging by, among others, the increasing extent to which various newspaper and periodical articles pay attention to the elderly worker, their importance is no longer ignored. This may reflect an attitude change in society. Also, the

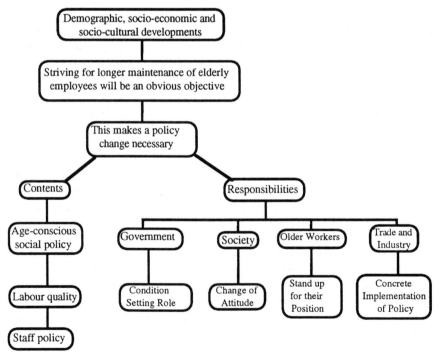

Figure 25.1. Which responsibilities are at what level.
Source: IKOL, Leeftijsbewust Personeelsbeleid, p. 18, Roermond 1992.

elderly worker's emancipation itself is highly relevant. They will have to fend for themselves to gain acceptance on the labour market.

The government

In order to increase the labour participation of older workers (now and in the future), the government will have to take a number of concrete policy measures. These measures must set pre-conditions so that the policy alteration pleaded for will become possible. For that purpose, a policy must be realized that consists of preventive, stimulating and discouraging elements.

(1) Preventive. The quality of labour should improve. Preventive elements are designed to improve the quality of labour, that is, labour contents, circumstances, relations and conditions must change in such a way that the employees' participation in employment will last longer. In addition, the government, as an employer, can function as an example.

(2) Stimulating. Legal measures and information provision aimed at encouraging older workers to continue working as long as they can and want to, fall under the heading of a stimulating policy. The stimulating elements focus on promotion of employment for those who want to and are able to continue to work, in the hope that this has a positive effect on others. This could be accomplished, for example, by taking measures that oppose stereotyping of elderly workers, by emphasizing their positive qualities, and by subsidizing their participation in employment. Loss of return could be compensated by using public resources. Age discrimination should be unacceptable during recruitment, reorganizations and training courses. Furthermore, a direct information campaign may be considered. Examples of themes for such a campaign are: age discrimination, demographic developments and (dis)qualification processes.

(3) Discouraging. Discouraging elements of the policy should aim at the restriction of unjust use of legal measures, such as early retirement rules. Over the past few years, many elderly workers, often relatively low-skilled and, as a result, with a bad labour market position, have been able to retire early rather easily on a large scale. Eventually, a structure should be realized in which it is common practice that elderly workers can go on doing paid work until their retirement age. Early retirement should only be possible in incidental, individual cases. Improving and creating more part-time jobs, flexible retirement, a pension scheme in accordance with years of employment, and solving the issue of pension breach, ought to be realized before society can expect elderly workers to be active longer.

The measures to be taken by the government are no more than preconditions. The social partners will have to agree on concrete measures to stimulate the labour participation of elderly workers. For that purpose, the age-conscious social policy offers possibilities. This policy will be discussed in the next section.

Age-conscious social policy

An age-conscious social policy not only aims at the older worker, but starts at the beginning of one's working life. It involves an integral social policy. If a specific policy is carried out for older workers only, there will be new generations of employees again and again who get into trouble at a later age. As a result healthy retirement is not possible for them either. The social partners should, therefore, promote a 'human resources' policy.

In a human resources policy, human capital is the main starting-point. The main objective of the policy is: maintaining and increasing the employees' capacities, workload, knowledge, creativity, responsibility and motivation.

Within companies the 'human resources' policy can be further shaped by a concrete implementation of an age-conscious social policy. This is a policy that intends to deal with the varying situations and needs of persons of different ages.

For each employee, including the elderly, an individual balance must be found between actual and maximum workload. In other words, this policy does not force age-related qualities of youth on older workers. Rather, attention and scope for age-related qualities of elderly workers is created (see table 25.2).

Age-conscious social policy: policy divisions

The implementation of policy divisions (proactive, preventive and corrective, see table 25.3) takes place in two complementary ways:

(1) by measures in the field of staff policy; and
(2) by measures that promote the quality of labour.

Table 25.2. Social policy per age and policy dimension.

Policy	Age 25+	40+	55+
Proactive	×	+	+
Preventive	+	×	+
Corrective	+	+	×

×: Policy emphasis; +: Part of normal policy.

Source: Discussion report in favour of the provisional council for the policy of elderly people, W. H. C. Kerkhoff and C. R. Tenhaeff, Amsterdam/The Hague 1989.

Table 25.3. Core elements of an age-conscious social policy

Objective social policy (policy divisions)	Target group	Objective: labour quality	Objective:staff policy
proactive	under 40 years	flexiblity of placement	career planning
preventive	40–55 years	prevention of early aging	prevention of problems concerning performance, health and motivation
corrective	55 plus	maintenance of staff for employment as possible	selectively dealing with people who wish to stop working

Source: IKOL, Leeftijdsbewust Personeelsbeleid, p. 21, Roermond 1992.

Table 25.3 also illustrates what objectives and general measures need to be central within the various policy divisions.

Proactive policy division. For labour quality the proactive policy division emphasizes flexibility of placing employees under 40. Staff policy for this age group concentrates on career policy. This means that it mainly focuses on promoting horizontal, vertical and diagonal mobility and mobility regarding contents. Figure 25.2 shows these aspects of internal mobility.

Preventive policy division. The preventive policy division in view of labour quality aims to prevent early aging. It targets employees between 40 and 55. Staff policy is directed at the prevention of problems concerning performance, health and motivation.

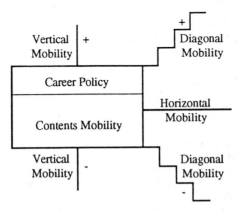

Figure 25.2. Internal mobility.
Source: Congressmap 'Leeftijdsbewust personeelsbeleid', Nederlands Studie Centrum, January 1993.

Corrective policy division. The corrective policy division focuses on the group of employees aged 55 and over. For both labour quality and staff policy the objective is to keep this group of employees employed as long as possible.

An age-conscious social policy as proposed here is innovative in that it emphasizes the necessity for the creation of a work situation that enables an employee's working life to last as long as possible. In the following section the separate policy divisions will be described in more detail.

Concrete implementation of age-conscious policy

When working out the age-conscious policy to be carried out, it will be necessary, with respect to labour quality and staff policy, to pay attention to measures described below. It should be added that a specific implementation will be needed per business or sector, as it depends on the labour organization's own situation.

Proactive policy dimensions

(a) *Labour quality objectives.* The proactive policy division with regard to labour quality mainly focuses on:

—Ergonomic adjustments of the production process. The production process should be designed on the basis of the employee's physical and psychological possibilities.
—Job rotation. For example, change the employees' duties every 5 years, while at the same time taking into account the physical and psychological capacities of the person concerned.
—Early awareness of problems concerning performance and adjusting the contents of the job by enriching, broadening, developing, or lightening the duties, and by adjusting and structuring the work.
—Setting up a horizontal organizational structure. This structure is aimed at horizontal decentralization. That is, a structure is striven for in which every employee formally has equal power, on account of his or her membership of the organization.
—Career planning and remuneration on the basis of skill and capacity assessment. Career planning means that for each employee it is assessed what may be his or her highest position within the company and how he or she can get there (for instance, what training is necessary for that purpose). The starting-point of skill and capacity assessment is maximum remuneration. On the basis of the physical and psychological possibilities of the employee, it is decided what remuneration he or she is entitled to.

(b) Staff objective. Staff policy is aimed at:

—Flexible retirement and using retirement means for periods of refresher leave.
—Setting up a central authoritative personnel department (organized per sector) which looks after employees' interests.
—Talent management designed to increase flexibility and motivation.

Preventive policy objectives

(a) *Labour quality objectives.* The preventive policy division, in view of labour quality, mainly aims to:

—Adjust working conditions in such a way that employees function optimally, both physically and psychologically.
—Adjust duties and work based on experience and training.
—Strengthen work consultation.
—Prevent early aging by retraining and refresher courses.

(b) Staff objectives. As regards staff policy, the emphasis wil be on:

—Staff planning on the basis of the strategic scheme of the organization. Staff planning means the planning of those activities within the framework of staff policy that are necessary to maintain adequate personnel. For that purpose, the existing staff is assessed, both quantitatively and qualitatively. Moreover, it will be evaluated what the development possibilities of the existing staff are and how the external staff supply will develop.
—Career counselling aimed at quality improvement of employees.

Corrective policy division

(a) *Labour quality objectives.* Labour quality within the corrective policy division should concentrate on:

—Preventing disablement and illness by taking into account physical and mental workload factors.
—Adjusting a position through allocation of tasks and return adjustment. Return adjustment means that an employer gets a subsidy from the government for an employee who can no longer fulfil his or her duties optimally.
—Strengthening counselling.
—Determining that age discrimination is unacceptable.
—Continuing to learn activities.
—Enabling flexible retirement without major changes of labour conditions (prevention of pension breach).
—Stringent, periodical medical control by health services.

(b) Staff objectives. The starting-points for staff policy are:

—Strengthening the role of corporate welfare work.
—Social counselling through individual staff counselling.

Practice

In The Netherlands the proportion of older workers within companies has been investigated both by the Ministry of Social Affairs and Employment and by the Institute for Categorial Consultation (IKOL).

IKOL has investigated a representative sample of companies. Figures from the Ministry show that some 6 per cent of the total number of employees within companies consists of employees over 55 (Ministerie van Sociale Zaken en Werkgelegenheid 1991). IKOL's examination has concentrated mainly on companies with fewer than 100 employees (small and medium-sized enterprises). In this sector, some 3.5 per cent of the total number of employees proved to be between 55 and 65 years in Limburg (see table 25.4). This contrasts sharply with the proportion of 55–64 year-olds in the entire population, which is well over one-third, and with the work-force of 50 years and over with respect to total work-force, some 13 per cent.

To show how, in practice, the age-conscious social policy is shaped, two examples are given: a bank and a car factory. Both make an effort to form such a policy. The bank particularly emphasizes prevention, whereas the car factory mainly takes corrective measures.

The bank points to demographic developments as one of the reasons to change their own policy. For the car factory, reduction of disablement is one of the main reasons. When setting up the policy, the bank did not opt for a specific policy aimed at elderly workers, but for setting up a policy that benefits all ages and is preventive. Consequently, most attention was paid to

Table 25.4. *The absolute number of workers of 55 years and over in small and medium-sized enterprises in the province of Limburg (The Netherlands) and as a percentage of the total number of workers in the enterprises, in 1992.*

Region	Number of workers of 55 years and over	Percentage of workers of 55 years and over
Venlo	36	3.1
Roermond	20	3.1
Heerlen	31	4.6
Maastricht	26	3.3
Total	113	3.5

Source: IKOL, Leeftijdsbewust Personeelsbeleid, p. 6, Roermond, 1992.

prevent constraints on social mobility, narrowing of experience and training. These elements proved to be highly influential on the employees' performance.

The car factory has set up a kind of social work provision to achieve speedy rehabilitation of their own staff. It provides for a great variety of jobs. An admission committee has set a number of criteria a person has to meet. This approach is mainly aimed correctively since it is not aimed at prevention, but at solving actual problems.

The main objective for a further investigation in the province of Limburg was to get an impression of the ideas managements hold on the elderly workers (van der Velden 1993). The conclusions are based on 16 interviews with managers of enterprises, labour organizations and employers' organizations. All recognize the problem of the elderly worker within the enterprise. They are aware that they have to change their policy, but they don't know how. What they really want is concrete information about the measures they have to take to keep the elderly workers employed. IKOL is, in cooperation with organizations for small and medium-sized enterprises, giving this information. So far the first results are very satisfactory.

References

Brink, R. H. S. van den 1991. Sociale vaardigheden van ouderen, een cross-sectioneel onderzoek (Social Skills of the elderly, a cross-sectional study), *Tijdschrift voor Gerontologie en Geriatrie*, February number 2.

Cosbo 1989. Themadag arbeid 4 april 1989 (Seminar on labour, 4 April 1989), Cosbo-info, Utrecht.

Dekkers, R. 1989. *Verjonging van ouderenbeleid . . . noodzaak? (Rejuvenation of the policies for the elderly . . . a necessity)*, Katholieke Universiteit Brabant, Tilburg.

Doup, A. C. B. W. (ed.) 1990. Leeftijdscriteria in het arbeidsbestel (Age criteria in the labour system), Alphen aan den Rijn, Sinzheimer Cahier nummer 1.

Haaren, P. W. M. and de Jong, G. R. A. (eds) 1990. Management van mensen (Managing people), Tijdschrift voor Organisatiekunde en Sociaal Beleid (themanummer M and O), July/August.

Kamps, B. J. H. and de Kort, E. C. G. *Oudere medewerkers: leren, belasting, beleving (Elderly workers: learning, burdening, experiencing)*, Katholieke Universiteit Brabant, Tilburg.

Kerkhoff, W. H. C. and Kruidenier, H. J. 1991. *Bedrijfsleven en de vergrijzing (Companies and rising age)*, Nederlands Instituut voor Arbeidsomstandigheden. Amsterdam.

Kerkhoff, W. H. C. and Tenhaeff, C. R. 1989. De plaats en positie van de oudere werknemers (Place and position of the elderly employee), Voorstudie nummer 3 ten behoeve van de Voorlopige Raad voor het Ouderenbeleid, The Hague.

Kerkhoff, W. H. C. 1987. Minder flexible en minder vitaal. Uitgerangeerd met 35? (Less flexible and less vital. Rejected at 35?), *Gids voor Personeelsmanagement*, number 4.

Kerkhoff, W. H. C. 1988. Demografie in arbeidsorganisaties (Demographis in labour organizations). In A. J. M. Knaapen (ed.) *Methoden, technieken en analyses voor personeelsmanagement* Deventer.

Kerkhoff, W. H. C. (ed.) 1987. *Ondernemingsbeleid, kwaliteit van de arbeid en voortijdige uittreding van oudere werknemers (Company policy, quality of work and early retirement of elderly workers)*, Stuurgroep Onderzoek op het gebied van de Oudere Mens (SOOM), Nijmegen.

Lammeren, B. van 1986. Leeftijdsbewust personeelsbeleid, vergrijzing of yuppocreatie (Age-conscious personnel policy, rising age or yuppie-creation). PW, number 10.

Ministerie van Sociale Zaken en Werkgelegenheid 1991. Rapportage Arbeidsmarkt 1991, 's Gravenhage.

Nederlands Studie Centrum 1991. Leeftijdsbewust personeelsbeleid (Age-conscious personnel policy). Conferentiemap naar aanleiding van studiedagen op 17 en 18 January 1991, Utrecht.

Noomen, J. L. 1991. Integraal Personeelsmanagement (*Comprehensive personnel-management*) (Nelissen, Baarn).

Sociaal Cultureel Planbureau 1990. *Social Cultureel Rapport 1990 (Social and cultural report)* (Social Cultureel Planbureau, Rijswijk).

Spikman, J. M. and Brouwer, W. H. 1991. Planningsvermogen van oudere en jongere volwassenen (Planning capacity of older and younger adults), *Tijdschrift voor Geronotologie en Geriatrie*, February number 2.

Studiecentrum voor Bedrijf en Overheid 1991. Loopbaanperspectieven voor de ouder wordende werknemer (Career policies for employees getting of age), Bundel van lezingen gehouden op studiedag 31 mei 1991, Eindhoven.

Tweede Kamer, Zitting 1990–1991, stuk nr. 21814 nrs. 1 en 2 1990. Ouderen in tel, Beeld en Beleid rond Ouderen 1990 − 1994 (The elderly do count, image and policy for the elderly 1990–1994, Tweede Kamer, The Hague.

Velden, J. van der 1993. *Implementatie Leeftijdsbewust Sociaal Beleid (Implementation of an age-conscious social policy)*, Roermond.

Verschuur, A and Wollaert, C. 1991. VUT of flexible pensioen? (Early retirement or flexible retirement?), Leeftijd, april 1991.

Wetenschappelijke Raad voor het Regeringsbeleid 1990. Een werkend perspectief (A workable perspective), Rapportnummer 38, The Hague.

26

Aged and disabled workers—successful business and rehabilitation: experiences from Samhall AB

K. Brzokoupil

Summary. The number of jobs available and the age structure in the work-force change. Work itself and access to work are important factors in rehabilitation and a way back to an active role. Disability may occur at any age and many elderly workers usually share similar obstacles to finding employment. The objectives and measures are mainly the same. In Sweden some of the disabled find employment at Samhall AB. Samhall's operating concept is to provide meaningful and vocational employment to persons with occupational disabilities. The production of goods and services are the means through which Samhall (the second-largest business group in Sweden) attempts to realize the operating concept: 27 900 of its employees are disabled; many are elderly. Knowledge of the fundamental values in the company is important on every level. The role of the management and the future for Samhall towards the year 2000 is presented.

Keywords: sheltered work, work rehabilitation, habilitation, Samhall.

Introduction

The number of jobs available is shifting and, at the same time, the age structure in the work-force is changing. The number of aging employees is higher than ever before. This situation varies throughout the world. Measures to handle the situation differ between countries and most governments are aware of the complex situation. At the same time, today's and tomorrow's generations have other job demands than those entering the labour market 10 to 20 years ago. This is relevant for all, with or without disability. Employment security, responsibility, personal development and the right to co-determination are desirable and priorities among young people in their thoughts of future working life.

There are groups among disabled and aged employees who are more skilled and trained than ever. Especially the young disabled expect great things from society. They ask for jobs in which they can pursue their career according to their knowledge. They don't have the competence and experience yet, which some of the older workers already have, but they have other qualities, for example solidarity with the employer as well as society.

The number of workers who retire before pensionable age is enormous in most European countries. High unemployment figures are an almost permanent condition. In some countries a generation never had employed parents. A few years ago, Sweden had very low unemployment figures. The economic situation was better and the official policy was 'work for all'. Today, unemployment figures are also rising in Sweden.

The percentage of unemployment in Sweden in January 1994 was approximately 14 per cent. Of these 4 per cent are in labour market policy measures like education, training and subsidized jobs. One-fifth of the total available work-force is outside the labour market.

Employers tend to hire a temporary work-force as a response to fluctuations on the market instead of employing a permanent number of employees. This is important for both young disabled and older workers. There is a trend towards alternative employment relations, and the concept of lifetime employment has gone. In future, the individual must have a large potential to adapt on a labour market that seems ever-changing.

All these factors, separately or combined, reduce the job opportunities both for older employees and disabled in the private as well as in the public sector.

The importance of work

In several ways the loss of a job either because of disability or for other reasons could mean that one is excluded, or partially excluded, from sharing human rights. Cut off from working life, one runs the risk of being passive and isolated. This will increase hazards like early retirement, hospital care, social services, and thus increase the costs of society and the individual (Stambrook et al. 1990, Jacobs et al. 1992, Skrobonja et al. 1992).

Work in a broader sense is important to the quality of life, not only because of income but also as a means to participate in society, for a meaningful life and personal development. Seen as such, work is of more significance to the disabled than to others, as disability often segregates. Work itself and access to work are important factors of rehabilitation and a way back to an active role in society (Jenkins et al. 1990, Kinney and Coyle, 1992, Hanson and Walker 1992, Vetter and Citovska 1990).

Generally, hours a day are spent at the workplace—8 usually well-planned, active and productive hours. With appropriate tasks, suitable adjustments, good work organization and a positive social climate, these hours could be stimulating, educational and developmental.

Having a job will give someone a sense of identity and participation, irrespective of the degree of disability or impairment. Identity—to accept and identify yourself as an employee and a professional—is one of the criteria which could be used as a measure of the maturity of a rehabilitation process. For instance, the psychiatric diagnosis, symptoms during treatment, and test of functional capacity performed in another environment are all of very little prognostic value in the work career of a disabled person who intends to 'go back' into society (Anthony et al. 1990).

There are realistic constraints as to how well one can expect to be treated at the workplace. The border between those seen—and seeing themselves—as 'disabled' and those who are 'normal' workers, diminishes. This is in itself rehabilitating. You are brought face to face with realistic demands. The questions and demands are about your capability to fit into the process of production, not about your diagnosis (Åberg 1991, Dauwalder and Hoffmann 1992).

At a workplace, external and internal environmental factors and routines are often apt to change more easily than in institutions. Even when one of the goals in clinical and institutional institutions is to support the restitution of identity, infantilization is often an unintended aspect of the relation between caregiver and the individual cared for. But, at a workplace, you are looked on as an adult from the very beginning.

An honest ambition in most places of work is that employees should gain in competence and be more skilful. In a workplace there is a structure—for example, to be punctual and accomplish the tasks. You will be judged by your capability as a worker, not by your difficulties. If work is arranged in a proper way, the environment will support not only social rehabilitation: it can also, if well adapted and continuously evaluated, be used in medical rehabilitation.

If you are disabled, for instance because of a rheumatic disease, a proper design of the workplace can result in being as good a worker as any other without this impairment. The technical adaptation made for a worker because of physical or psychological impairment could be beneficial for almost everybody. The adaptation that is needed to compensate for the functional impairment of a worker in the process of production, normally is to the benefit of others and for the quality of production. The disabled worker could be seen as a model where the demand on technical and/or physiological support is more obvious.

Disabled people of any age and many elderly workers share some of the same obstacles for employment, such as suitable education, lack of professional skill and needs for adjustment. They need employment in which they, according to their capacity, could achieve or keep a position. From a pragmatic point of view the objectives and measures are similar.

There exist myths and prejudices connected with 'aging and work' as well as with 'handicap and work'. As long as these ideas are not openly notified, admitted, discussed and disproved, they remain obstacles for both disabled and elderly employees.

The process of aging is an individual one. In some cases, aging starts as early as the age of 40. Finding the exact individual time is of great importance both for theory and practice in management (Ilmarinen 1991).

The pathological process sometimes has a different course and quicker pace in disability. For instance, individuals with Down's syndrome already have signs of Alzheimer's disease in their 40s. This should be kept in mind as the life expectancy for these groups is very much longer today than in the past. Other functional impairments in this syndrome will also influence their capability as they grow older. To what degree the consequences of this can be postponed is not yet fully understood (ILO 1992, Americans with Disabilities Act 1990, Bennett, 1992, Devenny et al. 1992).

There are reasons to believe that the aging process starts earlier in other groups of disabled as well when the demands at work are higher than their capacity permits.

The number of disabled demanding jobs will increase

The fact that elderly employees are more numerous now than before does not necessarily mean that the problems associated with aging will increase. Among elderly employees impairment and disabilities are more frequent. However, the capability to take care of diseases and impairments is very much better today than in the past and older workers will in a way be 'younger' as they are in a much better condition. Every new generation will be more vital and healthy than the generations before them, and be more skilled and educated.

In Sweden, as in other European countries, specialization in society is increasing and simpler job assignments are disappearing. Sometimes economic benefits are the underlying reasons, such as when more labour-intensive productions are located abroad where wages are lower. The number of jobs in the service and maintenance sectors has increased instead. Thirty-three per cent of the Swedish work-force today is in industry, and 67 per cent in the service sector. The forecast for the year 2000 is that 23 per cent will be in industry, and 77 per cent in the service and distribution areas. The same situation will arise in other European countries to the detriment of the disabled and elderly workers.

The number of immigrants with disabilities will increase in Europe during the 1990s. These groups will compete for employment as well. Some of them are very well educated and skilled.

Continued deinstitutionalization as well as development of medical and social rehabilitation will require jobs for people who previously were excluded from the ordinary labour market. Such individuals' clinical status is much better today and attitudes in society are changing in favour of integration instead of segregation and isolation.

Many of those who were formerly excluded from work are elderly. Some of them need pharmaceutical treatment which can interact with chemicals used in the production process. Therefore adjustment of upper limits for these chemicals may be necessary in the work environment. These requirements emphasize the need of experts to support the employers to detect risks in the working environment, as neglecting these risks will make the employers economically responsible for the injured employee (Elbe and Pohl 1986, Bennett 1992, Rubin 1992, Forteja 1991).

Those who were considered to be cut off from working life forever will increasingly demand to have their job back. They will realize that technical and organizational adaptations are possible that, in combination with social and economical support, can reassure the employer to adjust the workplace to accommodate the disabled.

The occupational health services

Public policy should support partners in the market to establish good working conditions and service by means of the Occupation Health Service (OHS). This will enable workers to realize a full and productive working life. In some countries the OHS is called Occupation and Environmental Health Services (OEHS). This is because the importance of combining measures in both the (work) environment and the human, social environment for creating healthy work is recognized today.

The OHS should evaluate the worker's performance in the production process. In the case of deteriorating health it should suggest suitable adjustments. The OHS should use its general knowledge of 'people at work', and understand that the development of working life lies in the successful development of the production of services and goods. In order to improve the working environment it is essential that experts collaborate more often in the reconstruction of work—not only in redesigning it. This entails that measures are not only taken for the disabled and/or elderly workers—those who most obviously are in need of a workplace and/or tasks adaptations—but that they are for everyone. Reconstruction of work has to do with social, economic and political processes, and also with the acknowledgement that the design and organization of work is based on social decisions and the technological level of development (Karasek and Theorell 1990).

An OHS that is well acquainted with a company knows the corporate culture, the means and values of the management and the potential for work environment changes. This knowledge includes both the physical and psychosocial components of the work environment, and is important for quality in the production of goods and services and for the suitability of work as a place for rehabilitation (Scheer 1990).

In most countries in Europe there is a tradition regarding the worker's safety and health. Especially in Sweden, good collaboration between employees and employers is well known. The adjustment of the Swedish Act of Working Law will result in reducing the incidence of persons who are excluded from the job market. It will also improve the situation for those persons with disabilities who already are in the work-force. However, there are obvious risks that the trend towards increased economic responsibility for rehabilitation for the employer will result in obstacles for those disabled who want to enter the job market. A job is the result of a negotiation between an employer and an employee. The employer is interested in getting someone to do the job. The employee's interest is to get a job and an income. Employers will increasingly focus their attention on retaining their current employees, while being extremely cautious in recruiting new employees.

Undoubtedly, these developments will affect those groups outside the labour force. Work capability, not age, nor impairment, nor disability, should be the criterion for hiring and retaining employees, as emphasized in the Americans with Disabilities Act. This is also expressed in the recommendation of an expert group commissioned by the WHO regarding aging and work (personal communication with J. Ilmarinen).

The measures to provide work for persons in need of special care

During the 1980s awareness of the needs and rights of persons with disabilities increased world-wide. A number of specific labour market policies for the disabled were developed in the form of legislative action, vocational rehabilitation, wage subsidized jobs and sheltered jobs. The Swedish labour market policy is unique in that it encompasses many of these measures for the disabled. One or several of these measures exist in most other countries (Neubert 1989, Wöhrl 1990, Reker et al. 1990).

In the early 1900s many countries established special job opportunities for the disabled. Today, sheltered jobs for the disabled exist world-wide. Organizations for the handicapped, governments and the parties on the labour market are all involved in efforts to increase the status of sheltered jobs. There are different regulations in different countries. Some have quota systems which compel the employer to hire a specific number of disabled employees. In others, special subsidies are given to organizations providing jobs for the disabled (Lösener 1990, Black and Meyer 1992).

To be able to take a job on the ordinary labour market, 'supported employment' has been developed. In supported employment the supporters both assist disabled persons to find a job, and support them for a specified time at the new workplace until they are able to manage the job. This support can include simple support such as transportation to and from the workplace, or aid with more complicated matters such as living and social contacts that are as important as the job itself in (re)habilitation.

In Europe the organization of providing work for disabled has developed in miscellaneous directions. The differences lie in issues such as how financial support is organized, the selection of disabled persons and eventually legislation that compels an employer to hire a certain percentage of disabled employees. These differences are reflected in the composition of disabled employed in sheltered work. In some countries the main workforce in sheltered workshops are mentally retarded, while in others a mixture of impairments exists. The demands on a successful business are shifting as well.

Today there is a trend to consider work, as arranged in sheltered workshops, accomplished by the disabled as a part of the labour market. During the 1990s, particularly from an international perspective, it can be expected that the widespread attempts to integrate the disabled into the labour market will influence the structure and direction of sheltered jobs in various ways.

The demand for economic efficiency in sheltered workshops has increased, especially as these have to compete on the market. That this can be accomplished according to the rules of good business behaviour is not fully understood. In shrinking economies where every public expenditure is scrutinized, subsidies given to an organization competing on the open market will course be questioned.

As a result, a broad international interest in how sheltered workshops can develop management methods, competence in business orientation and improve the often poorly developed marketing activities as a means for personal development for employed disabled persons in a strained economic situation, has evolved (McCaughrin et al. 1993, Schuster 1990).

In many areas, the possibilities of achieving a better use of resources through coordination are examined. The International Organization for Provision of Work for Handicapped People (IPWH) aims at those organizations that arrange, or are responsible for, sheltered workshops. One of the important tasks of the IPWH is to develop and share strategies in business and personnel care.

Samhall—meaningful and developmental workplaces for the disabled

When a person in Sweden is unable to obtain employment in the open labour market owing to any kind of disability, he or she is categorized as 'occupationally disabled'. Occupational disability can be described as the relationship between the individual's physical, mental, intellectual and socio-medical disabilities and the composition of the work environment in the broadest sense.

In Sweden the occupationally disabled have two main ways of getting a job via the local employment exchange. One way is to find subsidized employment, which means that the salary is partially covered by subsidies. Apart from that, the employer could get economic and practical support for technical adaptations of the workplace. Since 1994 there are also possibilities to get more extended support by personal assistance even in the workplace. The other way is to get employed at Samhall (the former sheltered workshops).

Samhall's jobs are available to the public employment services for referral of unemployed occupationally disabled persons who, with or without various forms of support, cannot find employment in the open labour market in spite of other forms of assistance from the authorities.

The employment exchange authorities categorize disabled applicants by means of eight 'categories'. An occupational disability cannot, however, be described in terms of diagnosis. The criteria are rather rough and, as the decision is made by a regional officer, the categorization can only be a provisional one. It is not possible to categorize an occupational disability in such rough terms that it could be grouped in all eight categories, but there will be overlap in the categorization. The vocational coordinators must have a good knowledge of the potential of work as a rehabilitation/habilitation process and the potentials of sheltered work, as provided by Samhall, as a tool for personal development.

Operating concept, objectives and methods

Samhall's operating concept is to provide meaningful and vocational employment to persons with occupational disabilities, wherever the need exists. The production of goods and services are the means through which Samhall attempts to realize the operating concept. Sales are conducted in accordance with normal business practices. Samhall units are located anywhere where there is need for job opportunities for occupationally disabled persons (Samhall 1990).

Table 26. .1. The categorization according to the Employment Exchange of the disabled in Samhall, 1993.

Codes of Handicap	Samhall all employees (%)
Disease in cardiocirculatory system	4
Impaired hearing	2
Ocular impairments	1
Locomotor disability	27
General somatic disability or disease	17
Psychiatric or psychological impairment	18
Intellectual impairment	16
Social medical impairment	15

Samhall is the second largest business group in Sweden. In January 1994 Samhall had approximately 31400 employees and almost 700 production units in more than 320 locations throughout the country. A total of 27 900 employees are disabled. Staff such as managers, salesmen, personal officers and supervisors totalled 3500. Operations are conducted by business units that have overall responsibility for special business areas and special sales companies, which are subsidiaries to the parent company, Samhall AB. The number of elderly disabled workers in Samhall is comparatively larger in the elderly groups than in corresponding groups in the general work-force (Fig. 26.1).

At Samhall approximately 20 per cent (6000) of the disabled workers were born in a country other than Sweden. The profile of the disabilities is almost the same in these immigrant groups as for Swedish-born employees.

It is important to emphasize that a job offered by Samhall AB is arranged like any other job in society and business. Samhall's companies must act and behave as such because they grow as an integral part of Swedish business and commerce. On 1 July 1992 Samhall was converted from a state-owned foundation to a limited liability company with the Swedish government as the sole owner. By this means Samhall continued to develop its role as an accepted partner on the market, which serves to enhance confidence in Samhall AB. It also increases the confidence of the disabled worker and the pride of the management, since both groups are equally responsible for Samhall's reputation and success.

Samhall's turnover is approximately US$1 billion per year. Thirty million working hours are spent yearly by disabled workers in different business areas.

The government subsidizes the company, covering salary only up to a certain, annually decided, number of working hours. Half of Samhall's income is from production, half is from governmental compensation for the additional

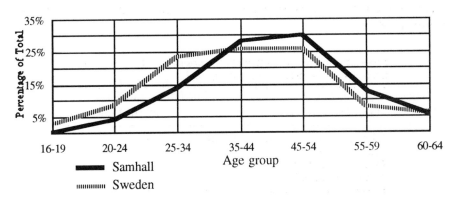

Figure 26.1. Age groups in Samhall and of the general work-force in Sweden: distribution of employees in Samhall AB (n = 28 529) and the total number of employees in the Swedish work-force (n = 3 952 000) on 1 January 1992.

costs of hiring disabled people, running workshops all over Sweden, and the need to adapt work and production so that it suits the working capacity of the disabled.

Target areas

The results of the Samhall companies are assessed in terms of the following four objectives:

(1) Meaningful and suitable jobs for occupationally disabled persons.
(2) A given share of recruitment from prioritized disability groups. The person in these prioritized groups are either intellectually (mentally) impaired, or have multiple disabilities.
(3) Transitions to employment outside the Samhall group.
(4) Economic results.

Within each of these target areas, objectives are established for both the short and long term, using the group's four objectives as target levels. None of the target areas is more important than any of the others. At all levels within the group, satisfactory results mean that the objectives have been successfully achieved in all four target areas.

In addition to fulfilling its assignment of providing job opportunities for occupationally disabled persons, Samhall sells rehabilitation services in the form of job training and expert assistance in the areas of rehabilitation and work environment.

Fundamental values

Samhall's operating concept is based on the importance of work for the individual's perception of life quality and sense of participation in society. Samhall should provide opportunities for the occupationally disabled who are not able to obtain employment on the open labour market. Samhall is thus an instrument of a labour market policy which has full employment as its objective.

All Samhall's operations focus on the individual. This means that the individual's abilities, desires and need for assistance and support are the guiding principles for the composition of the job opportunities offered by Samhall. The premise is that the right of all people to work and to attain self-fulfilment through their work can and should be maintained. Employment with Samhall does not focus on the impairment or functional incapacity. It is the relation between disability and work which is the starting-point for a placement in Samhall.

The individual's potential for work always depends on how the work assignments are related to his or her abilities. It is not meaningful to establish any general or specific lower limit of work ability for employment within Samhall. A general requirement is that the employee must be able to fulfil and actively participate in half-time employment. This is the minimum level covered by collective wage agreements currently in effect.

At the start of employment, however, the employee should be given a generously allocated introductory period, which may entail working less than half time. The kind of job may change during the employment time as the workshop changes its products and production. The objective is to combine training in workmanship with enlargement of social capacity.

Employment at Samhall should provide a step in the transition to the open labour market. The period of transition will vary according to the individual's needs and abilities. For some individuals, Samhall must provide a more long-term alternative to the open labour market. Despite the recession in the Swedish economy, more than 900 transitions of disabled people were performed in 1993.

Employees who go to another employer after having been at Samhall for some time are guaranteed re-entry to Samhall up to 12 months later. This will give the employee confidence to try a new job which in a way will be less secure than a job with Samhall. Being employed by Samhall implies that an employee does not run the risk of being fired in an economic depression.

Supported employment is also a possibility for transitions to the market. This means that the new employer is guaranteed that the job is done properly as the applicants get support from Samhall during their first period on the new job.

A job at Samhall assumes that the individual receives normal terms of employment and the agreed wage. Job content shall be meaningful. This presupposes that the job is based on the production of goods and/or services asked for on the market. Furthermore, the individual is assumed to be an active participant in the organization of work and the production process.

To adapt the organization to business and commercial life and to be able to offer a variety of work forms, Samhall has different forms of operations as well. Samhall's business programme encompasses a total of 14 business groups.

A disabled employee at Samhall who starts his/her career in one profession will probably not end or continue with the same tasks. The available tasks must change according to the production selected.

Concentration of specific tasks in workshops that are more capable of that production is necessary. It is not possible to let every single workshop develop its own line of production, even if these 'local' ideas at first sight seem to be more suitable to the capacity of disabled workers. To be able to compete on the market the entire capacity of Samhall must be used. This encompasses, among other things, the concentration of cooperative production, better use of the total capacity, and priority for economy in sales, customer contacts and in purchasing raw material. This must be done without any breach of the commission undertaken by the owner—as expressed in the business plan.

Job enlargement and job enrichment, development of jobs, education and training are means by which employees will broaden their skills. This will

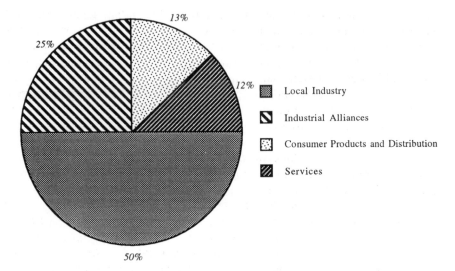

Figure 26.2. The business areas in which Samhall is active on the market.

enhance their chances of getting a job outside Samhall. The opportunity to acquire new knowledge and workmanship will give the disabled an advantage in the competition for jobs outside Samhall. Samhall also supports employees to start and run their own business as a way of transition to the open market, for instance as the owner of a kiosk or a small restaurant.

The adequate use of experts from the OHS is important to realize the business idea. Rehabilitation in and through work comprises the core of Samhall's operations. This means, among other things, that personnel and business development are linked. Personnel development is dependent on favourable business development and vice versa. In this respect Samhall's operations are characterized by a holistic view of work as a means to rehabilitation. This means that issues in areas such as personnel development, work environment and work organisation should be integrated with operational decisions such as business area, business development, production methods and related matters.

The work organization as a tool for rehabilitation and success in business

The work organization in Samhall is based on a process in which the working group has the dominant role. The procedures are variable and flexible. Such an organization of work is based on the following criteria:

(1) *Group dynamics.* This means actively and consciously making up a work group so that the team members are complementary to each other. The composition of the team must be based on age, sex, handicap, nationality, skilful-

ness in profession, experience and so on. As the individual has different experiences, an extensive competence is created in a team of workmen. The team should consist of no more than 5–9 employees. With an appropriate composition of the team, social interaction will thrive more fluently, because of the possibility of close social contacts, solidarity, security and communication, than in a larger group of workers. However, the team must be large enough to be able to deal with, for instance, absentees, internal and external training and transitions, without disturbances in production.

(2) *The content of work.* An expanded and better content of work is necessary to create meaningfulness and development. This can be achieved by assigning the responsibility for a whole process to the team. The organization must support possibilities to develop decision-making competence as well as the freedom of action in the framework of tasks and objectives of work. The group of workers will then be accountable—as far as possible—for a completed component or product. The team will also take control of earlier support, and functions of service, such as maintenance, cleaning, transportation and control. Also they are responsible for the quality of the process.

(3) *'Learning by doing'.* 'Learning by doing' will be the natural way to enlarge the worker's qualifications. It will give the individual the potential to strengthen his/her talents and skilfulness, and inspire workers to take responsibility.

(4) *Enhanced responsibility for the team.* Enhanced responsibility for the team will make it easier to fulfil the goal of production. As the group contributes actively to the organization of production, adequate selection of tools for the production process and the correction of any problems that arise will become progressively easier. The ease of attaining the production goal is further supported by feedback of the customer.

(5) *Planned job rotation.* Planned job rotation in the team will enlarge the competence of the individual as well as the group and minimize monotonous labour from a physical and psychological point of view.

The role of the management

The criteria used for clarifying the aims, measures and objectives of a work group are regarded as parts in the aspects of total quality management of Samhall.

The disabled individuals who are recommended and willing to take a job at Samhall must be aware of the possibilities of the employment and of a possible later transition to another employer. This puts a large strain on the worker, as the situation outside Samhall may be quite different. The attitude in society used to be that employment in Samhall was the last resort for the applicant.

Everything else should have been tried and failed. Today, these attitudes have changed and employment officers, society, client organizations, applicants and others look on Samhall as a means for work rehabilitation, and a phase in the process of habilitation.

To succeed with this, individual development plans for managers of Samhall are as necessary as for the disabled. These plans are an important means of retaining managers. In education and the system of support, knowledge of handicap and principles of rehabilitation must be combined with business-oriented tasks to produce successfully. It is not the objective to educate the management to be professional in the area of rehabilitation. The aim of education and training is to support them to see the possibilities in the operating concept. They must achieve success in business as well as maintaining a network for rehabilitation, such as institutions, organizations in the community and social services.

The managers are not caretakers. This specific role has changed remarkably since Samhall was established. The workers are respected. Employment at Samhall must provide the disabled worker with the opportunity to be seen as a respected citizen who contributes to society as a taxpayer, without the need of extra financial support from society.

The management role will involve increased attention to results, keyed to the objective of the total operation, with clear and measurable goals that can be broken down and followed up at each level. The main emphasis in this role will be on the capacity to create ideas and clarify them for the organization. In addition, there will be increased demands for a dialogue with employees and for responsiveness to their needs and reactions.

An important strategy involves concentrating on a market-oriented approach that is sensitive to both the employee's conditions and needs, and the demands and requirements of customers.

The availability of skilled and highly motivated managers is one of Samhall's most important prerequisites for meeting its objectives. Accordingly, the procurement and development of executives is one of the continuing high-priority core problems.

Demands on the 'profile' and leadership qualities of managers have been further clarified over the years. The natural background for this is the focus on business development and industry, but also an emphasis on human qualities, and a good capacity to develop an organization heavily based on the needs of employees. (Heaney 1991).

The rotation of executives between management and 'specialist' responsibilities is a natural part of planning. The company needs to develop a corporate culture that stimulates increased mobility of managers.

The emphasis on executive development will increasingly be on teaching conceptualization and the management of change. Management training will focus more on the needs and conditions of the individual employee.

The importance of individual introduction programmes on employment, 'development conferences' and development planning as tools for management for rehabilitation, have a prominent role in the management function (Reisman and Reisman 1993).

Samhall towards the year 2000—the near future

New recruitment, conducted in cooperation with the public employment offices which focus on groups of persons with severe disabilities, will be one of the prioritized objectives of Samhall in the near future. These groups are disable who are intellectually impaired, mentally disabled or have multiple disabilities. These groups of disable have great problems in obtaining a job. The owner of Samhull has decided that 40 per cent of new employees should belong to one of these groups. It is easy to understand that employing these groups is a great challenge for Samhall.

Samhall has now entered the new phase of development towards the year 2000. The first phase of the group's development started in 1980, when a foundation was established in each county to take over the sheltered workshops operated by the municipalities and other organizations. The second phase was initiated in 1987 with the integration of all companies and workshops under the common Samhall name. The conversion to a limited liability company on 1 July 1992 marked the start of the third phase of development. One characteristic of the last phase that the separate companies can act as a group with all the potential for coordination and development that entails.

The long-term strategies in the group's third phase has its origins in the report *Samhall towards the year 2000*, published in 1991. These strategies, which are based on the retention of Samhall's business concept, can be summarized by three key concepts:

(1) the developmental workplace;
(2) increased business orientation; and
(3) increased ability to cope with changes.

(1) *The development workplace* concentrates efforts on the individual's personal development, and supports requirements of his or her work and of the small development work groups, which are at the centre of group-oriented work methods. Good work within the group is characterized by independence, influence, learning, fellowship, participation and security.

(2) *Increased business orientation* signifies that a market- and customer-oriented approach is the primary foundation for the selection and development of business areas, organization and concepts. In the selection of a business area, care must always be taken to provide goods and services that are sufficiently diverse, so that the abilities and experience of both employees and recruits can be utilized within the framework of Samhall's operating concept. Samhall combines small- and large-scale operations. Samhall will

provide many small units close to the people, while simultaneously exploiting the benefits of large-scale operations in production development, marketing and purchasing in cases where this is cost-effective. The group's business development during the 1990s is characterized by increased participation in the service areas, as well as an adjustment to conditions in the future Europe.

(3) *An increased ability to cope* with changes assumes that Samhall will follow the increasingly rapid pace of change in society with respect to both the business climate and the labour market for occupationally disabled persons. Revision and renewal of business, company and work organizations are therefore essential and form constant components in Samhall's operations. The development of leadership and expertise is an important tool for the process of change in the group.

References

Åberg, J. (ed.) 1991. Rehabilitering i respekt. Samhall Skriftserie nr 21.

Americans with Disabilities Act 1990.

Anthony, W., Cohen, M. and Farkas, M., 1990. *Psychiatric Rehabilitation*, (Boston University, Center of Psychiatric Rehabilitation, Boston).

Bennett, D. H. 1992. Was muss die Gemeinde zur Versorgung ihrer chronisch psychisch Kranken tun?—Uberlegungen aus England, *Psychiatr. Prax.*, **19**: 212–216.

Black, J. W. and Meyer, L. H. 1992. But . . . is it really work? Social validity of employment training for persons with very severe disabilities, *Am. J. Ment. Retard.*, **96**: 463–474.

Dauwalder, J. P. and Hoffmann, H. 1992. Ecological vocational rehabilitation, *New Dir. Ment, Health Serv.*: 79–86.

Devenny, D. A., Hill, A. L., Patxot, O., Silverman, W. P. and Wisniewskike, E. 1992. Aging in higher functioning adults with Down's syndrome: an interim report in a longitudinal study, *J. Intellect. Disabil. Res.*, **36**: 241–250.

Elbe, A. and Pohl, G. 1986. *Lernen und Handeln: zur Verbesserung der Arbeitsbedingungen im Gastgewerbe* (Schriftenreihe Humanisierung des Arbeitslebens 69) (Campus Verlag, Frankfurt).

Forteja A. 1991. Operating a member-employing therapeutic business as part of an alternative mental health center, *Health Soc. Work*, **16**: 213–223.

Hanson, C. S. and Walker, K. F. 1992. The history of work in physical dysfunction, *Am. J. Occup. Ther.*, **46**: 56–62.

Heaney, C. A. 1991. Enhancing social support at the workplace: assessing the effects of the caregiver support program, *Health Educ. Q.*, **18**: 477–494.

Ilmarinen, J. 1991. *Myths and Facts about the Development of the Capacities of Aging Individuals*, (Institute of Occupational Health, Aavaranta Series 29) 226–236.

ILO 1992. *The ILO and the Elderly* (ILO Geneva).

Jacobs, H. E., Wissusik, D., Collier, R., Stackman, D. and Burkeman, D. 1992. Correlations between psychiatric disabilities and vocational outcome, *Hosp. Community Psychiatry*, **43**: 365–369.

Jenkins, C. D., Jono, T., Stanton, N. B. A. and Stroup-Benham, C. A. 1990. The measurement of health-related quality of life: major dimensions identified by factor analaysis, *Soc. Sci. Med.*, **31**: 925–931.

Karasek, R. and Theorell, T. 1990. *Healthy Work—Stress, Productivity and the Reconstruction of Working Life*, (Basic Books, New York).
Kinney, W. B. and Coyle, C. P. 1992. Predicting life satisfaction among adults with physical disabilities, *Arch. Phys. Med. Rehabil.*, **73**: 863–869.
Lösener, M. L. 1990. Legal contract of workshops for handicapped people—a wish or reality? *Rehabilitation*, (Stuttg), **29**: 182–185.
McCaughrin, W. B., Ellis, W. K., Rusch, F. R. and Heal, L. W. 1993. Cost-effectiveness of supported employment, *Ment. Retard.*, **31**: 41–48.
Neubert, D. A., Tilson Jr. G. P. and Ianacone, R. N. 1989. Postsecondary transition needs and employment patterns of individuals with mild disabilities, *Except. Child.*, **55**: 494–500.
Reisman, E. S. and Reisman, J. I. 1993. Supervision of employees with moderate special needs, *J. Learn. Disabil.*, **26**: 199–206.
Reker, T., Mues, C. and Eikelmann, B. 1990. Perspektiven der Arbeitsrehabilitation psychisch Kranker und Behinderter -ein Uberblick über den Stand und die Probleme im Landesteil Westfalen, *Offentl. Gesundheitswes.*, **12**: 691–695.
Rubin, A. 1992. Is case management effective for people with serious mental illness? A research review, (erratum in) *Health Soc. Work*, **17**(3): 238; *Health Soc. Work*, **17**: 138–150.
Samhall Policy document no 1.
Scheer, S. J. (ed) 1990. *Multidisciplinary Perspectives in Vocational Assessment of Impaired Workers* (Aspen, Rockville, MD).
Schuster, J. W. 1990. Sheltered workshops: financial and philosophical liabilities, *Ment. Retard.*, **28**: 233–239.
Skrobonja, A. Backov-Kolonic, M. and Fŏkovic, A. 1992. Evaluation of the quality of life of disabled workers based on the most frequently used type of health care, *Arh Hig Rada Toksikol*, **43**: 185–191.
Stambrook, M., Moore, A. D., Peters, L. C., Deviane, C. and Hawryluk, G. A. 1990. Effects of mild, moderate and severe closed head injury on long-term vocational status, *Brain Inj.*, **4**: 183–190.
Vetter, P. and Citovska, M. 1990. Rehabilitation of psychiatrically handicapped patients at halfway houses: the effect of work performance and other variables on the duration of hospitalization. A catamnestic study, *Psychiat. Prax.*, **17**: 78–84.
Wöhrl, H. G. 1990. Eingliederungschancen von Absolventen des BFW Heidelberg mit einer psychischen Behinderung (Intergration possibilities of graduates of the Vocational Retraining Center Heidelberg with mental handicaps), *Rehabilitation* (Stuttg), **29**: 84–92.

27

Personnel policies of an aging company: a Dutch case study

V. M. A. Gerardu and M. J. Schabracq

Summary. In this chapter we give an outline of the personnel policies focused on preventing and solving problems of older employees of the Dutch branch of a European industrial company of Dutch origin (NV Koninklijke Sphinx). It is meant to be a typical example of such an approach, with its main strong and weak points. As we do not have comprehensive and systematic longitudinal data of a quantitative nature at our disposal this is just a description of the state of affairs at the moment, not a critical evaluation study of its outcomes.

Keywords: personnel policies, aging company, case-study.

Introduction

Founded as a local glass and crystal grindery in 1834, NV Koninklijke Sphinx is one of the oldest still existing industrial companies in The Netherlands. It has now developed into one of Europe's leading ceramics manufacturers with plants in several European countries. It employs some 3000 workers and its total sales add up to 500 million Dutch guilders a year (£185 million). Core activities consist of the development, production and sales of ceramic and synthetic sanitary facilities, ceramic floor and wall tiles, and luxury bathroom equipment. The organizational structure is a strongly decentralized one: a staff department, based in Maastricht (The Netherlands), and eight autonomous business units, which are each responsible for their own profits.

In the next section we pay attention to some characteristics of the Dutch branch of the company. We go on to outline the general personnel policy, focusing on elderly personnel, and then describe some specific measures taken to operationalize this policy. In conclusion, we go into some results of this approach and discuss some inherent problems.

The Dutch situation

The Dutch branch of NV Koninklijke Sphinx consists of the staff department and three business units, all based in Maastricht. It employs 1421 workers. Their age and sex distribution is given in table 27.1. The table shows a picture of a predominantly male distribution of employees. The modal age category is the one from 30 to 40 (32 per cent). Forty-one per cent of the employees are over 40, 19 per cent over 50 and considerably less than 1 per cent over 60. The largest proportion of the employees have had some form of lower-level occupational education (52 per cent), a proportion which is probably considerably higher in the older age groups.

Forty per cent have had a medium-level occupational education, and 8 per cent a high-level occupational or university education. This age and education distribution is more or less typical for Dutch industrial enterprises.

Though we lack specific data about the mean period of time that elderly personnel at Sphinx hold on to their last jobs, it can be inferred from research in other companies that this period consists of a considerable number of years, probably more than 10 (Balemans 1992, Boerlijst et al. 1993, Broersen et al. 1993, Deltenre 1993, Haasnoot 1993). This would imply that several plateauing effects are operational, resulting in a diminished potential for horizontal mobility in the organization.

The overall sick-leave percentage is 9.36 per cent. Though this, at least by Dutch standards, is not an extremely high figure, it is still not a desirable figure when one takes into account the diminished feasibility of early retirement schemes and the changes in Dutch social legislature.

Up until now the main way of preventing problems of older personnel in The Netherlands consisted of the application of early retirement schemes. At

Table 27.1. *Age and sex of the Dutch employees.*

age	total	%	male	%	female	%
<21	6	0	6	0	0	0
21–30	382	27	313	22	69	5
31–40	449	32	378	27	71	5
41–50	308	22	274	19	34	2
51–60	273	19	263	19	10	1
> 60	3	0	3	0	0	0
Total	1421	100	1237	87	184	13

NV Koninklijke Sphinx employees aged 62 and over can retire with full pay, employees aged 60 and 61 with pay of 85 per cent of the gross salary and full pension rights. This has been, and still is, a very popular scheme with the elderly: 99 percent of all elderly workers make use of it. Owing to demographic developments (more elderly employees, less well-educated younger ones), however, this has become an extremely expensive way of dealing with the problem. Also, it leads to an unacceptable waste of human talent and possibilities, resulting in vacancies which are difficult and expensive to fill. These considerations make early retirement schemes a decreasingly feasible approach to prevent problems of older personnel. Consequently, one of the main tasks of the personnel and organization department at the moment consists of designing and implementing alternative ways to prevent and solve problems of older employees.

Moreover, some changes have been made in the Dutch social legislation. These changes aim at increasing the participation of (elderly) workers in the labour market. One consequence of these changes is that, in case of a reorganization, the company is no longer allowed to lay off older personnel solely on the criterion of their age. Another consequence is that organizations can look forward to a substantial legal fine when a worker is no longer able to fulfil his duties owing to illness or disablement. A third consequence is that companies have become financially responsible for sick-leave allowances during the first 6 weeks of the sick-leave period.

Together, these two factors imply that aging of the working population will become a more and more serious matter during the next 10 to 20 years. It is to be expected that this will have serious consequences for companies like NV Koninklijke Sphinx. First, older employees show, on average, a higher sick-leave percentage (Johns 1978, Keller 1983, Grosfeld and Schalk 1991). Second, one should take into account the relatively great proportion of (elderly) workers with a low level of education in the company. By now it is a well-known datum that level of education and sick-leave percentage are negatively correlated (Grosfeld and Schalk 1991). This can be considered to be a consequence of the fact that people with a lower level of education frequently hold jobs which are more harmful to their well-being and health. In the light of these facts it becomes clear that the present sick-leave percentage of 9.36 is not a very reassuring one, particularly because of the company's increased financial responsibility for the health and well-being of its employees.

Two years ago, a study into the well-being of the employees was conducted. The cause for this study was the high level of absenteeism in the sanitary ware division, which appeared to have more reasons than only physical ones. The employees had to complete a questionnaire anonymously, including questions concerning the functioning of their immediate superior. As is often the case in this kind of study (Haasnoot 1993, Broersen et al. 1993), the elderly employees especially were very positive. The work was considered demanding, but the employees' well-being was rather high.

However, positive results in this kind of study among elderly employees have often been reasoned away as results of several artefacts. First, the position can be defended that these results are to be attributed to differential selection: only the most satisfied workers would stay with the company while the others leave (Ilmarinen 1991). Another factor is that these results may stem from intergenerational differences in work ethics: members of the older generations would have a more positive view of their work based on their value orientation (Becker 1992). A third factor here is the phenomenon of 'response shift': people show a tendency to adjust their standards when their reality changes in a more favourable or a more unfavourable direction (a phenomenon which especially is studied in the context of cancer research, for example Breetvelt and Van Dam 1991).

However, this view that the higher labour satisfaction of older personnel as a consequence of artefacts is only valid if one holds the somewhat naïve opinion that labour satisfaction is directly and exclusively determined by biological age in itself. Only then would it make sense to label the effects described above as undesirable artefacts, and not as effects that are important in their own right. So, we hold the position that, notwithstanding the considerations about the possible role of artefacts, the finding that the elderly employees were relatively satisfied about their work is important in its own right. The elderly workers in the study did report a rather positive attitude towards their work and this may well be a good point of departure for preventive and curative interventions.

A last problem we would like to mention here is that Sphinx personnel has had to deal with many reorganizations over the past 15 years. Consequently, many of the employees have become somewhat weary and mistrustful of further change. This implies that the introduction of new policies has to be conducted in a most careful way. Good information about objectives, procedures and possible implications are necessary. Preferably, there also should be a formal possibility of participation in the decision making concerning the development and implementation of measures for the employees involved.

The general policy

Generally speaking, the board of Sphinx has a very positive attitude towards elderly employees. Plans proposed by the personnel department are usually adopted and money is made available, because it is felt that elderly employees have to be treated well. However, there is no explication of the general intentions: a mission statement that is part of the general organization philosophy is lacking. Such a statement often is useful as a point of departure and as a guideline for designing a comprehensive policy. In addition, it can be used as a touchstone for the evaluation of the consequences of other organizational changes for the elderly.

The main agent for the development of the general policy and its operationalizations is the personnel department. This department is in close contact with the operational departments. Every 2 weeks, there is a meeting in which employees encountering problems, or about to encounter problems, are discussed. The committee on safety, health and welfare is also involved in these meetings. The task of the company physician and social service is to point out possible problems of employees. The company physician, social workers, and the personnel department consult each other regularly. They act as advisers to the line manager of the employees concerned, who is responsible for the actual implementation of the developed policy.

Job rotation is considered a promising and powerful tool to prevent many of the problems of aging employees. An obvious possibility is to give elderly employees jobs of a more supervisory nature. One of their more important tasks then consists of supervising younger personnel. In this way, optimum use is made of the knowledge and experience of elderly employees.

In line with the considerations about job rotation, new Sphinx employees are, in principle, not engaged for a particular job: they have to be employable in various positions. This may create the right conditions for exchanges in position between elderly and younger employees, especially when jobs become too demanding for elderly workers while younger ones occupy less demanding positions.

Furthermore, in 1988 management development was introduced. In the future, this concept has to be elaborated more systematically, resulting in an overall human resource management approach. It is intended to establish a work-force that is able to deal with radical developments and changes, both technically and socially. Some of the projects mentioned form part of this management development approach (for example, training of middle management). As such, the introduction of management development can be considered as a first step toward a more general form of human resource management.

A separate department has been established for people who functioned less well in their previous work and for partially disabled employees. This department is concerned with assembly and packaging, which was formerly contracted out, but which is now executed by Sphinx personnel. In order to make such a department a success, it is crucial that its jobs are carefully designed. For instance, sufficient decision latitude (to arrange and execute tasks in a preferred way and at a manageable pace) has proved to be an important determinant of health, especially for older employees (Karasek and Theorell 1990). How important this consideration is can be inferred from the finding that, in general, a higher degree of automation is related to a higher sick-leave percentage (Smulders 1984). Another important point to attend to here consists of the prevention of the rise of stereotypes and stigmas attached to working in such a department.

Nonetheless, the automation of labour processes can have a preventive effect—at least, when it is implemented in an adequate way—since the work becomes less physically demanding. In principle, another effect of automation can be the enrichment of tasks, since control tasks are added to the execution of the work. Linked with automation are training projects, such as those focusing on handling new devices and enhancing the employee's knowledge of ceramics to improve his or her ability to evaluate products. The rationale for these training programmes is to be found in increased emphasis on quality control as a part of the new job.

At the moment most plants have been automated. However, in the sanitary ware division the current process will run for another 2 years. This implies that older employees in this department still have to do physically demanding work which can be harmful to their health. A serious problem in this respect is that the present level of training of this group of employees is insufficient for the new positions which will become available when automation will have been completed. At that moment it will be too late for (extra) training. Consequentially, a more prevention-oriented approach is needed.

Specific measures

Diagnostic measures

In order to develop specific measures, one has to have a clear picture of the possibilities and limitations in the organization for different possible measures in this respect. Some form of systematic organizational assessment is needed. Such an assessment can, for example, focus on the inventory of:

—the distribution of age, sick leave and personnel turnover in the different departments;
—functions which are especially adequate for elderly employees;
—the effectiveness of the internal communication;
—factors and mechanisms in jobs which are harmful to individual health and well-being;
—typical critical incidents in the working context that elderly employees find difficult to cope with;
—the organizational climate towards elderly employees, including the occurrence of stereotypes about elderly employees; and
—the applicability of specific measures.

At the moment a study of the workplace is in preparation. It will be executed by a psychological agency and it will focus on aspects such as atmosphere in the departments, working conditions, terms of employment and so on. This study is part of a planned 'performance appraisal' project. This planned project will be presented to the board soon in order to be implemented, in one division initially. It is intended to have performance appraisal discussions with all the employees on a yearly basis. In this way, it will be possible to compare and evaluate on a yearly basis whether plans for improvement of the situation have been realized. The results of the study will be discussed with the employees.

At this moment the figures about quantity and frequency of absenteeism are used as the main warning signals in NV Koninklijke Sphinx. Besides, the employees themselves are considered to have the highest expertise on their problems. When they indicate that the work is becoming too difficult or too demanding, this is discussed with the company health service.

Lastly, there is a 'periodical medical examination' for employees aged over 40 and working in shifts. This examination takes place every 2 years.

Preventive and curative measures

Every department has its less demanding jobs. For about 10 years now, a system of reserving specific positions for elderly employees who experience difficulties in performing their current job has been applied. Elderly employees do not occupy these new positions for a long time, because they soon retire. The system is based on an analysis of all jobs with regard to the physical strain they impose on the worker. In general, the system is intended to facilitate transitions from a physically demanding position (e.g. in the production of sanitary ware) to a position entailing less heavy work.

Exchange of positions may, however, cause problems. Elderly employees often have been holding their positions for a long period of time. This makes it very difficult for them to change jobs. Moreover, some of them regard the transition as a degradation. It is also very difficult for them to admit that, while getting older, they have problems with their job demands. For those in middle and higher positions, the status of their last position also plays an important role. In such a case, it can be wise to let them keep the title of their former position in their new position. Generally speaking, the employees who go through such a transition need some extra time, information and attention from the organization, and line manager, in order to adapt to their new circumstances (Doeglas and Schabracq 1993). Often some form of guidance or coaching is needed.

For middle management, which includes a lot of elderly employees, there are courses in 'managing in a changing society'. Managing independent, young employees is a difficult task for a lot of elderly managers. At first, these courses encounter some resistance, especially when they entail several days away from home. Afterwards, however, most of the participants are enthusiastic.

The project 'internal communication' aims at the improvement of the information flow towards the shop-floor. By providing better information one hopes to increase the motivation for change, especially among elderly employees who often have an aversion to change. In this way it may be possible to prevent the development of informal circuits spreading distorted information on certain subjects.

Lastly, a number of various measures have been developed:

- —employees aged 55 and over who are transferred to a less important position as a result of company circumstances or their personal medical situation keep their former salary.
- —employees older than 55 are exempted from shiftwork.
- —employees aged 59 and over are allowed to work 8 hours less a week.
- —employees aged 50 and over obtain additional holidays.
- —the oldest employees can follow a course, 'Preparation for retirement'.
- —various activities are organized for pensioners such as bingo, day trips, staff magazine, and so on.
- —for elderly employees and pensioners there is a social fund which pays for medical aid not included in the health insurance.
- —social service offers special guidance to employees aged 55 and over and to pensioners.

Conclusion

It is difficult to estimate the precise effects of the measures taken so far. In due course, when the study of the workplace has been carried out, more information will be available, including details on the elderly employees. However, the interpretation of this information will be difficult. This is partly a matter of the chosen research design: the study starts out somewhere in the middle of a process of change as a one-shot approach (Cook and Campbell 1979), and is, as such, not an appropriate way of determining causal factors. Partly also, this is a consequence of interference from other factors which make it impossible to draw clear conclusions about the causal structure of the problems. Examples of such factors are the proceeding process of automation, the developments around early retirement schemes and the changes in the Dutch Health Act, the Disablement Insurance Act and the Working Conditions Act. The study can, however, be a useful method of taking stock of the present situation. In order to be successful in this respect, it should have a broad scope and should be as comprehensive as is required.

In general, there are some indications of positive results of the policy so far. The percentage of recipients of disablement insurance benefits has decreased. There is, however, a shift in age categories. At the moment, special efforts are made to prevent elderly from falling under the Disablement Insurance Act. Because of the smaller percentage of elderly employees becoming disabled, the average age of recipients of disablement insurance benefits is decreasing. Moreover, respondents on average indicate that they are satisfied with the present preventive policy for elderly employees.

One issue which has become conspicuously clear is the need for more systematic research. A more systematic form of research, designed to identify causal factors behind individual and collective problems, could be useful here.

Its results would add greatly to a more comprehensive diagnosis of the difficulties. At the same time, it is advisable to make the implementation of policies and measures subject to evaluation research, with proper pre- and post-tests, and—if possible—some quasi-experimental design (Cook and Campbell 1979). Only then may it become clear which measures work and which do not. Results of such studies would provide a meaningful basis for improvements of measures and policies. The measures are important and expensive enough to justify this kind of research.

In conclusion, it can be said that the presented case is a typical one nowadays: a changing production organization trying hard to achieve a form of human resource management in order to deal with the problems it experiences with its (older) personnel. As it is often the case, however, the chosen approach might have been orchestrated more systematically (an adequate diagnosis, development of tailor-made measures, followed by a clearly phased implementation and evaluation). However, such a statement is essentially a cheap one. In practice, the development of this kind of personnel policies in complicated and turbulent surroundings is determined in large part by all kinds of sudden contingencies. A clearly phased process of diagnosis, design of policies, implementation and evaluation often are not achievable objectives. Things should have been done yesterday and phases are (partly) skipped: no clear diagnosis, no well-tailored measures, no terms for a thorough evaluation study. Those in charge find themselves in the middle of things and simply do not have the time to devise a more integrated approach. In many of these cases, some expert help from the outside is needed.

References

Balemans, M. 1992 *Naar een lijftijdsbewust personeelsbeleid (Towards an age-conscious personnel policy)* (University of Amsterdam, Amsterdam).

Becker, H. A. 1992. *Generaties en hun kansen (Generations and their chances)* (Meulenhoff, Amsterdam).

Boerlijst, J. G., Heijden, B. I. J. M. van der, Assen, A. van 1993. *Veertig-plussers in de onderneming (Over-forties in organizations)* (Van Gorcum, Assen).

Breetvelt, I. S. and Van Dam, F. S. A. M. 1991. Underreporting by cancer-patients: the case of response shift, *Social Science and Medicine*, **32**: 981–987.

Broersen, J. P. J., Zwart, B. C. H. de, Meijman, T. F., Dijk, F. J. H. van, Veldhoven, M. van, Schabracq, M. J. 1993. *Veroudering, werk en gezondheid (Aging, work and health)* (Study Centre on Work and Health, University of Amsterdam, Amsterdam).

Cook, T. D. and Campbell, D. T. 1979. *Quasi-experimentation* (Houghton Mifflin, Boston).

Deltenre, V. 1993. *Leeftijdsbewust personeelsbeleid (Age-conscious personnel policy)* (University of Amsterdam, Amsterdam).

Doeglas, J. D. A. and Schabracq, M. J. 1993. Transitie-management (Transition management), *Gedrag en Organisatie (Behaviour and Organization)*, **5**: 448–466.

Grosfeld, J. A. M. and Schalk, M. J. D. 1991. Verzuimfactoren afzonderlijk belicht. (Absence factors individually considered). In P. G. W. Smulders and T. J. Veerman (eds) *Handboek ziekteverzuim* (*Handbook of sick leave*) (Delwel, The Hague), 75–100.

Haasnoot, M. 1993. *Leeftijd en arbeidstevredenheid* (*Age and labour satisfaction*) (UvA, Amsterdam).

Ilmarinen, J. 1991. The aging worker, *Scandinavian Journal of Work, Environment and Health*, **17**, (suppl. 1): 1–141.

Johns, G. 1978. Attitudinal and non-attitudinal predictors of two forms of absence from work, *Organizational Behavior and Human Performance*, **22**: 431–444.

Karasek, R. A. and Theorell, T. 1990. *Healthy Work: Stress, Productivity and the Reconstruction of Working Life* (Basic Books, New York).

Keller, R. T. 1983. Predicting absenteeism from prior absenteeism, attitudinal factors and non-attitudinal factors, *Journal of Applied Psychology*, **68**: 536–540.

Smulders, P. G. W. 1984. *Balans van 30 jaar ziekteverzuimonderzoek* (*Report on 30 years of sick leave research*) (NIPG/TNO-Ministerie van Sociale Zaken (Ministry of Social Affairs, Leiden/The Hague).

Conclusion

The past decade, under the influence of a world-wide recession and growing competition from developing countries, has seen an increasing trend to stress the need to economize.

It is well known that the economic forces of trade and industry have the potential to profoundly shape behaviour and even thought processes and social values (Elias 1982). But there may be limits to the adaptability of man, and more so when man grows older.

The contributors to part IV all seem to agree on one issue: under growing pressure the first ones to suffer and drop out of the working process are those that are most vulnerable: the elderly and the disabled. It is in these groups that the balance between workload and capacity is stretched to its limit most easily. In a way, these groups are partially comparable with respect to work. They share several problems in keeping a job in the face of major changes in a market economy where money is short. Seemingly less profitable employees, they are thought to deplete the company they work for of badly needed resources, and are laid off first when reorganizations are necessary.

This strategy, at first sight, seems to be a solution to the problem. Over the past few years, however, it has become clear that this is merely a manoeuvre that serves to delay solving a socio-economic dilemma. In societies that seriously want to take care of all people, money inevitably has to be spent on social services that sustain a certain level of (financial) welfare for the whole population. The result of this is high social costs that eventually have to be paid for by the companies that laid off the people they felt were no longer cost-effective. Thus, taxes rise, labour becomes more expensive. Now we have gone full circle and have to face the problem in a more fundamental way. Employers and employees now have to deal with a situation in which a common interest is shared in reforming an existing socio-economic system in such a way that a humane society is able to survive.

The authors in this part of the book recognize the problem of the decreasing labour market participation of the elderly and stress the need for a restructuring of labour and greater participation in training in order to prevent the dropping out of the elderly employee. Also, all of them stress that if such a restructuring is going to be successful, policy makers need to take into account the specific needs and demands of the elderly.

However, this implies that old strategies, for instance making all employees work under tougher conditions—regardless of age-related capacities—are not going to work if one wants to keep the elderly at work and healthy. It also implies that new personnel policies, whether they are called age-dynamic or age-conscious, should make an attempt to be tailor-made: more attention must be paid to the *individual* worker and his or her specific problems. This entails a fundamental departure from personnel policies that try to get a maximum level of production output by applying the same rules to each employee: if employees could not keep it up, social services or some form of early retirement was available.

Lately, in several European countries, new legislation has made the employer financially more responsible for those employees that were laid off, in order to unburden public expenditure. It is hoped that these measures will counteract the increasing costs of unemployment, early retirement and several disability regulations that the state pays for. These measures can be seen as a first step in an attempt to force companies to take better care of personnel themselves. By this more pressure is put on companies to develop personnel policies by which they are able to keep all employees at work—and for as long as possible.

Thus, political parties, the government and the media can play an important role in helping to form a basis on which an age-conscious policy can be realized, as is stressed in van der Velden's contribution.

However, it is not only the government or the workplace that should be given attention. As Bruyn-Hundt has shown, it is not only the work at work that makes people burn out and have problems at work. The necessary activities outside the workplace can put a considerable strain on people as well, a strain that can have repercussions at work as well. This burden is different along lines of age and gender and must be accounted for in a well-balanced evaluation of the actual workload people have to carry.

Two of the chapters in this section (chapters 25 and 27) specifically focus on policies for the elderly within companies. These policies take the individual needs of employees into account, and aim at the prevention of potential later problems, diseases and eventual involuntary retirement. There are possible similarities with personnel policies that have evolved in helping the disabled keep or get a job. For instance, the important role management can play in finding ways to help an employee stay at work, participate and stay healthy while facing age-related problems, implicitly assumes a completely new role for management. If a management not only thinks along lines of efficiency, in the short term, but also stresses important other aspects of the production process and is able to incorporate new technological developments, this might result in a greater suitability of the workplace for the elderly worker. As stressed by Brzokoupil: 'Reconstruction of work has to do with social, economic and political processes, and also with the acknowledgement that the design and organization of work is based on social decisions and the technological level of development.'

References

Elias, N. 1982. *Het civilisatieproces. Sociogenetische en psychogenetische onderzoekingen* (*The process of civilization. Sociogenetic and psychogenetic studies*) (Spectrum, Aula no. 705/706, Utrecht/Antwerp).

Epilogue

T. F. Meijman

This epilogue is based not only on the book that you may just have read, but in particular on the closing statement to the symposium 'Work and Aging' which was organized in January 1993 by the Study Centre of Work and Health of the Faculty of Medicine, the Department of Work and Organization Psychology and the Department of Psychonomics of the Faculty of Psychology, all of the University of Amsterdam.

This symposium on Work and Aging, and also the publication that was composed of most of its contributions, was not a strictly scientific effort. The symposium was organized in order to stimulate the discussion of practitioners in the field of human resource management and occupational health services with academic scientists from various research areas. Outstanding academic experts from such divergent disciplines as work physiology, social economics, psychophysiology, occupational medicine and work sociology were brought together in one and the same conference, and they were stimulated to present to the audience what their specialized discipline could contribute to our knowledge of the aging worker and to the development of practical theories and tools for the personnel manager, the social worker and the occupational health practitioner.

I think that the organizers have been successful in this endeavour not only for the symposium itself but also with the published book that resulted from it. Highly specialized scientists mostly feel some uneasiness when they are forced to step aside from the idiosyncrasies of their own scientific community in order to discuss their ideas with experts from other disciplines and practitioners. Practitioners, on the other hand, have to overcome some shyness when they are forced to confront scientific experts with their practical problems and to discuss possible solutions. Yet, during the 2 days of this symposium both groups succeeded in meeting each other in fruitful and sometimes highly stimulating discussions, not only 'in the corridors' but also in the conference room itself. I am glad that it is possible to read most of the contributions and those of a number of invited authors in the present book.

My first conclusion is that mono-disciplinary research on aging can gain a lot from interdisciplinary discussions and from the discussions with practitioners from other fields.

My second conclusion, however, is a more pessimistic one. After attending to all the scholarly and highly stimulating contributions on 'aging and physical work capacity' or 'aging and information processing' and 'aging and social interactions', my feeling is that we do not know very much about these subjects and, in particular, are unaware of their relevance for the aging worker. Let me illustrate this statement with two remarks.

The first is that we do know many of the differences in physical and in cognitive performances between students in their 20s and residents of elderly homes. But we know much less of the gradually changing capacities of active workers of 45 or 55 years of age. And, more important, we do not know much about the meaning of this with respect to the (re)design of work and working situations, or to the restructuring of social relations in the work organization.

My second remark pertains to personal development in work. In this field, in my opinion, our knowledge is even more underdeveloped. Intuitively, everyone knows that motivations, emotions, interests and attitudes towards work change over the years through working life. But where are the longitudinal studies on vocational careers which may provide specific information on these processes, and, more importantly, which may provide tools for the implementation of social policies with respect to the aging worker? From the research on workload and work stress, we know that work places demands on the worker, and that the impact of these demands changes with the changing capacities of the elderly worker.

This symposium provided many, and excellent, examples of studies in these fields. But what do we know of the 'coin's other side'? Workers require things from their work which may motivate them and satisfy their interests and ambitions, which may enable them to develop new skills or stimulate them to develop their abilities and proficiencies. How are these processes changing during a vocational career? What is the impact on well-being and health, and what is the impact on the work performance of the frustration of such personal goals of the aging worker? The scientific literature is not very abundant on these subjects. Contributions from practice were, unfortunately, almost lacking in this symposium. It is not my task to comment on every contribution of this symposium, but I want to mention one study that interested me very much and that is relevant in this context. The results of the study that were presented by Boerlijst and his colleagues (chapter 18) on educational programmes for elderly workers show that such programmes are practically absent, at least in working organizations in The Netherlands. One gets the impression from this study that, in contrast to the general opinion that 'life begins after 40', working life stops after this age. At least, it is not considered worthwhile to invest in the education, training and development of the worker over 40.

Demographic changes and changes of the work-force with respect to its educational and its cultural level, however, will render this lack of personnel development policies—based on serious research into the changing capacities, interests, ambitions and personal goals of the aging worker—in work organizations more and more obsolete and counterproductive. In my opinion—and this is my final, optimistic, conclusion—the discussions during this symposium have broadened our views on the tasks which are ahead of us in the future, both in science and in practice. There is a need for research and the development of personnel management tools for the changing needs and the demands of aging workers. Changing capabilities and changing ambitions of aging workers will be more and more precious resources which will determine the success of modern work organizations. It is our duty as scientists and practitioners to provide these organizations with the knowledge and with the means to meet these new challenges. This symposium and the contributions to this book provide a stimulating basis for these tasks.

T. F. Meijman
Department of Psychology,
Section of Experimental and Work Psychology,
University of Groningen

Index